Berdimurodov Elyor Tukhliyivich and Dakeshwar Kumar Verma (Eds.)
**Carbon Dots in Biology**

# Also of interest

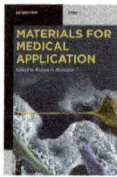

*Materials for Medical Application*
Robert B. Heimann (Ed.), 2020
ISBN 978-3-11-061919-5, e-ISBN 978-3-11-061924-9

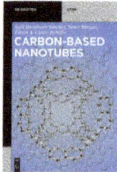

*Carbon-Based Nanotubes*
Raúl Hernández Sánchez, Saber Mirzaei and Edison
Arley Castro Portillo, 2022
ISBN 978-1-5015-1931-4, e-ISBN 978-1-5015-1934-5

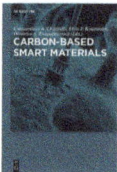

*Carbon-Based Smart Materials*
Constantinos A. Charitidis, Elias P. Koumoulos
and Dimitrios A. Dragatogiannis (Eds.), 2020
ISBN 978-3-11-047774-0, e-ISBN 978-3-11-047913-3

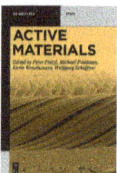

*Active Materials*
Peter Fratzl, Michael Friedman, Karin Krauthausen
and Wolfgang Schäffner (Eds.), 2022
ISBN 978-3-11-056181-4, e-ISBN 978-3-11-056220-0

# Carbon Dots in Biology

Synthesis, Properties, Biological
and Pharmaceutical Applications

Edited by
Berdimurodov Elyor Tukhliyivich
and Dakeshwar Kumar Verma

**DE GRUYTER**

**Editors**

Berdimurodov Elyor Tukhliyivich
Faculty of Chemistry
National University of Uzbekistan
Tashkent 100034
Usbekistan
elyor170690@gmail.com

Dr. Dakeshwar Kumar Verma
Govt. Digvijay Autonomous
Postgraduate College
Chhattisgarh
Rajnandgaon 491441
India
dakeshwarverma@gmail.com

ISBN 978-3-11-079992-7
e-ISBN (PDF) 978-3-11-079995-8
e-ISBN (EPUB) 978-3-11-086033-7

**Library of Congress Control Number: 2022951129**

**Bibliographic information published by the Deutsche Nationalbibliothek**
The Deutsche Nationalbibliothek lists this publication in the Deutsche Nationalbibliografie;
detailed bibliographic data are available on the internet at http://dnb.dnb.de.

© 2023 Walter de Gruyter GmbH, Berlin/Boston
Cover image: RED_SPY/iStock/Getty Images Plus
Typesetting: Integra Software Services Pvt. Ltd.
Printing and binding: CPI books GmbH, Leck

www.degruyter.com

# Contents

# About the editors

**Berdimurodov Elyor** is an associate professor at the National University of Uzbekistan. He has studied in the PhD programme at Tianjin University and the National University of Uzbekistan. He was a researcher at Tianjin University, Karshi State University, and Changchun Applied Chemistry Institute (Chinese Academy of Science). He was a participant in several international and domestic study and research programmes (Chinese Government Scholarships, Chinese great belt programme, National Fundamental, and practical grants). He has published around 22 articles in Scopus and Web of Science journals. H-index is 7 in Scopus, 7 in Web of Science, and 11 (i-10 index of 11) in Google Scholar. He has contributed over 12 book chapters to edited books. He was also a participant in over 15 international conferences from China, Russian, India, and other developed countries. His research areas are corrosion science, electrochemistry, quantum chemistry, material science, green chemistry, biology, medicine, pharmacy, and nanochemistry.

Berdimurodov Elyor Tukhliyivich Faculty of Chemistry, National University of Uzbekistan, Tashkent 100034, Uzbekistan, elyor170690@gmail.com

**Dakeshwar Kumar Verma's** research is mainly focused on the preparation and designing of organic and inorganic materials, nanomaterials that are useful for several industrial applications and material science. Dr Verma is the author of more than 60 research papers, review articles, and book chapters in peer-reviewed international journals published by ACS, RSC, Wiley, Elsevier, Springer, Taylor & Francis, etc. Additionally, he has published 8 edited/authored books with Elsevier, Wiley, and De Gruyter. He is also serving as an editor/author for various books that will be published by Elsevier, Wiley, Taylor & Francis, and De Gruyter. He has more than 850 citations with an H-index of 17 and an i-10 index of 24. Recently, two full-time PhD research scholars are working under his guidance. Dr Verma received the Council of Scientific and Industrial Research Junior Research Fellowship award in 2013. He also availed MHRD national fellowship during his PhD in 2013.

Dakeshwar Kumar Verma, PhD Assistant Professor, Department of Chemistry, Govt. Digvijay Autonomous Postgraduate College, Rajnandgaon, Chhattisgarh 491441, INDIA, dakeshwarverma@gmail.com

https://doi.org/10.1515/9783110799958-203

# List of contributors

**Elyor Berdimurodov**
Faculty of Chemistry
National University of Uzbekistan
Tashkent, 100034
Uzbekistan

**Khasan Berdimuradov**
Faculty of Industrial Viticulture and Food
Production Technology
Shahrisabz branch of Tashkent Institute of
Chemical Technology
Shahrisabz, 181306
Uzbekistan

**Kholmurodov Bahodir**
Faculty of Industrial Viticulture and Food
Production Technology
Shahrisabz branch of Tashkent Institute of
Chemical Technology
Shahrisabz, 181306
Uzbekistan

**Abduvali Kholikov**
Faculty of Chemistry
National University of Uzbekistan
Tashkent, 100034
Uzbekistan

**Khamdam Akbarov**
Faculty of Chemistry
National University of Uzbekistan
Tashkent, 100034
Uzbekistan

**Omar Dagdag**
Centre for Materials Science
College of Science
Engineering and Technology
University of South Africa
Johannesburg 1710

**Mohamed Rbaa**
Laboratory of Organic Chemistry
Catalysis and Environment, Faculty of Sciences
Ibn Tofail University
PO Box 133
14000, Kenitra
Morocco

**Brahim El Ibrahimi**
Department of Applied Chemistry
Faculty of Applied Sciences
Ibn Zohr University
Morocco

**Dakeshwar Kumar Verma**
Department of Chemistry
Government Digvijay Autonomous Postgraduate
College
Rajnandgaon
Chhattisgarh 491441
India

**Rajesh Haldhar**
School of Chemical Engineering
Yeungnam University
Gyeongsan, 712749
South Korea

**Pramod Kumar Mahish**
Department of Biotechnology
Government Digvijay Autonomous Postgraduate
College
Rajnandgaon
Chhattisgarh 491441
India

**Seong-Cheol Kim**
School of Chemical Engineering
Yeungnam University
Gyeongsan 38541
Republic of Korea

https://doi.org/10.1515/9783110799958-204

**Ekemini D. Akpan**
Institute for Nanotechnology and Water
Sustainability
College of Science
Engineering and Technology
University of South Africa
Johannesburg 1710
South Africa

**Eno E. Ebenso**
Institute for Nanotechnology and Water
Sustainability
College of Science
Engineering and Technology
University of South Africa
Johannesburg 1710
South Africa

**Vinayak Sahu**
National Institute of Technology Raipur;
And
Governmental Model College Raipur
Chhattisgarh
India
Email: vinayaksahu2208@gmail.com

**Abhinay Thakur**
Department of Chemistry
School of Chemical Engineering and Physical
Sciences
Lovely Professional University
Phagwara, Punjab
India

**Ashish Kumar**
NCE, Department of Science and Technology
Government of Bihar
India
Email: drashishchemlpu@gmail.com

**Palesa Seele**
MINTEK
South Africa

**Penny Mathumba**
MINTEK
Advanced Materials Division (AMD)
200 Malibongwe Drive
Randburg 2125
South Africa
Email: pennym@mintek.co.za

**Manoj Kumar Banjare**
MATS School of Sciences
MATS University
Pagaria Complex
Pandri, Raipur (C.G.), 492004
India

**Kamalakanta Behera**
Department of Chemistry
University of Allahabad
Prayagraj
Uttar Pradesh 211002
India

**Ramesh Kumar Banjare**
MATS College
MATS University
Aarang (C.G.), 492004
India

**Siddharth Pandey**
Department of Chemistry
Indian Institute of Technology Delhi
Hauz Khas
New Delhi, 110016
India

**MUHAMMAD ALAMGEER**
Khwaja Fareed University of Engineering and
Information Technology
Rahim Yar Khan
Pakistan

**Sonali Loya**
Govt. Nehru P.G. college
Dongargarh
Chhattisgarh
INDIA

**Swati Chandravanshi**
Govt. Nehru P.G. college
Dongargarh
Chhattisgarh
INDIA

**Mahdie Matin**
Endocrinology and Metabolism Research Center
Endocrinology and Metabolism Clinical Sciences
Institute
Tehran University of Medical Sciences
Tehran
Iran

**Mahtab Mirhoseinian**
Endocrinology and Metabolism Research Center
Endocrinology and Metabolism Clinical Sciences
Institute
Tehran University of Medical Sciences
Tehran
Iran

**Alireza Alikhanian**
Endocrinology and Metabolism Research Center
Endocrinology and Metabolism Clinical Sciences
Institute
Tehran University of Medical Sciences
Tehran
Iran

**Golnar Bayatani**
Endocrinology and Metabolism Research Center
Endocrinology and Metabolism Clinical Sciences
Institute
Tehran University of Medical Sciences
Tehran
Iran

**Mohammad Nazari Montazer**
Endocrinology and Metabolism Research Center
Endocrinology and Metabolism Clinical Sciences
Institute
Tehran University of Medical Sciences
Tehran
Iran

**Mohammad Mahdavi**
Endocrinology and Metabolism Research Center
Endocrinology and Metabolism Clinical Sciences
Institute
Tehran University of Medical Sciences
Tehran
Iran

**Burak Tüzün**
Plant and Animal Production Department
Technical Sciences Vocational School of Sivas
Sivas Cumhuriyet University
Sivas
Turkey

**Parham Taslimi**
Department of Biotechnology
Faculty of Science
Bartin University
74100 Bartin
Turkey

**Saima Ashraf**
Institute of Chemical Sciences
Bahauddin Zakariya University
Multan 60800
Pakistan

**Fahmida Jabeen**
Institute of Chemical Sciences
Bahauddin Zakariya University
Multan 60800
Pakistan

**Sabeen Iqbal**
Institute of Chemical Sciences
Bahauddin Zakariya University
Multan 60800
Pakistan

**Muhammad Salman Sajid**
Institute of Chemical Sciences
Bahauddin Zakariya University
Multan 60800
Pakistan

**Muhammad Naeem Ashiq**
Institute of Chemical Sciences
Bahauddin Zakariya University
Multan 60800
Pakistan

**Muhammad Najam-ul-Haq**
Institute of Chemical Sciences
Bahauddin Zakariya University
Multan 60800
Pakistan

**Shokoh Parhama**
School of Advanced Medical Technology
Isfahan University of Medical Sciences
Isfahan
Iran;
And
Centre for Sustainable Nanomaterials
IbnuSina Institute for Scientific and Industrial
Research
Universiti Teknologi Malaysia
81310 UTM Skudai, Johor
Malaysia

**Seyed Shirin Parham**
Department of Veterinary
ShahreKord Branch
Islamic Azad University
ShahreKord
Iran

**HadiNur**
Centre for Sustainable Nanomaterials
IbnuSina Institute for Scientific and Industrial
Research
Universiti Teknologi Malaysia
81310 UTM Skudai, Johor
Malaysia
And
Central Laboratory of Minerals and Advanced
Materials
Faculty of Mathematics and Natural Science
Universitas Negeri Malang
Malang
Indonesia

**Nicole RemaliahSamantha Sibuyi**
Department of Science and Innovation (DSI)//
Mintek Nanotechnology Innovation Centre (NIC)
Advanced Materials Division
Health Platform
Mintek, Randburg, South Africa
And,
DSI/Mintek NIC Biolabels Node
Department of Biotechnology
University of the Western Cape
Bellville
South Africa

**Anelisiwe Mbengashe**
DSI/Mintek NIC Biolabels Node
Department of Biotechnology
University of the Western Cape
Bellville
South Africa

**Zimkhitha Bianca Nqakala**
Organometallics and Nanomaterials
Department of Chemical Sciences
University of the Western Cape
Bellville
South Africa

**Antoinette Alliya Ajmal**
DSI/Mintek NIC Biolabels Node
Department of Biotechnology
University of the Western Cape
Bellville
South Africa

**Tswellang Mgijima**
Organometallics and Nanomaterials
Department of Chemical Sciences
University of the Western Cape
Bellville
South Africa

**Cate Malope Mashilo**
DSI/Mintek NIC Biolabels Node
Department of Biotechnology
University of the Western Cape
Bellville
South Africa

**Aluwani Matshaya**
DSI/Mintek NIC Biolabels Node
Department of Biotechnology
University of the Western Cape
Bellville
South Africa

**Samantha Meyer**
Department of Biomedical Sciences
Faculty of Health and Wellness Sciences
Cape Peninsula University of Technology
Bellville
South Africa

**Mervin Meyer**
DSI/Mintek NIC Biolabels Node
Department of Biotechnology
University of the Western Cape
Bellville
South Africa

**Martin OpiyoOnani**
Organometallics and Nanomaterials
Department of Chemical Sciences
University of the Western Cape
Bellville
South Africa

**Abram MadimabeMadiehe**
DSI/Mintek NIC Biolabels Node
Department of Biotechnology
University of the Western Cape
Bellville
South Africa

**Adewale Oluwaseun Fadaka**
DSI/Mintek NIC Biolabels Node
Department of Biotechnology
University of the Western Cape
Bellville
South Africa
And
Department of Anesthesia
Division of Pain Management
Cincinnati Children's Hospital Medical Center
Cincinnati, Ohio 45229, USA

**N.B. Iroha**
Department of Chemistry
Federal University
Otuoke
Bayelsa State
Nigeria

**C.O. Ezenwaka**
Department of Biology
Federal University Otuoke
Bayelsa State
Nigeria

**C.N. Opara**
Department of Microbiology
Federal University Otuoke
Bayelsa State
Nigeria

**F.E. Abeng**
Department of Chemistry
Cross River University of Technology
Calabar
Nigeria

Elyor Berdimurodov*, Khasan Berdimuradov, Kholmurodov Bahodir, Abduvali Kholikov, Khamdam Akbarov, Omar Dagdag, Mohamed Rbaa, Brahim El Ibrahimi, Dakeshwar Kumar Verma, Rajesh Haldhar and Pramod Kumar Mahish

# Chapter 1
# Recent trends and developments in carbon dots

**Abstract:** The carbon dots are new materials in modern chemistry. The modern development ways for carbon dots were discussed in this chapter. Currently, the carbon dots are synthesized by the top-down and bottom-up methods. The electrochemical methods, ultrasonic treatment, laser ablation method, and arc discharge method were mostly used in the top-down methods. The bottom-up methods have some advantages such as convenient methodology, precise control, easy instrumentation, cost-effectiveness, involvement of non-toxic precursor molecules, practical applicability, and green materials. The carbon dots are synthesized from green sources such as carbohydrates, biomass, and bio-waste. The carbon dots are modified with the supramolecular hosts to obtain the unique carbon dots in the biometric elements, catalysts, and sensor applications. The carbon dots are modified to follow the 2D materials to enhance their unique properties.

**Keywords:** Carbon dots, top-down syntheses, bottom-up syntheses, green materials, biomass

---

*Corresponding author: Elyor Berdimurodov,** Faculty of Chemistry, National University of Uzbekistan, Tashkent 100034, Uzbekistan
**Khasan Berdimuradov, Kholmurodov Bahodir,** Faculty of Industrial Viticulture and Food Production Technology, Shahrisabz Branch of Tashkent Institute of Chemical Technology, Shahrisabz 181306, Uzbekistan
**Abduvali Kholikov, Khamdam Akbarov,** Faculty of Chemistry, National University of Uzbekistan, Tashkent 100034, Uzbekistan
**Omar Dagdag,** Centre for Materials Science, College of Science, Engineering and Technology, University of South Africa, Johannesburg 1710, South Africa
**Mohamed Rbaa,** Laboratory of Organic Chemistry, Catalysis and Environment, Faculty of Sciences, Ibn Tofail University, PO Box 133, 14000 Kenitra, Morocco
**Brahim El Ibrahimi,** Department of Applied Chemistry, Faculty of Applied Sciences, Ibn Zohr University, Agadir 86153, Morocco
**Dakeshwar Kumar Verma,** Department of Chemistry, Government Digvijay Autonomous Postgraduate College, Rajnandgaon, Chhattisgarh 491441, India
**Rajesh Haldhar,** School of Chemical Engineering, Yeungnam University, Gyeongsan 712749, South Korea
**Pramod Kumar Mahish,** Department of Biotechnology, Government Digvijay Autonomous Postgraduate College, Rajnandgaon, Chhattisgarh 491441, India

https://doi.org/10.1515/9783110799958-001

## 1.1 Introduction

### 1.1.1 Importance of carbon dots in material and engineering science

The size of carbon dots is lower than 10 nm. Carbon dots contained mainly $sp^2$-hybridized graphitic carbon. Their important properties depend on the structural, optical, physical, chemical, and electronic performances. These materials are ease of functionalization, good chemical inertness, thermal stability, high water solubility, unique luminescence properties, and low toxicity [1, 2]. Their unique performance significantly depends on the synthesis sources, which may be organic, polymer, green source, and inorganic sources. Some obtained results confirmed that the surface structure, functional groups, heteroatoms, doping agents, and size are also reasons for their unique properties [3, 4].

Carbon dot-based nanomaterials are the new trend in material and engineering science. They are 0D materials as a new trend in drug delivery, sensing, catalysis, and bioimaging. They have good performances such as their low-cost synthesis methodology, high biocompatibility, low toxicity, and good optical properties. These properties make them become more effective materials in modern science. The carbon dots were synthesized by the cost-effective and easy-operation methods [2, 5, 6]. Additionally, the sonolysis of carbon precursors, thermolysis, electrochemical and chemical oxidations, and laser ablation methods were also used in the carbon dot synthesis. Currently, the mostly synthesized carbon dots have the following unique properties such as exceptional

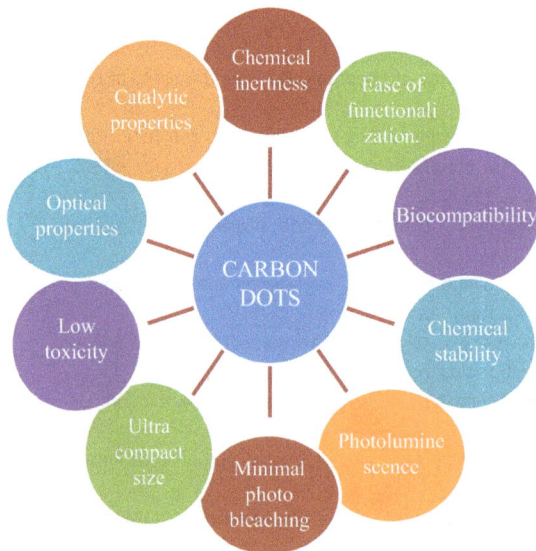

**Figure 1.1:** Unique properties of modern carbon dots [7].

productivity, superior photostability, high quantum yield, biocompatibility, electrical properties, and excellent optical performances (Figures 1.1 and 1.2) [7, 8].

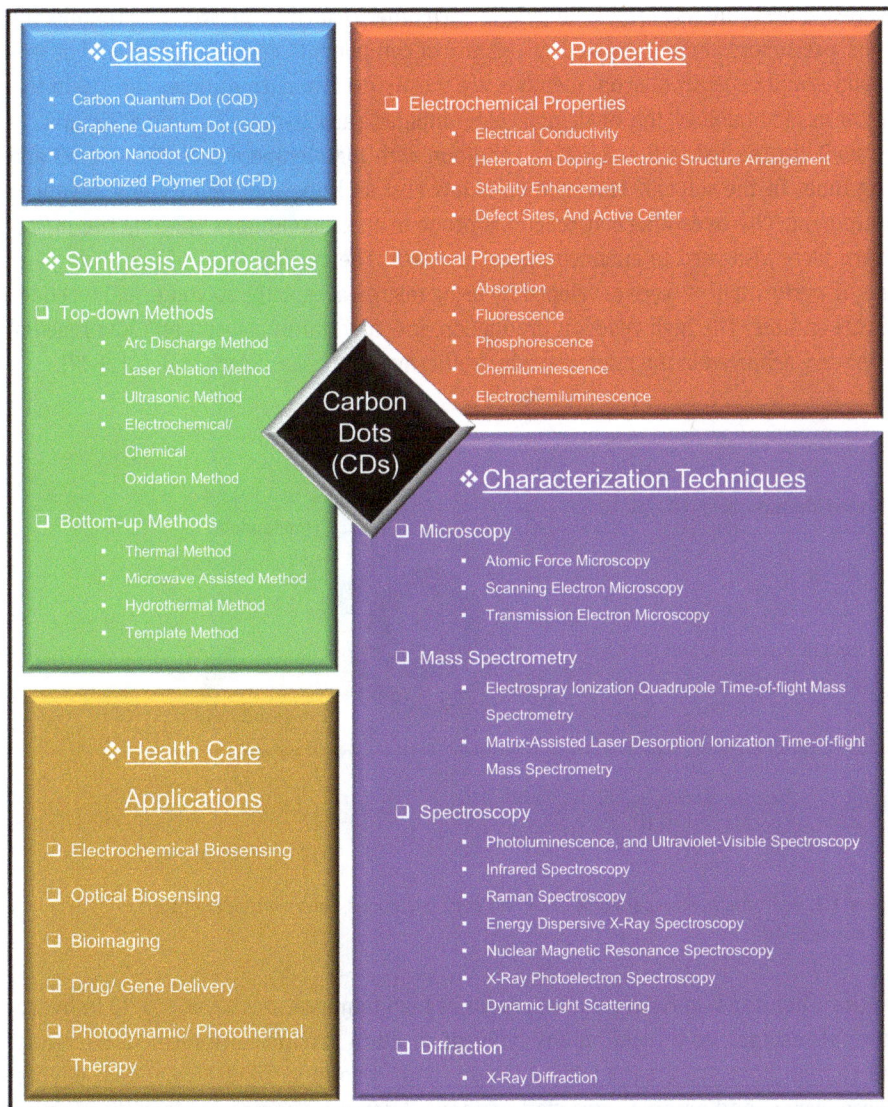

❖ Classification

- Carbon Quantum Dot (CQD)
- Graphene Quantum Dot (GQD)
- Carbon Nanodot (CND)
- Carbonized Polymer Dot (CPD)

❖ Synthesis Approaches

❑ Top-down Methods
- Arc Discharge Method
- Laser Ablation Method
- Ultrasonic Method
- Electrochemical/ Chemical Oxidation Method

❑ Bottom-up Methods
- Thermal Method
- Microwave Assisted Method
- Hydrothermal Method
- Template Method

❖ Health Care Applications

❑ Electrochemical Biosensing

❑ Optical Biosensing

❑ Bioimaging

❑ Drug/ Gene Delivery

❑ Photodynamic/ Photothermal Therapy

❖ Properties

❑ Electrochemical Properties
- Electrical Conductivity
- Heteroatom Doping- Electronic Structure Arrangement
- Stability Enhancement
- Defect Sites, And Active Center

❑ Optical Properties
- Absorption
- Fluorescence
- Phosphorescence
- Chemiluminescence
- Electrochemiluminescence

Carbon Dots (CDs)

❖ Characterization Techniques

❑ Microscopy
- Atomic Force Microscopy
- Scanning Electron Microscopy
- Transmission Electron Microscopy

❑ Mass Spectrometry
- Electrospray Ionization Quadrupole Time-of-flight Mass Spectrometry
- Matrix-Assisted Laser Desorption/ Ionization Time-of-flight Mass Spectrometry

❑ Spectroscopy
- Photoluminescence, and Ultraviolet-Visible Spectroscopy
- Infrared Spectroscopy
- Raman Spectroscopy
- Energy Dispersive X-Ray Spectroscopy
- Nuclear Magnetic Resonance Spectroscopy
- X-Ray Photoelectron Spectroscopy
- Dynamic Light Scattering

❑ Diffraction
- X-Ray Diffraction

**Figure 1.2:** Main characteristics of carbon dots [9].

## 1.2 Current trends in the synthesis of carbon dots

At present, the carbon dots are synthesized by the top-down and bottom-up methods. These synthesis methodologies differ from each other related to the synthesis routine, solvents, precursors, temperature effects, nature of components, carbon sources, and others. Solvothermal or hydrothermal methods are mostly used in current times [10, 11]. In this synthesis methodology, the small molecules (anilines, carbohydrates, amino acids, and citric acid) are reacted in the auto-clap condition with high temperatures (150–270 °C) and long times in the solvent. This method is low cost and does not require any advanced equipment. The hydrothermal reactions occur in these conditions [12–14]. The microwaves were also used to enhance the carbon dots. The various types of precursors and unique performances were developed by using microwaves in the solvothermal reactions of carbon dots. The high reaction production, low amount of solvent, and low reaction time were achieved in the microwave-hydrothermal synthesis of carbon dots [15–17].

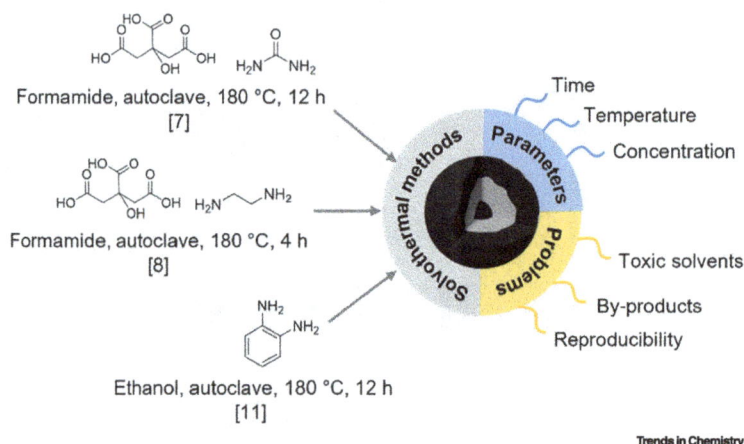

**Figure 1.3:** Hydrothermal preparations of carbon dots: limitations, controllable parameters, and examples [15].

On the other hand, various reactions occur at high temperatures (over 170 °C) in hydrothermal reactors; as a result, many types of reaction products are formed. Currently, the carbon dots are cleaned from the reaction products by dialysis, heat-drying, freeze-drying, and organic extraction methods. After the cleaning process, the reaction productivity maybe reduced (Figure 1.3). These limitations would be solved by the new syntheses methodologies such as machine learning, laser synthesis in the liquid phase, flow chemistry, and mechanochemistry. These ways make the synthesis of carbon dots become more controllable, more productive, and take unique properties [15, 18–20].

## 1.3 Developments in the synthesis of carbon dots

In modern times, various new strategies in the synthesis of carbon dots are being developed. The main factor of carbon dots is a surface character, which is controlled by various modern methods. The studies confirmed that the functional groups are attached to the surface of carbon dots. For example, the hydroxyl, carboxyl, carbonyl, ether, epoxy, and amine functional groups are mainly linked to carbon dot surface. Various heteroatoms such as boron, sulphur, phosphorous, oxygen, and nitrogen are doped on the surface by using organic, polymeric, and biological materials [21, 22].

The rise of quantum yield is an important factor in the development of the synthesis strategy of carbon dots. The obtained results confirmed that biological ways are more efficient techniques for the rise in quantum yield. The photoluminescence intensity and higher biocompatibility of carbon dots were increased with the biological synthesis ways. The biological synthesis ways required biological sources such as grass, tea leaves, coconut shell, soya beans, coffee beans, garlic, rice bran, egg, sugar beet molasses, leaves, pomegranate, banana, honey, and yoghurt [23, 24].

The top-down and bottom-up modern syntheses methods are mostly used in the development of carbon dots. The macroscopic carbonaceous materials such as graphite, carbon nanotubes, and activated carbon were employed in the top-down methods. Figure 1.4 shows the trends of developments of carbon dot syntheses. The electrochemical methods, ultrasonic treatment, laser ablation method, and arc discharge method were mostly used in the top-down methods. These methods used the following conditions: high energy, high potential, laser emission, and high acidity [25, 26].

The bottom-up methods have some advantages such as convenient methodology, precise control, easy instrumentation, cost-effectiveness, involvement of non-toxic precursor molecules, practical applicability, and green materials. However, the quantum yield is little. The reaction processes in this method required more time and high energies [27, 28].

## 1.4 Trends in green carbon dots

Green materials have an important role in modern chemistry. The carbon dots are synthesized from green sources such as carbohydrates, biomass, and bio-waste. The synthesis of carbon dots from environmentally friendly materials cannot require expensive methods and hazardous chemicals. All syntheses methodologies are green. These properties promote the future materials of carbon dots [7]. The precursors were prepared from natural and synthetic compounds. The modern methods named top-down and bottom-up preparations have less economic and environmental impacts. In these methodologies, high energy, expensive precursors, hazardous organic molecules, and large

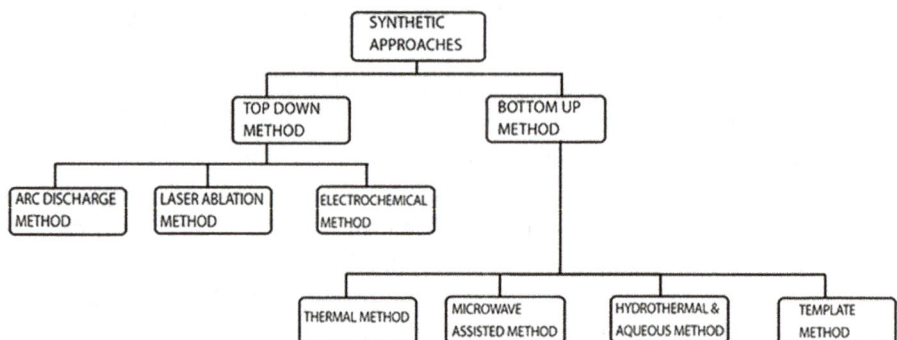

(a)

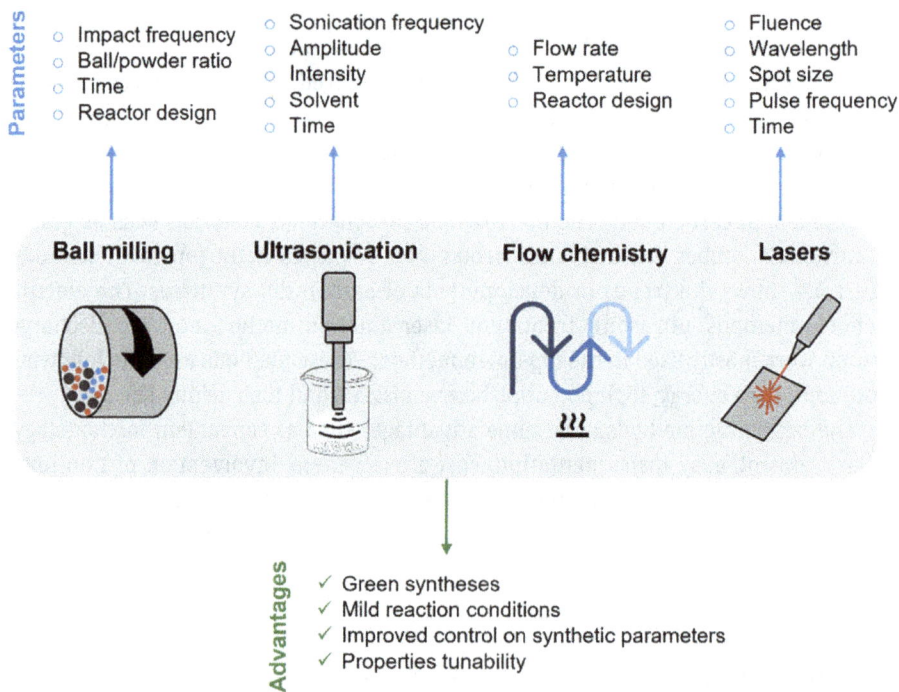

(b)

**Figure 1.4:** Developments in the synthesis of carbon dots: (a) strategies [29] and (b) advantages of modern methods [15].

amounts of toxic solvents were used. Therefore, green sources of carbon dots are required in modern chemistry [30–33].

The waste of biomass is a serious problem in modern times because the production of horticultural products were accumulated in the large amount. The recycling of waste biomass is very important in the modern era. The waste biomass is of low cost and is the raw material for the source of carbon dots. Many research works suggested that the carbon dots were effectively synthesized from the waste biomass [14, 34, 35]. For example, the low-size carbon dots were prepared from the coconut shell biomass. The obtained carbon dots are good blue-emitting carbon quantum dots and water-soluble highly fluorescent material. The carbon dots from the agricultural waste were used as an effective agent in cancer treatment. The various sources of biomass were used for carbon dot syntheses: silkworm cocoon, spent coffee grounds, pseudo-stem of banana plant, cat feedstock waste, tender coconut waste, durian peel, rice husk, waste tea residue, papaya waste, apple seeds, and so on. Pyrolysis, solvent-free carbonization, hydrothermal carbonization, thermal carbonization, oxidative pyrolysis, chemical oxidation, roasting, charring, sand bath-assisted method, ultrasonic wet method, and calcination method were mostly used to synthesize carbon dots from the biomass [7]. For example, Zhao et al. synthesized carbon dots from chitosan, cellulose, lignin, and hemicelluloses by hydrothermal methods at various temperatures (150–200 °C). The synthesis procedure and its main properties were shown in Figure 1.5. It is indicated that the obtained carbon dots have the following good properties: environmental friendliness, excellent biocompatibility, emission wavelength, tunable excitation, and high photostability. The fluorescence performance of the obtained carbon dots was employed in the detection of copper(II) ions at low concentrations. It is confirmed that these carbon dots are more effective agents in the analytical chemistry for metal detection [36].

## 1.5 Carbon dot modification with supramolecular compounds

The carbon dots are modified with supramolecular hosts to obtain unique carbon dots in the biometric elements, catalysts, and sensor applications. The host–guest interactions can promote the fluorescence performance of carbon dots (Figure 1.6). As a result, the modification of carbon dots with the supramolecular host–guest part is an effective agent in detection of biomolecules, organic compounds, anions, and metal cations. In the last 10 years, the following supramolecular host is attached to the carbon dots: the carboxyl esterase, calixarene, resorcinarene macrocycles, crown ethers, pillar[n]arenes, and cucurbit[n]urils. These hosts interacted with the carbon dots through the covalent and non-covalent bonds. The modification of carbon dots with the above host by the hydrothermal–carbonation methods. The structural and intrinsic properties of carbon dots are improved by the supramolecular hosts. These properties are mainly responsible

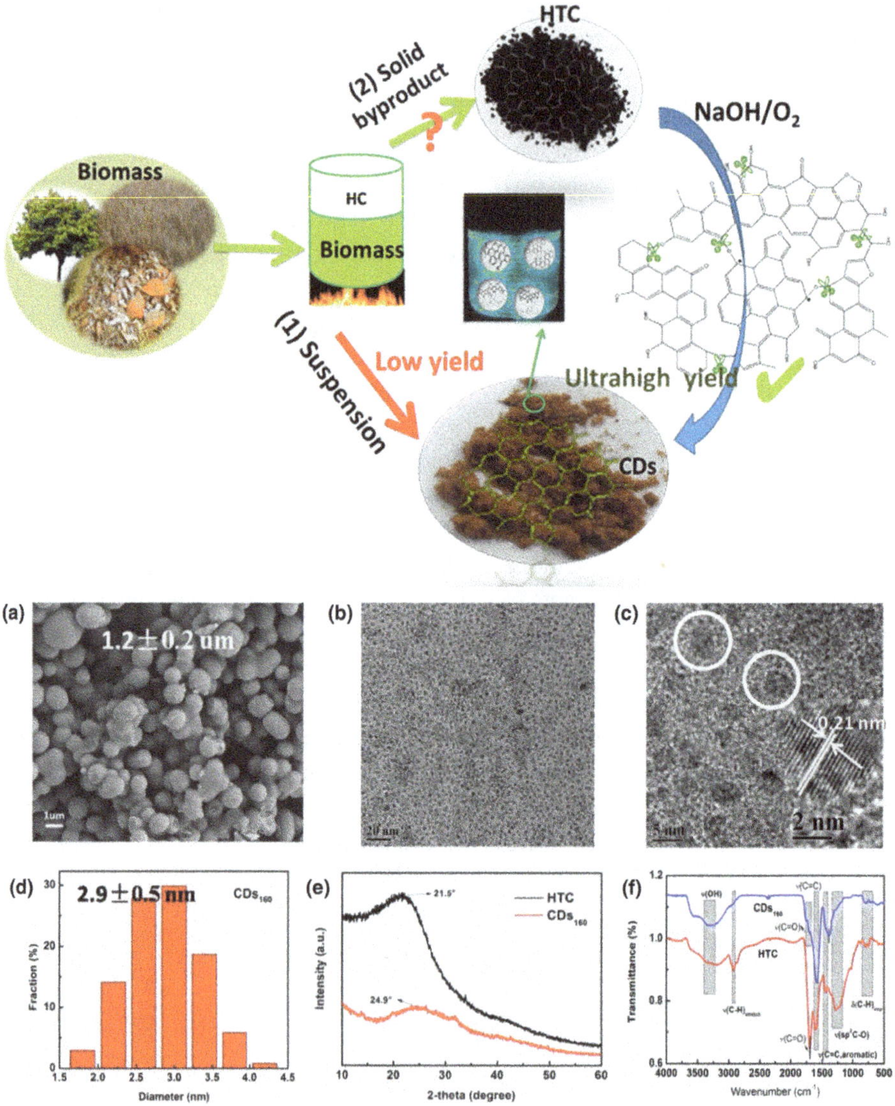

Figure 1.5: Synthesis of carbon dots from the waste biomass [36].

for metal detection in analytical chemistry. The host and guest interacted with each other through hydrogen bonds, electrostatic forces, and variable intermolecular interactions including π–π. The supramolecular host contained the hydrophobic cavity, in which the functional groups were on the surface of carbon dots. The host molecules are attached to the surface of carbon dots; as a result, the surface characteristics are changed to develop special performances. These modifications were widely used in environmental remediation, metal extraction, drug delivery, and battery materials.

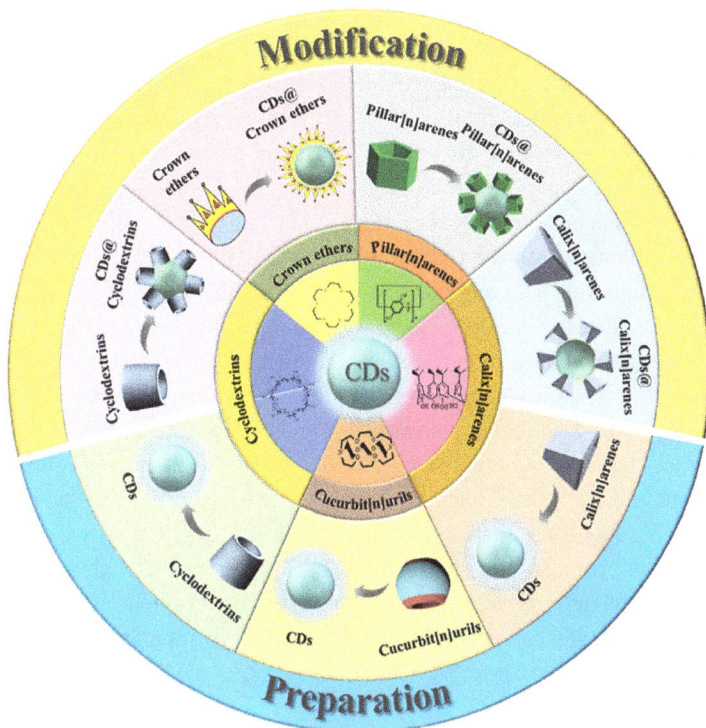

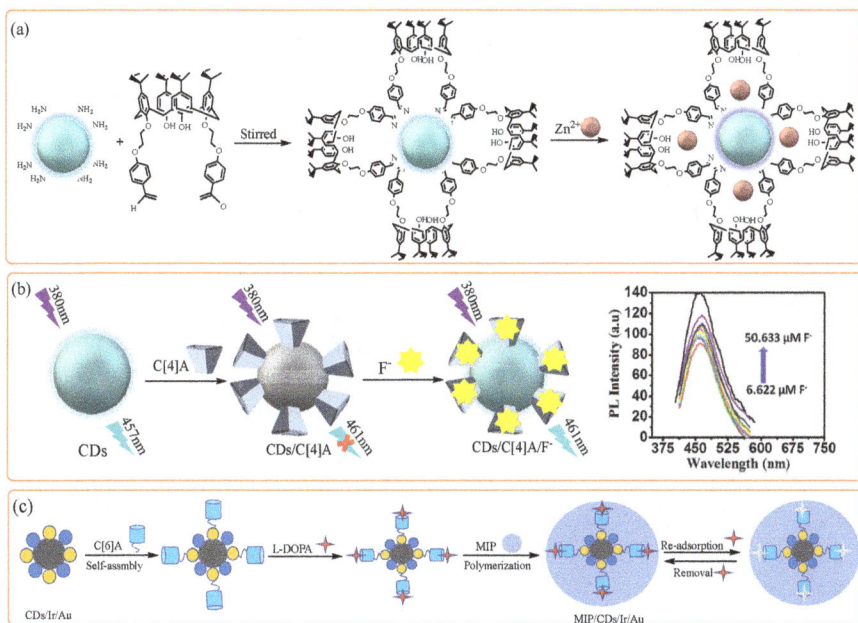

**Figure 1.6:** Carbon dot modifications with the supramolecular hosts and applications [37].

Currently, the supramolecular host with carbon dots was successfully used in the intramolecular charge transfer, internal filtration effect, Forster resonance energy transfer, dynamic quenching, static quenching, and photoelectron transfer. For example, the calix[$n$]arenes contained 3D scaffolds with internal cavities, which interacted with the surface of carbon dots. The amino functional groups are mainly attached to the surface of carbon dots. The calix[$n$]arenes interacted with the amino functional groups, which are attached on the surface (Figure 1.6). The fluorescent properties of carbon dots were enhanced with the calix[$n$]arene supramolecular complex with the carbon dots. The carbon dots have more active regions on the surface of carbon dots. The calix[$n$]arenes are adsorbed on these active regions by the inner cavity. Consequently, the fluorescence performance of carbon dots is enhanced. The free amino functional groups of carbon dots were synthesized from ethylenediamine and glycerol by hydrothermal reactions. The zinc ions are adsorbed into a cavity of carbon dots; as a result, the fluorescence intensity was increased at 418 nm. The $sp^2$ nitrogen atoms and hydroxyl functional groups are mainly responsible for the adsorption of zinc ions' inner cavity of carbon dots. Then, the electron-providing ability of these two groups was blocked, which can promote the rise of fluorescence performance. High fluorescence is very important in the determination of metal ions from the examples at low concentrations. The supramolecular complex with the carbon dots is a highly effective material in the determination of metal ions [37].

## 1.6 Recent trends in carbon dots/2D hybrid materials

The carbon dots/2D hybrid materials are new materials in energy storage, optoelectronic, bioimaging, photocatalysis, and sensing applications. The reason for this is that they are low toxicity, biocompatibility, photostability, photoluminescence, and small size. The carbon dots interact with the nanomaterials through the functional groups attached to the surface of the carbon dots. The carbon dots are modified to follow the 2D materials: layered double oxides (LDOs), layered double hydroxides (LDHs), layered transition metal oxides (LTMOs), transition metal dichalcogenides (TMDCs), graphitic carbon nitride (g-C$_3$N$_4$), and graphene-based materials. These 2D materials have anisotropic physicochemical properties, show large surface-to-volume ratio, and are small in size. These properties make the carbon dots more effective in various applications [38].

The g-C$_3$N$_4$ effectively interacted with the carbon polymerized dots, carbon nanodots, graphite quantum dots, and carbon quantum dots to enhance the unique properties in energy storage, photocatalysis, catalysis, optoelectronics, and sensing. The nature of functional groups on the surface of carbon dots is a key factor in the electrochemical sensing of carbon dots. For example, the carboxyl functional groups on the surface of carbon dots support copper detection (Figure 1.7). The metal ions effectively

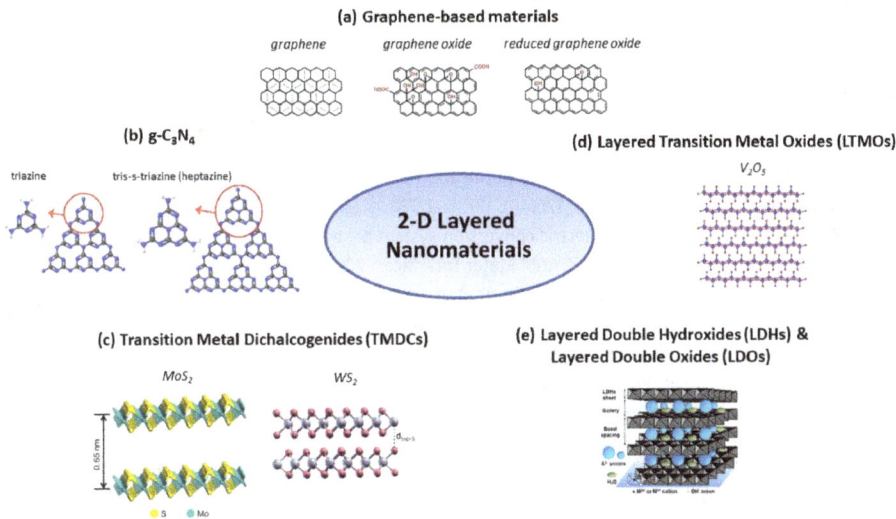

**Figure 1.7:** Various carbon dots/2D hybrid materials [38].

interacted with the metal surface to form covalent bonds between the carboxyl functional groups and metal d-orbitals [38].

# 1.7 Conclusions

In this chapter, the recent trends and developments in carbon dots were discussed and reviewed. At recent times, the carbon dots are synthesized by the top-down and bottom-up methods. The electrochemical methods, ultrasonic treatment, laser ablation method, and arc discharge method were mostly used in the top-down methods. These methods used the following conditions: high energy, high potential, laser emission, and high acidity. The bottom-up methods have some advantages such as convenient methodology, precise control, easy instrumentation, cost-effectiveness, involvement of non-toxic precursor molecules, practical applicability, and green materials. The carbon dots are synthesized from green sources such as carbohydrates, biomass, and bio-waste. The synthesis of carbon dots from environmentally friendly materials cannot require expensive methods and hazardous chemicals. The carbon dots are modified with the supramolecular hosts to obtain unique carbon dots in the biometric elements, catalysts, and sensor applications. The host–guest interactions can promote the fluorescence performance of carbon dots. The carbon dots are modified to follow the 2D materials: LDOs, LDHs, LTMOs, TMDCs, g-$C_3N_4$, and graphene-based materials. These 2D materials have anisotropic physicochemical properties, show large surface-to-volume, ratio, and are small in size. These properties make carbon dots more effective in various applications

# References

[1]    Shahshahanipour M. et al. An ancient plant for the synthesis of a novel carbon dot and its applications as an antibacterial agent and probe for sensing of an anti-cancer drug. Mater Sci Eng C. 2019, 98, 826–833.
[2]    Xu X. et al. Electrophoretic analysis and purification of fluorescent single-walled carbon nanotube fragments. J Am Chem Soc. 2004, 126(40), 12736–12737.
[3]    Ma Z. et al. Bioinspired photoelectric conversion system based on carbon-quantum-dot-doped dye–semiconductor complex. ACS Appl Mater Interfaces. 2013, 5(11), 5080–5084.
[4]    Ridha AA, et al. Carbon dots; the smallest photoresponsive structure of carbon in advanced drug targeting. J Drug Delivery Sci Technol. 2020, 55, 101408.
[5]    Wang D. et al. Facile and scalable preparation of fluorescent carbon dots for multifunctional applications. Engineering. 2017, 3(3), 402–408.
[6]    Zhao C, et al. Hydrothermal synthesis of nitrogen-doped carbon quantum dots as fluorescent probes for the detection of dopamine. J Fluoresc. 2018, 28(1), 269–276.
[7]    Kurian M, Paul A. Recent trends in the use of green sources for carbon dot synthesis–A short review. Carbon Trends. 2021, 3, 100032.
[8]    Chakraborty D, Sarkar S, Das PK. Blood dots: Hemoglobin-derived carbon dots as hydrogen peroxide sensors and pro-drug activators. ACS Sustain Chem Eng. 2018, 6(4), 4661–4670.
[9]    Mansuriya BD, Altintas Z. Carbon dots: Classification, properties, synthesis, characterization, and applications in health care – an updated review (2018–2021). Nanomaterials. 2021, 11(10), 2525.
[10]   Xia C. et al. Evolution and synthesis of carbon dots: From carbon dots to carbonized polymer dots. Adv Sci. 2019, 6(23), 1901316.
[11]   De Medeiros TV. et al. Microwave-assisted synthesis of carbon dots and their applications. J Mater Chem C. 2019, 7(24), 7175–7195.
[12]   Khayal A. et al. Advances in the methods for the synthesis of carbon dots and their emerging applications. Polymers. 2021, 13(18), 3190.
[13]   Chan KK, Yap SHK, Yong K-T. Biogreen synthesis of carbon dots for biotechnology and nanomedicine applications. Nano-Micro Lett. 2018, 10(4), 1–46.
[14]   Chahal S. et al. Green synthesis of carbon dots and their applications. RSC Adv. 2021, 11(41), 25354–25363.
[15]   Bartolomei B, Dosso J, Prato M. New trends in nonconventional carbon dot synthesis. Trends Chem. 2021, 3(11), 943–953.
[16]   Kumar R, Kumar VB, Gedanken A. Sonochemical synthesis of carbon dots, mechanism, effect of parameters, and catalytic, energy, biomedical and tissue engineering applications. Ultrason Sonochem. 2020, 64, 105009.
[17]   Ng HKM, Lim GK, Leo CP. Comparison between hydrothermal and microwave-assisted synthesis of carbon dots from biowaste and chemical for heavy metal detection: A review. Microchem J. 2021, 165, 106116.
[18]   Bag P, et al. Recent development in synthesis of carbon dots from natural resources and their applications in biomedicine and multi-sensing platform. Chem Select. 2021, 6(11), 2774–2789.
[19]   Zhang J, Yu S-H. Carbon dots: Large-scale synthesis, sensing and bioimaging. Mater Today. 2016, 19(7), 382–393.
[20]   Liu ML, et al. Carbon dots: Synthesis, formation mechanism, fluorescence origin and sensing applications. Green Chem. 2019, 21(3), 449–471.
[21]   Ji C. et al. Recent developments of carbon dots in biosensing: A review. ACS Sensors. 2020, 5(9), 2724–2741.
[22]   Sakdaronnarong C. et al. Recent developments in synthesis and photocatalytic applications of carbon dots. Catalysts. 2020, 10(3), 320.

[23]  Wang DM, Lin KL, Huang CZ. Carbon dots-involved chemiluminescence: Recent advances and developments. Luminescence. 2019, 34(1), 4–22.
[24]  Saleh-Mohammadnia M. et al. Fluorescent cellulosic composites based on carbon dots: Recent advances, developments, and applications. Carbohydr Polym. 2022, 294, 119768.
[25]  Yusuf VF. et al. Recent developments on carbon dots-based green analytical methods: New opportunities in fluorescence assay of pesticides, drugs and biomolecules. New J Chem. 2022, 46.
[26]  Miao S. et al. Hetero-atom-doped carbon dots: Doping strategies, properties and applications. Nano Today. 2020, 33, 100879.
[27]  El-Shabasy RM. et al. Recent developments in carbon quantum dots: Properties, fabrication techniques, and bio-applications. Processes. 2021, 9(2), 388.
[28]  Wu ZL, Liu ZX, Yuan YH. Carbon dots: Materials, synthesis, properties and approaches to long-wavelength and multicolor emission. J Mat Chem B. 2017, 5(21), 3794–3809.
[29]  Singh I. et al. Carbon quantum dots: Synthesis, characterization and biomedical applications. Turk J Pharm Sci. 2018, 15(2), 219.
[30]  Tejwan N, Saha SK, Das J. Multifaceted applications of green carbon dots synthesized from renewable sources. Adv Coll Interf Sci. 2020, 275, 102046.
[31]  Radnia F, Mohajeri N, Zarghami N. New insight into the engineering of green carbon dots: Possible applications in emerging cancer theranostics. Talanta. 2020, 209, 120547.
[32]  Shahraki HS, Ahmad A, Bushra R. Green carbon dots with multifaceted applications–Waste to wealth strategy. FlatChem. 2022, 31, 100310.
[33]  Ma P. et al. Application progress of green carbon dots in analysis and detection. Part Part Syst Charact. 2022, 39(9), 2200104.
[34]  Lin X. et al. Carbon dots based on natural resources: Synthesis and applications in sensors. Microchem J. 2021, 160, 105604.
[35]  Meng W. et al. Biomass-derived carbon dots and their applications. Energy Environ Mater. 2019, 2(3), 172–192.
[36]  Zhao Y. et al. Synthesizing green carbon dots with exceptionally high yield from biomass hydrothermal carbon. Cellulose. 2020, 27(1), 415–428.
[37]  Yan F. et al. Carbon dots modified/prepared by supramolecular host molecules and their potential applications: A review. Anal Chim Acta. 2022, 1232, 340475.
[38]  Falara PP, Zourou A, Kordatos KV. Recent advances in carbon dots/2-D hybrid materials. Carbon. 2022, 195, 219–245.

Omar Dagdag\*, Rajesh Haldhar, Seong-Cheol Kim,
Elyor Berdimurodov\*, Ekemini D. Akpan and Eno E. Ebenso\*

# Chapter 2
# Main properties and characteristics of carbon dots

**Abstract:** Carbon dots (CDs) are new composites in nanomaterials. CDs are new types of carbon allotropes such as carbon nanotubes, activated carbon, graphite, and many other carbon nanotubes. Scientists have recently created new CDs and are exploring their applications, including energy technology, optics, and biomedicine. This chapter describes the different types and characteristics of CDs, including specificity, scalability, and biocompatibility.

**Keywords:** Carbon dots, functional nanomaterials, materials, optical, dispersibility, biocompatibility properties

## 2.1 Introduction

Carbonated materials are used in many fields such as technology, chemistry, biomedicine, and other interdisciplinary fields [1]. There are many types of carbon compounds in nature, and carbon compounds play an important tool in the innovation of carbon-based advanced materials. From conventional three-dimensional (3D) graphite [2] to develop carbon-based nanomaterials such as fullerenes [3], one-dimensional carbon nanotubes (CNTs) [4], and two-dimensional graphene [5], the simple exploration of new carbon materials and their use is also an important topic in the fields of chemistry, materials, and physics. In recent years, there has been a search for graphene and CNTs with optical, electrical, and biocompatible properties.

---

\*Corresponding authors: Omar Dagdag, Centre for Materials Science, College of Science, Engineering and Technology, University of South Africa, Johannesburg 1710, South Africa, e-mail: dagdao@unisa.ac.za
\*Corresponding authors: Eno E. Ebenso, Centre for Materials Science, College of Science, Engineering and Technology, University of South Africa, Johannesburg 1710, South Africa, e-mail: ebensee@unisa.ac.za
\*Corresponding authors: Elyor Berdimurodov, Faculty of Chemistry, National University of Uzbekistan, Tashkent 100034, Uzbekistan, e-mail: elyor170690@gmail.com
Ekemini D. Akpan, Centre for Materials Science, College of Science, Engineering and Technology, University of South Africa, Johannesburg 1710, South Africa
Rajesh Haldhar, Seong-Cheol Kim, School of Chemical Engineering, Yeungnam University, Gyeongsan 38541, Republic of Korea

https://doi.org/10.1515/9783110799958-002

Recently, much attention has been paid to new types of 3D models in luminous carbon materials, such as carbon dots (CD), nanoparticles with non-spherical morphology, and nanometric energy (<10 nm).

Compared to quantum dots [2, 6], CDs are multifunctional, have low toxicity, have high fluorescence, and are cost-effective, making them suitable for many applications such as biomedicine and optoelectronics. In addition to this, during the synthesis of new nanomaterials, CDs can be converted into solar cells due to their injection/fission rate and stored as additional electrons (electron input/electron capture) [7, 8]. CD training has significantly improved (Figure 2.1).

CDs have complex, crystalline microstructures of various sizes, so an important relationship between CD structure and content needs to be explored. Many important aspects of CD have been described, including fluorescence, UV absorption, scattering, and biocompatibility. The following sections discuss about CD along with its contents.

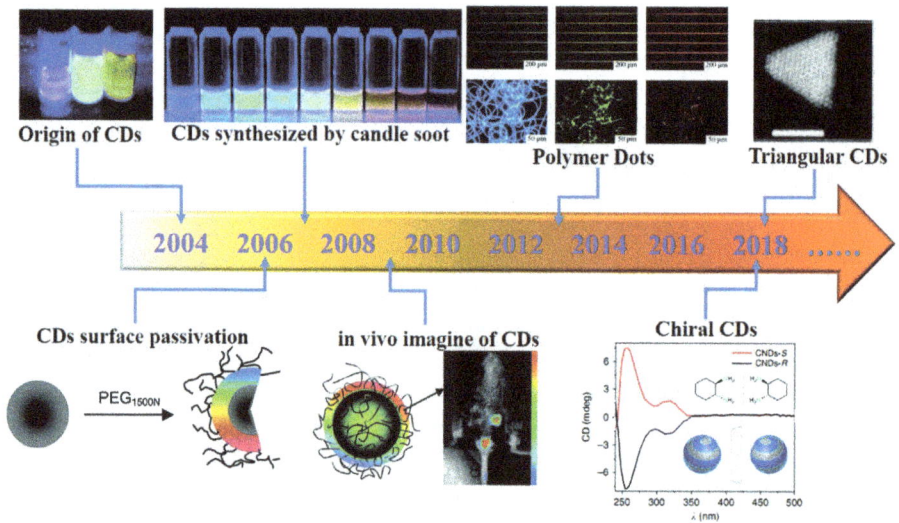

**Figure 2.1:** Milestones in the development of CDs [2].

# 2.2 Main properties and characteristics of carbon dots

## 2.2.1 UV–visible

Tang et al. [9] prepared microwave-generated graphene quantum dots (GQDs). This GQD was detected by deep UV radiation and double wavelength UV radiation at 282 and 228 nm, respectively. Additionally, the growth in solubility and microwave effect of

GQD can be seen by comparing UV absorbance. For example, if the reaction increases, the absorption edge turns red, but the maximum position does not change. The origin of the top is due to the cross of the edge, which contains oxygen GQD. The $\pi$–$\pi$* interactions of the carbon–carbon bonds and n–$\pi$* interactions of carbon–oxygen bonds are mainly responsible for the maximum UV absorbance at 228 and 282 nm, respectively. In the research work of Li et al. [10], green preparation methods were performed to obtain the photoluminescent CDs. Then, the formed CDs reacted with NaOH to form the solid CDs1. Later, CDs1 reacted with the HCl acidic solution to form CDs2. It is indicated that the maximum absorption depth of the UV–visible surface of CDs1 is 540 nm. CDs2 is also a light source similar to CDs1, although it has a longer wavelength. It can be seen that the absorption of UV radiation affects the working surface of the CD surface.

## 2.2.2 Fluorescence

The effect of other compounds on CD fluorescence has been studied in several recent studies. Dean et al. [11] hydrothermally synthesized CDs with 35% quantum yield (QY), which can be regulated by photoluminescence (PL) with locally available O atoms. The atomic material of O plays an important role in regulating the coupling bandwidth between HOMO (highest occupied molecular orbital) and LUMO (lowest unoccupied molecular orbital) [12]. Experimental results showed that when the oxygen concentration level increased in the system, the bandwidth gradually decreased, and a red-shift of the emission peak appeared at 625 nm (Figure 2.2a). Bao et al. [13] also confirmed the discharge mechanism of the surface condition made from carbon fibre concentrates oxidized by $HNO_3$. The size, shape, and molecular properties of CD depend on the synthesis methodology: temperature, time, and component concentrations. In the synthesis methodology, the following main factors were observed: (1) the amount of oxidation on the surface rose with the growth of synthesis time, which is responsible for the increase in the wavelength of the radiation; (2) higher acid concentrations and shorter reaction times lead to longer discharge wavelengths. Han et al. [14] studied the fluorescence properties of CDs, effects to these properties, and impacts of surface-state energy were studied. Self-healing and disease are manifested by three types of polymer CD (PCD). It is clear from the obtained X-ray photoelectron spectroscopy (XPS) results that the nitrogen and carbon atoms were accounted for 13.83% and 20.11%, respectively. The energy distinction between the HOMO and LUMO regions decreased when the degree of surface oxidation rises. As a result, the amorphous regions were decreased. The rise in the C = N content is responsible for the rise in red-shift of PL bands. The energetic position of the PCD changed with the rise in the C = N content, indicating that a large number of electrical changes were found, and a red change was found in the PL band (Figure 2.2b). Zhu et al. [15] suggested that CD fluorescence has a correlation effect between C = O and C = N. Three types of CDs were prepared and installed with three different solvents ($H_2O$, $C_2H_5OH$, and DMF) according to the hydrothermal application shown in Figure 2.2d. The obtained results were

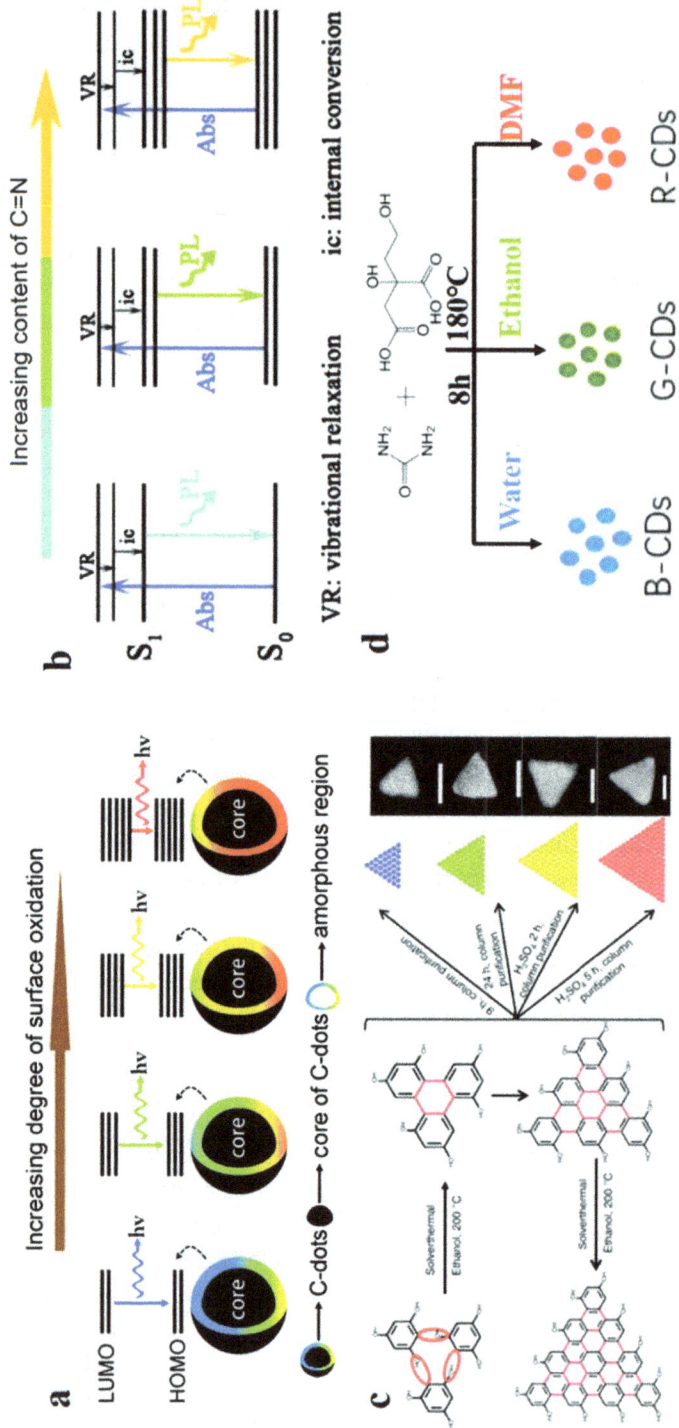

**Figure 2.2:** Procedures in changes of fluorescence properties of CDs: solvent, size, functional groups, and surface effects [2].

confirmed by fluorescence and XPS spectral data. The size of CDs is the next significant factor to identify stability by limiting the change in the fluorescence wavelength. For example, Fan et al. [16] have specially installed CDs with multi-coloured narrow band-width emissions. Transmission electron microscopic images show triangular CDs and QY CDs of about 54–72%. Symmetric phloroglucinol was selected for this research work. The active functional groups of amino and hydroxyl are mainly responsible for the formation of CDs. The blue, green, yellow, and red colours appeared on 1.9, 2.4, 3.0, and 3.9 nm, respectively (Figure 2.2c). The distance between the bonds decreases with increasing size, leading to constant friction [17].

## 2.2.3 Phosphorescence

In recent years, researchers developed CDs with higher RTP (room-temperature phosphorescent) content and longer shelf life [2]. Deng and his colleagues reported another RTP phenomenon by dispersing PVA (polyvinyl alcohol). CD was prepared by pyrolysis of ethylenediaminetetraacetic acid disodium salt at 400 °C under $N_2$. It is indicated that when CDs were coated with a PVA matrix, fluorescence was observed under the influence of UV light at room temperature. When comparing the sample images with CD-PVA and CD with $H_2O$, a peak at 220 nm and bandwidth between 300 nm were observed due to the change in the n–π* binding and π–π* transitions of C=O bonds, respectively. In the phosphorus band, strong binding at 260–340 nm was observed for C = O and another band at 500 nm, suggesting that phosphorescence may be induced by carbon dioxide, which binds to C = O. The obtained results suggested that the phosphorescence of CD is caused by the three phases of the gaseous carbonyl component and that the PVA molecule activates these products by adding hydrogen. The energy of vibration or rotation was importantly depleted by the presence of hydrogen [18].

Furthermore, Li et al. [19] suggested that the C = N interaction affects the RTP performance on CD. The C = N bonds of this system were mainly responsible for phosphorescence effects. It constructed the CD RTP complex by treating urea and nitrogen-doped CD (NCD) with the heat method. Because the doubling of the recrystallized urea and the hydrogen bonding between the burette and the NCD reduces the triple trigger distribution, the CD system has a longer lifetime than the single device in Figure 2.3a. Figure 2.3b and d illustrates the optical properties and impacts of hydrogen presence in H-NCD compounds. H-NCDs were formed by hydrogen bonding. The optical properties of H-NCDs were changed by the protonation processes. In Figure 2.3d, the green colour of H-NCDs was altered to blue by protonation. As observed from the obtained data that H-NCD had a longer lifetime and higher RTP utilization than the three CD types, it is confirmed that the third group was involved in the hydrogen bonding between H-NCD and the biuret.

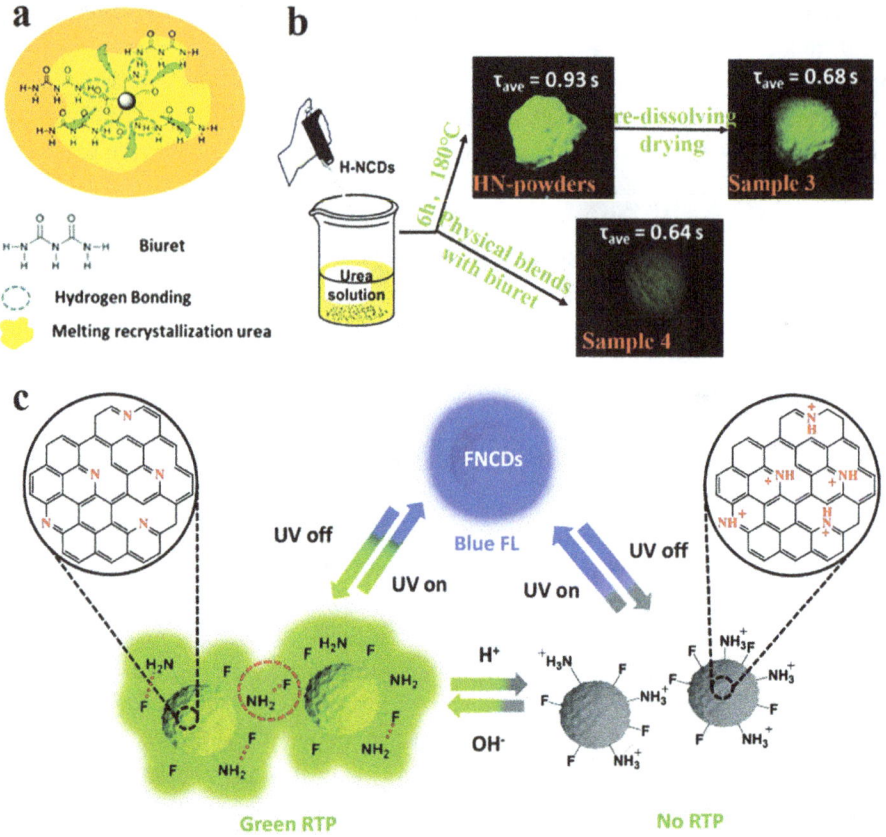

**Figure 2.3:** Pictorial representation of RTP and impacts [2].

## 2.2.4 Dispersibility and biocompatibility

Most CDs have good hydrophilicity and dispersion, so they form O-products in the first step of the process. Additionally, changes in the hydrophilicity and hydrophobicity of CD can be achieved by adjusting the surface coefficients. Hsu et al. [20] reported that hydrothermal deposition of hydrophilic CD has been described in boilers. It has been suggested that these groups play an important role in determining the hydrophilicity of CD derivatives. In addition, Mitra et al. [21] produced hydrophobic CDs obtained by microwave pyrolysis of co-polymer Pluronic F-68.

## 2.3 Conclusions

CDs are a new generation of carbon-based nanomaterials. Currently, CDs are mostly used in pharmacology, medicine, nanotechnologies, surface, drug determination, wastewater treatments, and removal processes from heavy metals. Many CDs have recently been tested for composition, material, and base. In this section, we introduce some of the important aspects of CDs discussed earlier, such as their biocompatibility, dispersibility, and optical properties.

## Abbreviations

**RTP**     room-temperature phosphorescent
**PCDs**    Polymer CDs
**PVA**     Polyvinyl alcohol

## References

[1]   Zhang B-T, Zheng X, Li H-F, Lin J-M. Application of carbon-based nanomaterials in sample preparation: A review. Anal Chim Acta. 2013, 784, 1–17.
[2]   Li Z, Wang L, Li Y, Feng Y, Feng W. Frontiers in carbon dots: Design, properties and applications. Mater Chem Front. 2019, 3, 2571–2601.
[3]   Thompson BC, Fréchet JM. Polymer–fullerene composite solar cells. Angew Chem Int Ed. 2008, 47, 58–77.
[4]   DRESSELHAUS, M. S. D. G.; Eklund, PC; Rao, AM, The Physics of Fullerene-Based and Fullerene-Related Material. Carbon Nanotubes, 2000, p. 331–379.
[5]   Feng W, Long P, Feng Y, Li Y. Two-dimensional fluorinated graphene: Synthesis, structures, properties and applications. Adv Sci. 2016, 3, 1500413.
[6]   Derfus AM, Chan WC, Bhatia SN. Probing the cytotoxicity of semiconductor quantum dots. Nano Lett. 2004, 4, 11–18.
[7]   Kaur M, Kaur M, Sharma VK. Nitrogen-doped graphene and graphene quantum dots: A review on synthesis and applications in energy, sensors and environment. Adv Coll Interf Sci. 2018, 259, 44–64.
[8]   Essner JB, Baker GA. The emerging roles of carbon dots in solar photovoltaics: A critical review. Environ Sci Nano. 2017, 4, 1216–1263.
[9]   Tang L, Ji R, Cao X, Lin J, Jiang H, Li X, et al. Deep ultraviolet photoluminescence of water-soluble self-passivated graphene quantum dots. ACS Nano. 2012, 6, 5102–5110.
[10]  Li D, Jing P, Sun L, An Y, Shan X, Lu X, et al. Near-infrared excitation/emission and multiphoton-induced fluorescence of carbon dots. Adv Mater. 2018, 30, 1705913.
[11]  Ding H, Yu S-B, Wei J-S, Xiong H-M. Full-color light-emitting carbon dots with a surface-state-controlled luminescence mechanism. ACS Nano. 2016, 10, 484–491.
[12]  Hu S, Trinchi A, Atkin P, Cole I. Tunable photoluminescence across the entire visible spectrum from carbon dots excited by white light. Angew Chem Int Ed. 2015, 54, 2970–2974.
[13]  Bao L, Liu C, Zhang ZL, Pang DW. Photoluminescence-tunable carbon nanodots: Surface-state energy-gap tuning. Adv Mater. 2015, 27, 1663–1667.

[14]  Han L, Liu SG, Dong JX, Liang JY, Li LJ, Li NB, et al. Facile synthesis of multicolor photoluminescent polymer carbon dots with surface-state energy gap-controlled emission. J Mater Chem C. 2017, 5, 10785–10793.

[15]  Zhu J, Bai X, Bai J, Pan G, Zhu Y, Zhai Y, *et al*. Emitting color tunable carbon dots by adjusting solvent towards light-emitting devices. Nanotechnology. 2018, 29, 085705.

[16]  Yuan F, Yuan T, Sui L, Wang Z, Xi Z, Li Y, et al. Engineering triangular carbon quantum dots with unprecedented narrow bandwidth emission for multicolored LEDs. Nat Commun. 2018, 9, 1–11.

[17]  Qu S, Zhou D, Li D, Ji W, Jing P, Han D, *et al*. Toward efficient orange emissive carbon nanodots through conjugated sp2-domain controlling and surface charges engineering. Adv Mater. 2016, 28, 3516–3521.

[18]  Deng Y, Zhao D, Chen X, Wang F, Song H, Shen D. Long lifetime pure organic phosphorescence based on water soluble carbon dots. Chem Comm. 2013, 49, 5751–5753.

[19]  Li Q, Zhou M, Yang Q, Wu Q, Shi J, Gong A, *et al*. Efficient room-temperature phosphorescence from nitrogen-doped carbon dots in composite matrices. Chem Mater. 2016, 28, 8221–8227.

[20]  Hsu P-C, Chang H-T. Synthesis of high-quality carbon nanodots from hydrophilic compounds: Role of functional groups. Chem Comm. 2012, 48, 3984–3986.

[21]  Mitra S, Chandra S, Kundu T, Banerjee R, Pramanik P, Goswami A. Rapid microwave synthesis of fluorescent hydrophobic carbon dots. RSC Adv. 2012, 2, 12129–12131.

Vinayak Sahu

# Chapter 3
# Synthetic strategies of carbon dots

**Abstract:** Since its discovery in 2004, carbon dots (CDs), an affluential in the field of carbon nanomaterials, have gotten a lot of attention. Many ways to synthesizing CDs have been explored to date, and many efforts have been made to leverage simple, cost-effective, and sizable-scale synthesis passage. Simultaneously, CD photoluminescence (PL) mechanisms have been extensively studied, with the primary causes being molecule state, surface state, carbon core state, and their synergistic effect. CDs have been widely used in catalysis, biomedicine, optoelectronic devices, lubrication, sensing, and other fields due to their advantageous properties such as brilliant optical, tremendous biocompatibility, good catalytic activity, tiny size, less virulent, and ecological. This chapter provides a comprehensive overview of the development of CD synthesis strategies. The potential and challenges for CD research are examined in depth based on the synopsis. The goal of this study is to motivate associated researchers to overcome different technical hurdles, fill in the gaps in conventional research, and fully use the prospects of CDs.

## 3.1 Introduction

Conventional quantum dots (QDs), which include heavy metals like Cd and Pb, have excellent ocular properties and vast applicability potential; nevertheless they lack biocompatibility and environmental friendliness due to their known high lethality and harmful for environment and severely restrict their applicability, particularly applications in the field of biology [1–3]. As a result, developing a new form of environmentally friendly QDs is critical. Luckily, carbon dots (CDs), a new class of zero-dimensional fluorescent carbon nanomaterial, were introduced by Xu et al. [4] in 2004 while purifying single-walled carbon nanotubes (CNTs). CDs, in contrast to semiconductor QDs, have reduced lethality, greater biocompatibility, and superior photostability, indicating that they have a strong prospect to be a safe and less-virulent alternative. CDs are nanoparticles of carbon with a diameter below 10 nm and having fluorescent properties [5, 6]. A single CD structure contains a core of carbon and heteroatoms like N, S, O, and functional groups like $-OH$, $-COOH$, and $-NH_2$ [7]. CDs with varied morphologies have non-uniform names due to their complex structure

**Vinayak Sahu,** National Institute of Technology Raipur, Chhattisgarh 492010, India;
Governmental Model College Raipur, Chhattisgarh 492099, India, e-mail: vinayaksahu2208@gmail.com

https://doi.org/10.1515/9783110799958-003

and countless variants. Yang's group [6] classified CDs into three categories based on their cores of carbon and chemical states of the surface:carbon nanodots (CNDs), graphene quantum dots (GQDs), and polymer dots (PDs). Anisotropic GQDs have quantum confinement and single/multiple films of graphene, crystalline also has lateral dimensions, whereas CNDs have a spherical morphology, resulting heights equal to lateral dimensions roughly, and CNDs have two subcategories into amorphous carbon nanoparticles (CNPs) and cross-linked CNDs with PDs with and polymeric creatures, quantum confinement effect not shown by polymerizing small molecule/polymer precursors. The distinction between carbon sources of synthetic and synthetic techniques is largely responsible for the diversity of CDs. CNTs, carbon nano-onions (CNOs), nano-diamond, graphene, fullerene, and their derivatives have many advantages: (1) Optical qualities that are favourable PL is rarely visible in pure CNOs, nano-diamond, CNTs, graphene, and fullerene due to the Dirac cone's zero optical bandgap. CDs, on the other hand, have controllable emission of PL and great resistant to photobleaching as well as high, near 100% PL quantum yields (PLQYs) [8] and (2) synthesis is simple and cost-effective. Various synthetic preparation methods for the above carbon nanomaterials have been developed, such as for CNTs, graphene, and nano-diamond chemical vapour deposition, for CNO high-temperature annealing, and for fullerene arc discharge, numerous shortcomings such as expensive, poor yields, tedious processes, and a large amount of raw material requirements must still to overcome. The vast majority of carbonaceous materials, including glucose, graphene oxide (GO), citric acid, CNTs, hair, coal, leaves, graphite, grass, durian, and even garbage of household, have been used as CD precursors [9–19]. In addition, a wide range of synthetic techniques, including hydrothermal synthesis, chemical oxidation, microwave treatment method, pyrolysis method, and others, have been used to create CDs. Consequently, getting CDs at inexpensive prices is simpler; (3) excellent stability and solubility. The aforementioned CDs have poor solubility and stability in aqueous or other solvents due to their restricted surface groups and chemical inertness. Although it is frequently expensive and ineffective, surface modification is meant to increase their solubility. Due to the large number of surface groups in CDs, they exhibit good long-term solubility and stability in aqueous solvent. Importantly, the polarity of CDs can be easily altered by utilizing the right precursors, making them amenable to dispersion in a wider range of solvents. Additionally, CDs have appealing catalytic activity due to their distinct electronic structures and energies. As a result, CDs are receiving more and more attention. Many attempts are being made right now to take advantage of simple, economical, and extensive preparation. Chen and co-workers [23] have recently developed a hyperthermia magnetic technique to quickly and easily produce multi-colour luminous CDs on a large scale. The specifics mechanism of the PL of CDs is still up for discussion despite the fact that the synthesis of CDs has advanced significantly. Carbon core state, surface state, molecule state, and their synergistic effect are currently the main perspectives for the PL genesis of CDs. In addition, their benefits include favourable optical properties, attractive catalytic performance, chemical inertness, tiny size, ecological, as well as ease of synthesis [20–31].

Numerous outstanding reviews on PL mechanisms and synthetic methods or the possible applicabilities of CDs have been recently described. For instance, Tan's group [32] both offered information on the development of CDs in light-emitting diodes (LEDs) and nanomedicine from the year 2020. In a different review, Lu's group described the mechanisms of PL of CDs, which included crosslink-enhanced emission, internal factors dominated emission, and external factors dominated emission. Additionally, Yang's group [33] and Ding's group [34] outlined developments in chiral CDs and solid-state CDs, respectively, by the year 2020. However, CDs continue to make amazing strides forward every day. More significantly, a thorough study that methodically presents synthesis procedures, PL roots, and prospective CD applications is currently lacking. As it is known, prospective CD applications, mechanisms of PL, and synthetic techniques interact and support one another rather than existing alone. For instance, the ability to prepare multi-colour-emitting CDs at controlled temperatures encourages the application in bioimaging and optoelectronic devices. In addition, the creation of CDs with clearly defined structures using the right precursors and reaction circumstances makes it easier to understand how their PL mechanism works. Meanwhile, the discovery of the mechanism of PL will direct the synthesis of CDs with controlled PL, for biomedicine applications need a preparation method to obtain CDs with stable PL, good PLQYs, and low lethality. Additionally, a deeper comprehension of the origin of PL of CDs should make it easier to build applications (such as sensing and bioimaging) that depend on PL alterations.

In order to carefully explain the synthetic processes, the source of fluorescence, and the prospective uses of CDs, a thorough examination is absolutely necessary. Due to this, this study focuses on thoroughly documenting the development of common preparation procedures, the most widely recognized theories for PL mechanisms, and prospective CD applications. The many procedures used to create CDs, such as electrochemical, pyrolysis reaction, chemical oxidation, microwave treatment, laser ablation, hydrothermal synthesis, and other methods, are described in depth.

## 3.2 Synthetic strategies

The synthesis processes for CDs are of two types: bottom-up and top-down route. Both routes are briefly described in Figure 3.1.

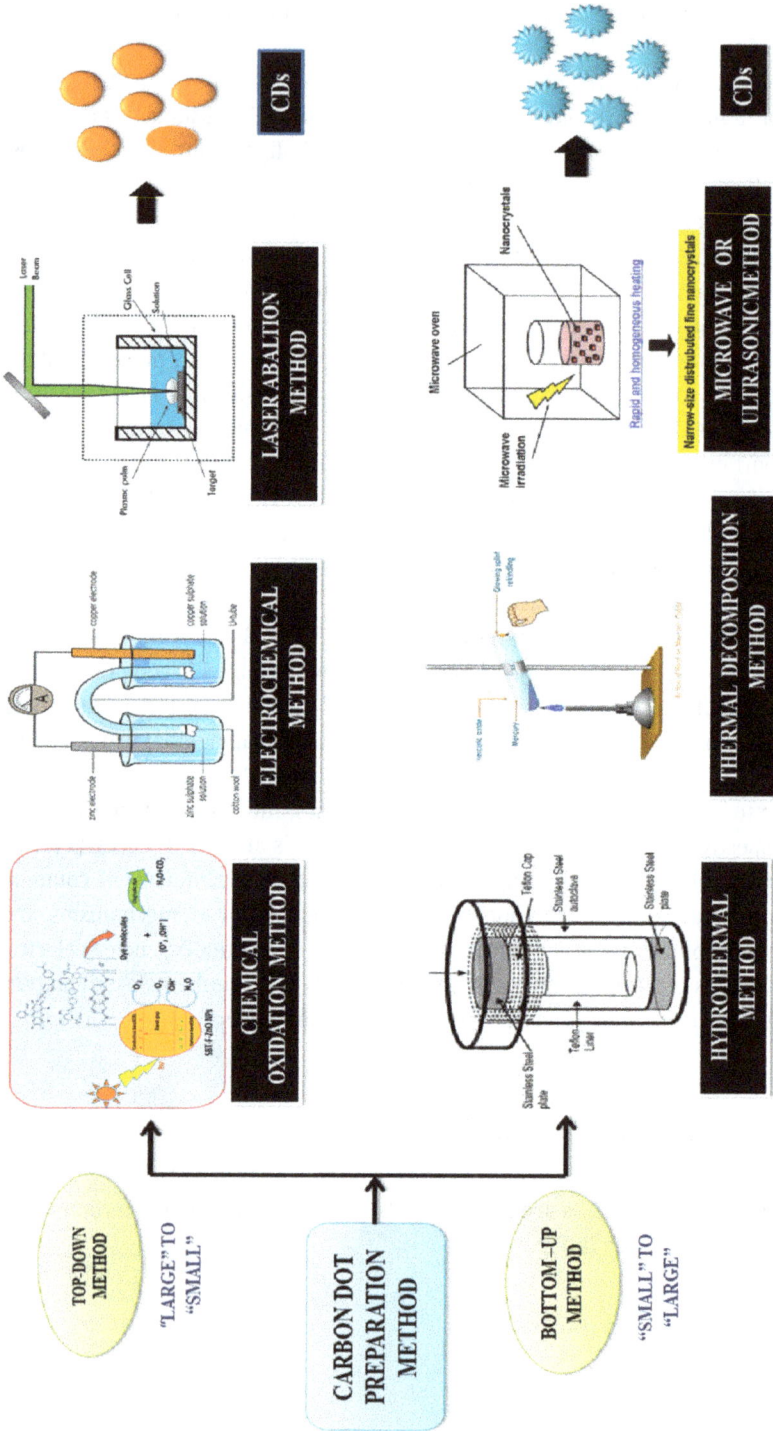

**Figure 3.1:** Synthetic strategies of CDs.

## 3.2.1 Chemical oxidation

Chemical oxidation involves oxidizing the reactants to produce CDs using strong acidic solutions like conc. $HNO_3$ + conc. $H_2SO_4$ or oxidants like $H_2O_2$ and $FeCl_3$. This method was used to the modification of CDs with the various functional groups; as a result, the CDs are become water soluble and effective for the bio-imaging. Xu et al. [4] in 2004 separated the suspension using gel electrophoresis after oxidizing arc-discharge soot with strong $HNO_3$. Finally, CDs that emit blue, orange, yellow, and green light were unintentionally acquired [4]. Even though the CDs highest PLQY was just 1.6% this groundbreaking breakthrough created a fascinating new area of research in carbon nanomaterials. Later, Mao and colleagues [28] used candle soot instead of arc-discharge soot to synthesize CDs while concentrated $HNO_3$ was refluxed for 12 h. The produced CDs have 3% PLQY and have good water solubility, and an average diameter of particles 2–6 nm may be used as a cell-imaging probe, then, by refluxing in conc. $HNO_3$, carbon soot [35], and carbon black [36] also served as carbon sources to synthesize CDs, but all of the resulting CDs had a low PLQY of less than 5%. Sun's group [37] published a two-step approach in 2010 to create CDs with a high PLQY of close to 60% utilizing PEG1500N as a functionalizing agent, based on the observation that surface modification was one of the most successful strategies to upgrade PLQYs. First, $HNO_3$ was used to reflux carbon soot, which served as the carbon source, to create the precursor CNPs. In order to create PEG-passivated CDs with good PLQY, the starting material was first mixed with $SOCl_2$ and then subjected to a reaction with PEG1500N [37], which held tremendous promise for bioimaging. Similar to this, Li and colleagues [38] and Huo and colleagues [39] merged surface passivation and chemical oxidation to create CDs with high PLQY using GO and activated carbon, conc. $HNO_3$ as oxidizing agent, and amino group containing compounds as surface functionalizing agents. This method is cheap and easy but produces CDs with poor yield and non-tuneable PL. Then, group of Zhu [40] and group of Zhang [41] used carbon fibres (CFs) as carbon sources to synthesize CDs by oxidative cutting. CDs were an efficient fluorescent sensor for the selective detection of $Cu^{2+}$ ions. Based on the optical properties, it is possible that a significant number of tiny $sp^2$ carbon by chemical oxidation, crystallites present in coal have been easily terminated and surface-passivated, which is the method by which coal-based CDs were formed [42–44]. The production method of CDs created using chemical oxidation utilizing powder of carbon as precursor carbon source was also attributed by Leblanc's group to the breaking off of particles of graphite buried in the huge carbon skeleton, which resulted in the creation of tiny particles with oxidized surfaces. Strong acids were utilized in all of the aforementioned literatures to create CDs, restricting the usage of chemical oxidation. $H_2O_2$ was used to effectively peel off GQDs from black carbon/GO or without side products in order to reduce the environmental risks associated with strong acids. However, this method needed auxiliary catalysts or high temperatures [45–47]. More importantly, bulk carbon compounds that were used as precursors had poor yields and low PLQYs. These limitations were anticipated to be alleviated by the tiny compounds used as precursors. Chen's team still used $HNO_3$ as the oxidant while synthesizing PDs which have 55.7% PLQY using o-phenylenediamine (o-PD) as the source of carbon [48]. Yu's team [43] was able to overcome this weakness. Low-temperature method using one pot to create red-emissive PDs by oxidative polymerizing p-phenylenediamine (p-PD) at 80 °C using $FeCl_3$ as the mild oxidant [49] and p-PD oligomer clusters, tiny dots with distinct boundaries, huge uniform dots, and monodispersed dots were all involved in the creation of CDs. In addition to avoiding the need of strong acid, this innovation made it simple to make long-wavelength-emissive CDs. One of the first methods for the synthesis of CDs, the chemical oxidation approach, allows for the quick and efficient creation of multi-colour-emitting CDs from soot, coal, GO, CFs, and so on. Centrifugation, neutralization, and dialysis are some of the time-consuming steps and difficult post-treatments that are often required by the procedure. Moreover, to get CDs with high PLQYs, passivation of the surface is a crucial phase. Furthermore, using acid oxidizing agents causes more environmental problems and falls short of the desire for green preparation. Even while the use of simple nonacid oxidizing agent and small precursor's molecule can reduce environmental risks and produce CDs good PLQY, it remains difficult to produce CDs with particular groups and intended emissions. As a result, chemical oxidation cannot be broadly applied.

## 3.2.2 Laser-abscission method

The typical laser-abscission method uses a laser beam of high energy to attack a target of carbon, which results in the exfoliation of nanoparticles of carbon that are then functionalized to create CDs. It is important to note that the laser-abscission method can directly create CDs without further passivation by carefully choosing the suitable carbon target or its medium. Sun's group [48] was the first to describe using a wavelength of 1,064 nm, Nd:YAG laser with a combination of cement and graphite powder as the carbon target to make CDs with 10% PLQY of in 2006 that emits various colours providing a novel technique for making CDs and demonstrating the viability of CDs for bioimaging, but with evident flaws such preparation at high temperatures and low pressures (900 °C and 75 kPa) and the requirement for organic species to passivate the CDs. Du's [51] team created single-step laser-abscission preparation of CDs by improving experimental settings, considerably streamlining the operations in order to skip the challenging preparation phases. In a nutshell, graphite powders dispersed in PEG200N were exposed to a Nd:YAG laser for 2 h while being helped by ultrasound. This resulted in combined preparation and passivation of CDs of 3.2 nm sizes and PLQY 5%. The pulsed laser's creation of a high-pressure, high-temperature in PEG200N solvent, which caused graphite particles to heat up to plasma state, may be the cause of CD generation under laser irradiation; pure CDs were created by the subsequent condensation of carbon plasma; their surfaces were immediately oxidized by O atoms that had been broken down; and subsequent reactions with PEG200N or its fragment molecules produced by the plasma ultimately led to the creation of CDs with carboxylate groups [49]. These discoveries clarified the intricate phases of the laser ablation method, but the issues of poor yields, low PLQYs, and non-tuneable morphology and PL of CDs persisted. Yang's [41] team created luminous CDs by laser-irradiating graphite flakes dispersed in PEG1500N solution. They then adjusted the laser pulse width to affect nucleation and growth, successfully creating CDs with average diameters of 3, 8, and 13 nm. This work was inspired by the findings shown earlier [50]. Similar to this, Li and colleagues [41, 42] confirmed that the input of laser fluence could readily adjust the sizes and PL emissions of CDs using a long-pulse-width laser conducive to research the link between optical performance and sizes of CDs. Instead of using bulk carbon materials, they used toluene as a precursor for the carbon, eliminating the need for passivators like PEG. By using a real-time detection system to track the PL changes of the solution, the formation process of CDs was revealed, showing that, first, large pieces of graphene were formed as an intermediate. As the irradiation continued, the graphene increased and superimposed to form large multi-layer grapheme. Finally, the huge multi-layer graphene was laser ablated to produce CDs [51]. Kang's group [52] also claimed that GQDs' surface flaws were more likely to become functionalized with shorter laser wavelengths, and that multiple GQD surface states could be created by adjusting the laser's wavelength. The discovery of the link between the laser's wavelength and the functionalization of CDs gave rise to a fresh idea for modifying their shape and PL [52].

The discovery of the connection between the laser's wavelength and the functionalization of CDs gave rise to a fresh idea for modifying their shape and PL. One of the early approaches to creating CDs was the laser ablation process, which has the advantage of being easy to use.

### 3.2.3 Electrochemical synthesis

Good conductivity carbon materials are typically employed as working electrode and sources of carbon in electrochemical synthesis. Oxidation reaction occurs in the anode once a certain voltage is applied, which involves the separation of CDs from carbon precursors. Additionally, the electrolyte may implement bottom-up electrochemical synthesis of CDs by serving as a carbon source [53]. Sham and colleagues [56] used the electrochemical method for the first time in 2007 to directly make emitting blue light CDs that have 6.4% PLQY.

For the preparation of water-soluble and luminescent CDs, three electrode systems were used: MWCNTs as carbon source covered carbon paper as the working electrode and as a counter electrode Pt-wire used, as a reference electrode Ag/AgClO$_4$, and acetonitrile solution-containing tetrabutylammonium perchlorate (TBAP) as the electrolyte. According to one theory, CDs were exfoliated into the electrolyte solution when TBA cations broke their structures near the flaws during electrochemical cycling after intercalating into the gap of MWCNTs [54]. This research introduced a novel strategy for the direct synthesis of water-soluble CDs, although yields, PLQYs, and fluorescence tuneability of CDs needed to be improved. After that, CDs with a diameter of 2.0 nm on an average were prepared using graphite rod as the working electrode and phosphate buffer solution as the electrolyte of a three electrode system, respectively [55]. A two-electrode system using ammonia solution as the electrolyte and the prepared disc electrodes as the working electrode and counter electrode was used to fabricate CDs in another instance, reducing the cost of production. In this case, petroleum coke was used to create disc electrodes and this was a cost-effective solution. In addition to these, ionic liquids (ILs) were employed as the electrolyte to complete the electrochemical synthesis of CDs, and graphene sheet and screen-printed carbon electrodes were used as electrodes. The hydroxylation or oxidation of graphite and subsequent release of CDs from the anode might be the results of the production of oxygen and hydroxyl radicals by anodic oxidation of water [56]. Cell imaging and photovoltaics can use the produced CDs because of their high degree of crystallinity, superior water solubility, and limited size distribution. Despite the literatures mentioned above, the issues of poor yields, low PLQYs, monotonous CD fluorescence, and essential dialysis operation persisted. Li et al. [57] prepared the GQDs were obtained by electrochemical oxidation method with NaOH aqueous solution, saline solution of $PO_4^{3-}$ buffer and aqueous solution of KCl as electrolyte. Alternatively, Pang and colleagues [58] showed a controlled electrochemical approach for the size-selective synthesis of monodisperse luminous CDs, although it was unable to significantly alter the PL emissions of CDs.

Fortunately, Kang's team [59] was able to adjust the alkali-assisted electrochemical system's current intensity and produce a series of CQDs whose PL is size-dependent that changed from brown to blue as diameters declined from 3.8 to 1.2 nm. Additionally, utilizing acids as the electrolyte resulted in no CQD production, demonstrating that the alkaline environment was the key element. In another instance, this group created various PL emissions and sizes of CQDs with photo increased catalytic activity by varying the parallel spacing of graphite rods, utilizing two graphite rods as both anode and cathode and ultrapure water as electrolyte [60]. However, the two aforementioned literatures produced little CQDs. Liu and colleagues [41, 44, 47] originally presented a simple bottom-up production approach of CDs, finishing collection of solid-state CDs after becoming aware of the difficulties and needs for CDs synthesis. The creation mechanism of CDs was hypothesized based on structural properties. Nitriles were polymerized and carbonized to create CDs; at the same time, functional groups with positive charge produced by ILs passivated the surface of CDs. Importantly, the produced CDs with a PLQY of more than 10% and exceptional solubility in water may be used for cell imaging and ion detection. Although the PLQYs and CD yields were not much increased by this study, it did offer a fresh notion for improving the electrochemical process.

The electrochemical method can produce CDs with water-solubility (mostly GQDs) with high crystallinity and uniform size that are suitable for catalytic applications [59, 60] and photovoltaic uses [55], and it is also environmentally benign. Importantly, by modifying the current density or potentials, CD diameters and PL hues may be altered. However, this technology has clear drawbacks, like poor PLQY, low yields, and a time-taking process, making it challenging to obtain CDs on large scale for a low cost.

## 3.2.4 Hydrothermal method

The main steps in the hydrothermal/solvothermal method are as follows: precursors are dissolved in water or other solvents, and then they are put in a tightly sealed container that is where they go through carbonization and polymerization or exfoliated to form CDs under relatively high pressure and temperature.

By slicing oxidized graphene sheets (GSs) into GQDs with a 5% PLQY, Wu's group [61] first revealed a hydrothermal method. The elimination of bridged oxygen atoms from the epoxy lines caused by certain ultrafine GS particles that broke off during the hydrothermal process might be the genesis mechanism for GQDs [61]. These results offered a new, simple method for producing GQDs, although poor PLQYs, non-uniform GQD sizes, and pre-treatment of source materials needed to be addressed. Zhu et al. prepared amino-functionalized GQDs using ammonia solution as a solvent by the solvothermal method, giving GQDs, up to 29% of PLQY and a narrow size distribution. By adjusting the temperature of the reaction, PL colours of GQDs were tailored from blue to yellow, allowing GQDs to be used in multi-colour LEDs and bioimaging [62]. However, this study continued to employ expensively oxidized GSs as the carbon source. In order to get out of the situation, sustainable and replenishable leaves [22, 63] were used as carbon sources to create CDs, greatly lowering the prices. Specifically, poplar leaves were used as a carbon source to obtain CDs with a 10.64% of PLQY and a high throughput of 1.4975 kg by hydrothermal processing, providing an environmentally friendly method for mass-producing CDs at a low cost. Notably, CDs were advantageous for electrocatalytic water splitting, sensing, and bioimaging due to their strong photostability and minimal cytotoxicity. Additionally, taxus leaves were used to create deep red emissive PDs that shown significant promise in the fields of optoelectronics and biomedicine because to their deep red PL, good PLQY of 31%, and narrow full width at half maximum [63]. Using a one-step solvothermal process, waste-expanded polystyrene was also transformed into high-value CDs in addition to leaves [64]. However, the difficult postprocessing to eliminate unreacted by-products was essential for the top-down hydrothermal/solvothermal approach, greatly increasing preparation time and expense. Dai's team [19, 39, 67] created a productive hydrothermal approach to synthesis GQDs that only uses starch as a precursor in order to overcome the drawback. In a nutshell, the starch dispersed in deionized water was heated at 463 K for 2 h to obtain the products only containing GQDs, water, and carbide precipitate. After centrifugation, the purified GQDs were collected [65], discarding the time-consuming post-treatment and providing a straightforward and affordable hydrothermal method to produce CDs. Small molecules as precursors have been shown to be more adaptable in altering the structural and optical characteristics of CDs during hydrothermal/solvothermal processes than bulk materials. The tiny chemical L-ascorbic acid was initially used to create monodispersed blue-emitting CDs with a uniform size but a low PLQY of 6.79% by heating it to 180 °C for 4 h [66]. Fortunately, it was discovered that N-doped CDs obtained from chitosan and chitin had higher PLQYs and larger sizes than N-free CDs synthesized from glucose under the same hydrothermal synthetic conditions, indicating that molecular precursors had a clear impact on the PLQYs and morphology of CDs. Yang's team [67] demonstrated that designing CDs with the usage of heteroatoms could change their physicochemical properties very effectively and especially enhance PLQYs using citric acid and ethylenediamine (EDA) as the respective carbon and nitrogen sources. To create N-doped CDs, anilines and their derivatives were often used as nitrogen sources [28, 68–73]. Using oPD, $m$-phenylenediamine (mPD), and pPD, respectively, with dithio salicylic acid in acetic acid via solvothermal reaction, red-, green-, and blue-emissive CDs with aggregation-induced PL and an optimal PLQY of 20.77% were created. These CDs were then applied to white LEDs (WLEDs) and fingerprint detection [69]. Additionally, CDs with adjustable PL from blue to near-infrared [68, 70, 71, 73], even CDs with dual-emission [28], have been created by modifying amine species, adding amide auxiliary [68], carbon sources [28, 71, 72], additional nitrogen sources [73], or adjusting amine species. The effective synthesis of triple-emission [72] for bioimaging and WLEDs shows how widely applicable the hydrothermal method is. In addition to aniline, other nitrogen sources used to create N-doped CDs included urea [74] and EDA [75, 76].

## 3.2.5 Pyrolysis route

The steps that make up the common pyrolysis method include carbonizing carbon sources at high temperatures. CDs are gathered following purification and after treatment process such chromatographic separation, dialysis, centrifugation, and filtration. However, the separating process can be stopped by encapsulating CDs with a suitable template to restrain their development and shape during pyrolysis [77]. In general, it is separated into liquid-phase and solid-phase pyrolysis depending on the various states of carbon sources. Liu's group [105, 110] developed a pyrolysis technique to create CDs which is soluble in oil using anhydrous citric acid and surface functionalization agent 1-hexadecylamine and octadecane as solvent 300, inspired by the synthesis of semi-conductor and magnetic nanocrystals. The resultant CDs had various optical and structural features depending on the duration of reaction or capping agent used to control the depolymerization, breakdown, and pyrolysis process [78]. This study demonstrated that the optical characteristics of CDs may be changed by only altering the pyrolysis parameters such as solvents, agents, or duration of pyrolysis. It also provided an easy approach to synthesize CDs having good PLQY. The development of very blue luminous CDs with a PLQY of 31.6–40.6% was later accomplished by Wu's group [79] using a one-step pyrolysis in solid phase. In this procedure, the disodium salt of EDTA was just igniting at 400 °C for 2 h without the use of any solvent or agent [79] facilitating the pyrolysis route. Similar to this, GQDs could also be made by citric acid through pyrolysis, and after increasing the period of pyrolysis, the result was GO because of a greater extent of carbonization [80].

The CDs produced by pyrolysis had simple manipulation and desirable PLQYs, but their application potential in catalysis was limited by their non-uniform size and propensity to aggregate [81]. Although it was time-consuming and expensive, separation by chromatographic could be employed to produce CDs with uniform size. Zaitsev [79] et al. coupled the pyrolysis process and templating method. To be more precise, CDs made by pyrolyzing citric acid were created using silica gel-containing immobilized aminopropyl groups as a template to control their form. The sizes and shapes of CDs produced by pyrolysis were also controlled using metal-organic framework materials as templates [81–83]. However, collecting CDs from templates required alkaline solution etching of the support, which was bad for the environment. Some publications have created N-CDs having excellent PLQYs by sensibly choosing surface passivation agents or precursors in order to enhance the structural and PL characteristics of CDs produced by pyrolysis [84–87]. For instance, GQDs with N-doping with 59.2% PLQY were prepared by one-pot pyrolysis using tris(hydroxymethyl)aminomethane as a doping material and surface functionalizing agent and for carbon source citric acid was utilized [88]. These GQDs also showed a strong detectability of 2,4,6-trinitrophenol.

Additionally, citric acid and EDA were pyrolyzed to enhance the N-CDs with PLQY of 80% [86], preparation of CDs desirable for biological imaging process. In a different instance, N-CDs were produced by pyrolyzing ammonium citrate and showed strong corrosion inhibitive properties [85]. A high PLQY CD can be made by pyrolysis, which has numerous advantages including easy operation, inexpensive sources of carbon, and preparation on large scale. Additionally, the hydrophobic/hydrophilic characteristics and doping of atom, kinds of CDs can be altered employing various surface functionalizing agents or suitable starting materials. Although the approach is labour-intensive, consumes a lot of energy, and is unable to avoid the tiresome purification and after-treatment it is required to produce high quality CDs of a consistent size. The preparation of long-wavelength CDs by pyrolysis is particularly challenging, severely restricting the use of CDs in biomedicine and optoelectronics.

## 3.2.6 Microwave treatment

The quick production of CDs is made possible by microwave treatment. It just takes a few minutes. Reactants with polar molecules have a tendency to absorb microwave energy during the therapy and then transform it internally. The internal energy provided by the microwave approach, in contrast to the solvothermal/hydrothermal method. In the year 2009, Yang's group [21, 31] developed using PEG200N and glucose by microwave treatment. CDs with high PLQYs and multi-colour emission were not formed. Doping with nitrogen was thought to significantly raise PLQYs and alter PL hues. As a result, N-CDs with emission of red colour and have 15% PLQY were created by microwave irradiating inexpensive urea and citric acid as the nitrogen and carbon, respectively. The PLQY of N-doped CDs could be increased to 40.2% when EDA was used in place of urea as the nitrogen source [89]. In order to create red-emitting N-CDs with a 15% PLQY, Wang et al. [34, 39] microwave-treated pPD in a solution of ethanol and water. Because of the red-colour PL and excellent PLQY, N-CDs can now be used as nanothermometers in living cells [90]. Magnetic-fluoroscent N-CDs, doped with Fe, were also prepared using citric acid and EDA with $FeCl_2$. These CDs, due to their superparamagnetic and fluorescent behaviours, showed promise in multi-modal cellular imaging [91]. The application of the aforementioned literatures in optoelectronics was, however, constrained by their inability to prepare multi-colour PL CDs. Liu et al. [41] created CDs using microwave irradiation and poly(ethylenimine) solution with glutaraldehyde (GA) added to it to provide an answer. They were able to create multi-colour luminous CDs with an adjustable wavelength ranging from 464 to 556 nm [92] by altering the GA and PEI ratio. CDs were prepared by the microwave methods, as a result, the multi-colour CDs were prepared. Additionally, it was demonstrated that the microwave approach contributed to prevent direct pep contacts or transfer of excessive resonance energy in order to produce resistant to self-quenching CDs in solid-state PL in a matter of minutes [89, 93–95]. Numerous studies have revealed that large-sized raw materials can also be used to create CDs using the microwave method in addition to small molecules. For instance, the cleaving generation mechanism of GQDs [61] has provided insight. GO as the source of carbon to create greenish-yellow luminous GQDs having 11.7% PLQY, using the microwave technique in an acid environment [96], but this study needed a 3-h reaction time and employed strong acids as the solvent, restricting its widespread use. In order to address these drawbacks, Wang's group [97] used microwave technology to manufacture B-GQDs having 21.1% PLQY, while avoiding the need of strong acids and cutting the 30 min of reaction time. However, it was still necessary to investigate low-cost sources of carbon. Based on this, Wang's team [98] developed a one-pot microwave process to make N-doped CDs with a high PLQY of 46% utilizing silkworm chrysalis as the natural carbon source, thus lowering the cost of sources of carbon. More recently, Zhang's team [26] produced narrow-dispersed CDs on a kilogramme scale at low cost without the need of time-consuming fractionalization techniques and just utilizing cheap industrial surfactants as carbon sources. The reaction media can be quickly and uniformly heated using a microwave approach, which drastically reduces the reaction time and increases product yields. The microwave way is more practical, cost-effective, and time-saving than previous approaches, showing great promise for widespread industrialization.

Furthermore, it has been demonstrated that a vast array of carbon-containing materials are readily available for use as carbon sources in microwave synthesized CDs. As a result, microwave processing enables the fabrication of multi-colour and high-PLQY CDs by choosing particular precursors. As a result, this method is frequently used to create CDs. microwave-derived CDs, in contrast to hydrothermal/

solvothermal CDs, have wilder size distributions and lower PLQYs, which causes them to perform badly in fields like optoelectronics and biomedicine.

## 3.2.7 Other chemical methods

In addition to the methods mentioned earlier, extensive research has been conducted to examine additional chemical techniques for the quick and effective manufacture of CDs. By employing inexpensive, natural eggs of chicken as the starting material and a quick one-step plasma-induced technique, Chen's group [99] was able to significantly reduce the reaction time. Yan et al. [100] converted benzoic acid in GQDs with blue-emitting light. The two aforementioned literatures introduced novel techniques for quickly synthesizing CDs, but they had one thing in common: they called for expensive specialized equipment. The ultrasonic method is unique in that it benefits from low-tech requirements, cheap equipment, and easy manipulation. A simple ultrasonic approach to create CDs was given by Lee and colleagues [101], Meral and colleagues [102], Liu and colleagues [103], Yan and colleagues [104], Leblanc and colleagues [7, 105]. They used a variety of carbon sources including food waste, graphene, blueberries, citric acid, application of the generated CDs to ion lubrication, detection, and bioimaging, and additive manufacturing. Orange-emission N-doped CDs, for instance, were prepared by ultrasonication technique [105] employing solvent-deionized water, oPD, and citric acid as nitrogen and carbon and sources, and acting as the fluorescent ink for 3D printing by embedding in sodium polyacrylate. More significantly, waste food was transformed into CDs with blue emission having uniform size and great photo stability on a large scale utilizing ultrasound irradiation [101]. However, the long wave length emission or adjustable PL of CDs could not be produced using the ultrasonic technique. There have also been other straightforward methods for making CDs. For instance, Zhang and colleagues [106] and Nandi and colleagues [107] employed the photo-fenton reaction and the sono-fenton reaction, respectively, to massively produce GQDs utilizing GO as a carbon source. These methods were proven to be simple, but they did not successfully address flaws like a lack of available raw materials. This is due to the fact that external irradiations like sonication and ultraviolet light enhanced the Fenton reaction. On the other hand, Liu's team [108] showed how to produce graphite-based GQDs with a size of 8 nm using a magnetron sputtering technique.

Additionally, a number of environmentally friendly and sustainable synthesis methods, such as the Schiff-base reaction [21, 109–111], self-exothermic synthesis [20, 24], co-polymerization reaction base catalysis, and reduction methods [16], have attracted a lot of interest due to their advantages, including energy conservation, ease of use, and lack of specialized equipment. For instance, Huang's group [20] report a large-scale one-pot synthesis of highly PL CDs utilizing Schiff-base condensation by only storing a combination of triethylenetetramine and *p*-benzoquinone at ambient temperature for 20 h. This group developed a method to prepare CDs in bulk scale by mixing hydroquinone with $H_2O_2$ and further adding EDA at room temperature [20]. This method was based on the fact that a self-exothermic reaction made the organic precursors (hydroquinone and EDA) carbonized to form the carbon cores of CDs. However, the sustainable synthesis technique is still in its infancy and requires further work to improve its effectiveness and decrease the use of hazardous materials in the production of CDs on a wide scale. Recently, Chen's group [23] introduced a new rapid magnetic hyperthermia strategy to produce multi-fluorescence CDs using carbamide and citrate with three different cations as precursors in order to achieve green, quick, and large-scale CD production. This method, which preliminarily possessed the capacity for industrial production [23], fully exploited the magnetocaloric effect to quickly and easily synthesize CDs with a yield of 60% and a production rate up to 85 g h$^{-1}$. However, there was still room for further cost reduction and productivity enhancement. In conclusion, a lot of work has been put into thoroughly researching the creation of CDs, and a lot of progress has been accomplished. Numerous methods, including chemical oxidation, laser ablation, electrochemistry, hydrothermal/solvothermal synthesis, pyrolysis reaction, and microwave treatment, have been used to create CDs. Regardless of the technique used, the ability to alter the PL hues, PLQYs, sizes, hydrophilicity/hydrophobicity, and surface groups of CDs makes them extremely promising in a number of fields, including biomedicine, sensing, and optoelectronics. The hydrothermal/solvothermal approach is regarded as the best preparation technique for efficiently fabricating CDs.

More significantly, it can easily prepare specifically made CDs with a high PLQY, multi-colour emissions, and a consistent size, making it widely used. Other chemical techniques, such as gaseous detonation [100], magnetic hyperthermia [23], self-exothermic reaction [24], and the aldol condensation method [27], have also been developed and paved the way for industrial preparation in order to achieve green, quick, one-pot, simple, and large-scale production of CDs. The vast majority of carbon-containing substances, such as glucose, citric acid, GO, CNTs, graphite, coal, leaves, grass, durian, hair, and even household garbage, have manifested as precursors [5, 6, 9–18], laying the groundwork for inexpensive CD preparation. However, there are still unavoidable drawbacks including ill-defined optical qualities and CD sizes, undesirable purity, and intermittent manufacture that must be overcome, substantially limiting the application potential of CDs. Therefore, it is imperative to improve the current processes and continue to research new avenues for the quick, effective, environmentally friendly, scaled-up, and manageable production of CDs. It is common knowledge that improved knowledge of the PL genesis of CDs is anticipated to fundamentally direct and improve their synthesis methods.

# References

[1]     Michalet X, Pinaud FF, Bentolila LA, Tsay JM, Doose S. Quantum dots for live cells, in vivo imaging, and diagnostics. Science. 2005, 307(5709), 538–544.

[2]     Kirchner C, Liedl T, Kudera S, Pellegrino T, Javier AM, Gaub HE, Stolzle S, Fertig N, Parak WJ. Cytotoxicity of colloidal CdSe and CdSe/ZnS nanoparticles. Nano Lett. 2005, 5(2), 331–338.

[3]     Derfus AM, Chan WCW, Bhatia SN. Probing the cytotoxicity of semiconductor quantum dots. Nano Lett. 2004, 4(1), 11–18.

[4]     Xu X, Ray R, Gu Y, Ploehn HJ, Gearheart L, Raker K, Scrivens WA. Electrophoretic analysis and purification of fluorescent single-walled carbon nanotube fragments. J Am Chem Soc. 2004, 126(40), 12736–12737.

[5]     Baker SN, Baker GA. Luminescent carbon nanodots: Emergent nanolights. Angew Chem Int Ed. 2010, 49(38), 6726–6744.

[6]     Zhu S, Song Y, Zhao X, Shao J, Zhang J, Yang B. The photoluminescence mechanism in carbon dots (graphene quantum dots, carbon nanodots, and polymer dots): Current state and future perspective. Nano Res. 2015, 8(2), 355–381.

[7]     Mintz KJ, Bartoli M, Rovere M, Zhou Y, Hettiarachchi SD, Paudyal S, Chen J, Domena JB, Liyanage PY, Sampson R, Khadka D, Pandey RR, Huang S, Chusuei CC, Tagliaferro A, Leblanc RM. A deep investigation into the structure of carbon dots. Carbon. 2021, 173, 433–447.

[8]     Chen X, Zhang X, Xia LY, Wang HY, Chen Z, Wu FG. One-step synthesis of ultrasmall and ultrabright organosilica nanodots with 100% photoluminescence quantum yield: Long-term lysosome imaging in living, fixed, and permeabilized cells. Nano Lett. 2018, 18(2), 1159–1167.

[9]     Hai X, Feng J, Chen X, Wang J. Tuning the optical properties of graphene quantum dots for biosensing and bioimaging. J Mater Chem B. 2018, 6(20), 3219–3234.

[10]    Yan Y, Gong J, Chen J, Zeng Z, Huang W, Pu K, Liu J, Chen P. Recent advances on graphene quantum dots: From chemistry and physics to applications. Adv Mater. 2019, 31(21), 180–186.

[11]    Kang Z, Lee ST. Carbon dots: Advances in nanocarbon applications. Nanoscale. 2019, 11(41), 19214–19224.

[12]    Fernando KA, Sahu S, Liu Y, Lewis WK, Guliants EA, Jafariyan A, Wang P, Bunker CE, Sun YP. Carbon quantum dots and applications in photocatalytic energy conversion. ACS Appl Mat Interf. 2015, 7(16), 8363–8376.

[13]    He C, Shuang E, Yan H, Li X. Structural engineering design of carbon dots for lubrication. Chin Chem Lett. 2021, doi: https://doi.org/10.1016/j.cclet.2021.03.026.

[14]    Hu C, Li M, Qiu J, Sun YP. Design and fabrication of carbon dots for energy conversion and storage. Chem Soc Rev. 2019, 48(8), 2315–2337.

[15]    Li Y, Xu X, Wu Y, Zhuang J, Zhang X, Zhang H, Lei B, Hu C, Liu Y. A review on the effects of carbon dots in plant systems. Mater Chem Front. 2020, 4(2), 437–448.

[16]    Liu ML, Chen BB, Li CM, Huang CZ. Carbon dots: Synthesis, formation mechanism, fluorescence origin and sensing applications. Green Chem. 2019, 21(3), 449–471.

[17]    Wang B, Song H, Qu X, Chang J, Yang B, Lu S. Carbon dots as a new class of nanomedicines: Opportunities and challenges. Coord Chem Rev. 2021, 442, 214010.

[18]    Yao B, Huang H, Liu Y, Kang Z. Carbon dots: A small conundrum. Trends Chem. 2019, 1(2), 235–246.

[19]    Wu Y, Cao M, Zhao Q, Wu X, Guo F, Tang L, Tan X, Wu W, Shi Y, Dai C. Novel high-hydrophilic carbon dots from petroleum coke for boosting injection pressure reduction and enhancing oil recovery. Carbon. 2021, 184, 186–194.

[20]    Chen BB, Liu ZX, Deng WC, Zhan L, Liu ML, Huang CZ. A large-scale synthesis of photoluminescent carbon quantum dots: A self-exothermic reaction driving the formation of the nanocrystalline core at room temperature. Green Chem. 2016, 18(19), 5127–5132.

[21]   Liu ML, Yang L, Li RS, Chen BB, Liu H, Huang CZ. Large-scale simultaneous synthesis of highly photoluminescent green amorphous carbon nanodots and yellow crystalline graphene quantum dots at room temperature. Green Chem. 2017, 19(15), 3611–3617.

[22]   Li W, Liu Y, Wang B, Song H, Liu Z, Lu S, Yang B. Kilogram-scale synthesis of carbon quantum dots for hydrogen evolution, sensing and bioimaging. Chin Chem Lett. 2019, 30(12), 2323–2327.

[23]   Zhu Z, Cheng R, Ling L, Li Q, Chen S. Rapid and large-scale production of multi-fluorescence carbon dots by a magnetic hyperthermia method. Angew Chem Int Ed. 2020, 59(8), 3099–3105.

[24]   Song SY, Sui LZ, Liu KK, Cao Q, Zhao WB, Liang YC, Lv CF, Zang JH, Shang Y, Lou Q, Yang XG, Dong L, Yuan KJ, Shan CX. Selfexothermic reaction driven large-scale synthesis of phosphorescent carbon nanodots. Nano Res. 2021, 14, 2231–2240.

[25]   Du XY, Wang CF, Wu G, Chen S. The rapid and large-scale production of carbon quantum dots and their integration with polymers. Angew Chem Int Ed. 2021, 60(16), 8585–8595.

[26]   Fang L, Wu M, Huang C, Liu Z, Liang J, Zhang H. Industrializable synthesis of narrow-dispersed carbon dots achieved by microwave-assisted selective carbonization of surfactants and their applications as fluorescent nano-additives. J Mater Chem A. 2020, 8(40), 21317–21326.

[27]   Li L, Li Y, Ye Y, Guo R, Wang A, Zou G, Hou H, Ji X. Kilogram-scale synthesis and functionalization of carbon dots for superior electrochemical potassium storage. ACS Nano. 2021, 15(4), 6872–6885.

[28]   Shuang E, Mao QX, Wang JH, Chen XW. Carbon dots with tunable dual emissions: From the mechanism to the specific imaging of endoplasmic reticulum polarity. Nanoscale. 2020, 12(12), 6852–6860.

[29]   Zhu P, Tan K, Chen Q, Xiong J, Gao L. Origins of efficient multiemission luminescence in carbon dots. Chem Mater. 2019, 31(13), 4732–4742.

[30]   Zhi B, Yao X, Wu M, Mensch A, Cui Y, Deng J, Duchimaza-Heredia JJ, Trerayapiwat T, Niehaus KJ, Nishimoto Y, Frank BP, Zhang Y, Lewis RE, Kappel EA, Hamers RJ, Fairbrother HD, Orr G, Murphy CJ, Cui Q, Haynes CL. Multicolor polymeric carbon dots: Synthesis, separation and polyamide-supported molecular fluorescence. Chem Sci. 2021, 12(7), 2441–2455.

[31]   Song Y, Zhu S, Zhang S, Fu Y, Wang L, Zhao X, Yang B. Investigation from chemical structure to photoluminescent mechanism: A type of carbon dots from the pyrolysis of citric acid and an amine. J Mater Chem. 2015, 3(23), 5976–5984.

[32]   Zhao B, Tan Z. Fluorescent carbon dots: Fantastic electroluminescent materials for light-emitting diodes. Adv Sci. 2021, 8(7), 2001977.

[33]   Ru Y, Ai L, Jia T, Liu X, Lu S, Tang Z, Yang B. Recent advances in chiral carbonized polymer dots: From synthesis and properties to applications. Nano Today. 2020, 34, 100953.

[34]   Xu A, Wang G, Li Y, Dong H, Yang S, He P, Ding G. Carbon-based quantum dots with solid-state photoluminescent: Mechanism, implementation, and application. Small. 2020, 16(48), 2004621.

[35]   Tian L, Ghosh D, Chen W, Pradhan S, Chang X, Chen S. Nanosized carbon particles from natural gas soot. Chem Mater. 2009, 21(13), 2803–2809.

[36]   Dong Y, Chen C, Zheng X, Gao L, Cui Z, Yang H, Guo C, Chi Y, Li CM. Onestep and high yield simultaneous preparation of single- and multi-layer graphene quantum dots from cx-72 carbon black. J Mater Chem. 2012, 22(18), 8764–8766.

[37]   Wang X, Cao L, Yang ST, Lu F, Meziani MJ, Tian L, Sun KW, Bloodgood MA, Sun YP. Bandgap-like strong fluorescence in functionalized carbon nanoparticles. Angew Chem Int Ed. 2010, 49(31), 5310–5314.

[38]   Shen J, Zhu Y, Chen C, Yang X, Li C. Facile preparation and upconversion luminescence of graphene quantum dots. Chem Commun. 2011, 47(9), 2580–2582.

[39]   Qiao ZA, Wang Y, Gao Y, Li H, Dai T, Liu Y, Huo Q. Commercially activated carbon as the source for producing multicolor photoluminescent carbon dots by chemical oxidation. Chem Commun. 2010, 46(46), 8812–8814.

[40]   Peng J, Gao W, Gupta BK, Liu Z, Romero-Aburto R, Ge L, Song L, Alemany LB, Zhan X, Gao G, Vithayathil SA, Kaipparettu BA, Marti AA, Hayashi T, Zhu JJ, Ajayan PM. Graphene quantum dots derived from carbon fibers. Nano Lett. 2012, 12(2), 844–849.

[41] Liu C, Xiao G, Yang M, Zou B, Zhang ZL, Pang DW. Mechanofluorochromic carbon nanodots: Controllable pressure-triggered blue- and red-shifted photoluminescence. Angew Chem Int Ed. 2018, 57(7), 1893–1897.

[42] Ye R, Xiang C, Lin J, Peng Z, Huang K, Yan Z, Cook NP, Samuel EC, Hwang C, Ruan G, Ceriotti G, Raji AR, Marti AA, Tour JM. Coal as an abundant source of graphene quantum dots. Nat Commun. 2013, 4, 2943.

[43] Hu C, Yu C, Li M, Wang X, Yang J, Zhao Z, Eychmuller A, Sun YP, Qiu J. Chemically tailoring coal to fluorescent carbon dots with tuned size and their capacity for cu(ii) detection. Small. 2014, 10(23), 4926–4933.

[44] Jia J, Sun Y, Zhang Y, Liu Q, Cao J, Huang G, Xing B, Zhang C, Zhang L, Cao Y. Facile and efficient fabrication of bandgap tunable carbon quantum dots derived from anthracite and their photoluminescence properties. Front Chem. 2020, 8, 123–129.

[45] Jiang F, Chen D, Li R, Wang Y, Zhang G, Li S, Zheng J, Huang N, Gu Y, Wang C, Shu C. Eco-friendly synthesis of size-controllable amine-functionalized graphene quantum dots with antimycoplasma properties. Nanoscale. 2013, 5(3), 1137–1142.

[46] Zhu C, Yang S, Wang G, Mo R, He P, Sun J, Di Z, Kang Z, Yuan N, Ding J, Ding G, Xie X. A new mild, clean and highly efficient method for the preparation of graphene quantum dots without by-products. J Mater Chem. 2015, B 3(34), 6871e6876.

[47] Lu Q, Wu C, Liu D, Wang H, Su W, Li H, Zhang Y, Yao S. A facile and simple method for synthesis of graphene oxide quantum dots from black carbon. Green Chem. 2017, 19(4), 900–904.

[48] Liu MX, Ding N, Chen S, Yu YL, Wang JH. One-step synthesis of carbon nanoparticles capable of long-term tracking lipid droplet for real-time monitoring of lipid catabolism and pharmacodynamic evaluation of lipidlowering drugs. Anal Chem. 2021, 93(12), 5284e5290.

[49] Xia J, Chen S, Zou GY, Yu YL, Wang JH. Synthesis of highly stable redemissive carbon polymer dots by modulated polymerization: From the mechanism to application in intracellular ph imaging. Nanoscale. 2018, 10(47), 22484–22492.

[50] Sun YP, Zhou B, Lin Y, Wang W, Fernando KAS, Pathak P, Meziani MJ, Harruff BA, Wang X, Wang H, Luo PG, Yang H, Kose ME, Chen B, Veca LM,, Xie SY. Quantum-sized carbon dots for bright and colorful photoluminescence. J Am Chem Soc. 2006, 128(24), 7756–7757.

[51] Hu SL, Niu KY, Sun J, Yang J, Zhao NQ, Du XW. One-step synthesis of fluorescent carbon nanoparticles by laser irradiation. J Mater Chem. 2009, 19(4), 484–488.

[52] Hu S, Liu J, Yang J, Wang Y, Cao S. Laser synthesis and size tailor of carbon quantum dots. J Nanopart Res. 2011, 13(12), 7247–7252.

[53] Yu H, Li X, Zeng X, Lu Y. Preparation of carbon dots by non-focusing pulsed laser irradiation in toluene. Chem Commun. 2016, 52(4), 819–822.

[54] Kang S, Ryu JH, Lee B, Jung KH, Shim KB, Han H, Kim KM. Laser wavelength modulated pulsed laser ablation for selective and efficient production of graphene quantum dots. RSC Adv. 2019, 9(24), 13658–13663.

[55] Niu F, Xu Y, Liu M, Sun J, Guo P, Liu J. Bottom-up electrochemical preparation of solid-state carbon nanodots directly from nitriles/ionic liquids using carbon-free electrodes and the applications in specific ferric ion detection and cell imaging. Nanoscale. 2016, 8(10), 5470–5477.

[56] Zhou J, Booker C, Li R, Zhou X, Sham TK, Sun X, Ding Z. An electrochemical avenue to blue luminescent nanocrystals from multiwalled carbon nanotubes (mwcnts). J Am Chem Soc. 2007, 129(4), 744–745.

[57] Zheng L, Chi Y, Dong Y, Lin J, Wang B. Electrochemiluminescence of watersoluble carbon nanocrystals released electrochemically from graphite. J Am Chem Soc. 2009, 131(13), 4564–4565.

[58] Li Y, Hu Y, Zhao Y, Shi G, Deng L, Hou Y, Qu L. An electrochemical avenue to green-luminescent graphene quantum dots as potential electronacceptors for photovoltaics. Adv Mater. 2011, 23(6), 776–780.

[59]  Xu Y, Liu J, Zhang J, Zong X, Jia X, Li D, Wang E. Chip-based generation of carbon nanodots via electrochemical oxidation of screen printed carbon electrodes and the applications for efficient cell imaging and electrochemiluminescence enhancement. Nanoscale. 2015, 7(21), 9421–9426.

[60]  Li Y, Liu X, Wang J, Liu H, Li S, Hou Y, Wan W, Xue W, Ma N, Zhang JZ. Chemical nature of redox-controlled photoluminescence of graphene quantum dots by post-synthesis treatment. J Phys Chem C. 2016, 120(45), 26004–26011.

[61]  Bao L, Zhang ZL, Tian ZQ, Zhang L, Liu C, Lin Y, Qi B, Pang DW. Electrochemical tuning of luminescent carbon nanodots: From preparation to luminescence mechanism. Adv Mater. 2011, 23(48), 5801–5806.

[62]  Li H, He X, Kang Z, Huang H, Liu Y, Liu J, Lian S, Tsang CH, Yang X, Lee ST. Water-soluble fluorescent carbon quantum dots and photocatalyst design. Angew Chem Int Ed. 2010, 49(26), 4430–4434.

[63]  Han Y, Huang H, Zhang H, Liu Y, Han X, Liu R, Li H, Kang Z. Carbon quantum dots with photoenhanced hydrogen-bond catalytic activity in aldol condensations. ACS Catal. 2014, 4(3), 781–787.

[64]  Pan D, Zhang J, Li Z, Wu M. Hydrothermal route for cutting graphene sheets into blue-luminescent graphene quantum dots. Adv Mater. 2010, 22(6), 734–738.

[65]  Liu J, Geng Y, Li D, Yao H, Huo Z, Li Y, Zhang K, Zhu S, Wei H, Xu W, Jiang J, Yang B. Deep red emissive carbonized polymer dots with unprecedented narrow full width at half maximum. Adv Mater. 2020, 32(17), 1906641.

[66]  Son H, Liu X, Wang B, Tang Z, Lu S. High production-yield solid-state carbon dots with tunable photoluminescence for white/multi-color lightemitting diodes. Sci Bull. 2019, 64(23), 1788–1794.

[67]  Chen W, Li D, Tian L, Xiang W, Wang T, Hu W, Hu Y, Chen S, Chen J, Dai Z. Synthesis of graphene quantum dots from natural polymer starch for cell imaging. Green Chem. 2018, 20(19), 4438–4442.

[68]  Chen W, Shen J, Wang Z, Liu X, Xu Y, Zhao H, Astruc D. Turning waste into wealth: Facile and green synthesis of carbon nanodots from pollutants and applications to bioimaging. Chem Sci. 2021, 12, 11722–11729.

[69]  Zhang B, Liu C, Liu Y. A novel one-step approach to synthesize fluorescent carbon nanoparticles. Eur J Inorg Chem. 2010, 28, 4411–4414.

[70]  Briscoe J, Marinovic A, Sevilla M, Dunn S, Titirici M. Biomass-derived carbon quantum dot sensitizers for solid-state nanostructured solar cells. Angew Chem Int Ed. 2015, 54(15), 4463–4468.

[71]  Zhu S, Meng Q, Wang L, Zhang J, Song Y, Jin H, Zhang K, Sun H, Wang H, Yang B. Highly photoluminescent carbon dots for multicolor patterning, sensors, and bioimaging. Angew Chem Int Ed. 2013, 52(14), 3953–3957.

[72]  Li H, Han S, Lyu B, Hong T, Zhi S, Xu L, Xue F, Sai L, Yang J, Wang X, He B. Tunable light emission from carbon dots by controlling surface defects. Chin Chem Lett. 2021, doi: https://doi.org/10.1016/j.cclet.2021.03.051.

[73]  Xu X, Mo L, Li W, Li Y, Lei B, Zhang X, Zhuang J, Hu C, Liu Y. Red green and blue aggregation-induced emissive carbon dots. Chin Chem Lett. 2021, doi: https://doi.org/10.1016/j.cclet.2021.05.056.

[74]  Jiang K, Sun S, Zhang L, Lu Y, Wu A, Cai C, Lin H. Red, green, and blue luminescence by carbon dots: Full-color emission tuning and multicolor cellular imaging. Angew Chem Int Ed Engl. 2015, 54(18), 5360–5363.

[75]  Wang B, Li J, Tang Z, Yang B, Lu S. Near-infrared emissive carbon dots with 33.96% emission in aqueous solution for cellular sensing and light-emitting diodes. Sci Bull. 2019, 64(17), 1285–1292.

[76]  Shuang E, Mao QX, Yuan XL, Kong XL, Chen XW, Wang JH. Targeted imaging of the lysosome and endoplasmic reticulum and their pH monitoring with surface regulated carbon dots. Nanoscale. 2018, 10(26), 12788–12796.

[77]  Liu B, Chu B, Wang YL, Hu LF, Hu S, Zhang XH. Carbon dioxide derived carbonized polymer dots for multicolor light-emitting diodes. Green Chem. 2021, 23(1), 422–429.

[78]   Qu S, Wang X, Lu Q, Liu X, Wang L. A biocompatible fluorescent ink based on water-soluble luminescent carbon nanodots. Angew Chem Int Ed. 2012, 51(49), 12215–12218.
[79]   Mikhraliieva A, Zaitsev V, Xing Y, Coelho-Júnior H, Sommer RL. Excitationindependent blue-emitting carbon dots from mesoporous aminosilica nanoreactor for bioanalytical application. ACS Appl Nano Mater. 2020, 3(4), 3652–3664.
[80]   Wang F, Pang S, Wang L, Li Q, Kreiter M, Liu CY. One-step synthesis of highly luminescent carbon dots in noncoordinating solvents. Chem Mater. 2010, 22(16), 4528–4530.
[81]   Pan D, Zhang J, Li Z, Wu C, Yan X, Wu M. Observation of ph-, solvent-, spin-, and excitation-dependent blue photoluminescence from carbon nanoparticles. Chem Commun. 2010, 46(21), 3681–3683.
[82]   Dong Y, Shao J, Chen C, Li H, Wang R, Chi Y, Lin X, Chen G. Blue luminescent graphene quantum dots and graphene oxide prepared by tuning the carbonization degree of citric acid. Carbon. 2012, 50(12), 4738–4743.
[83]   Gu ZG, Li DJ, Zheng C, Kang Y, Woll C, Zhang J. Mof-templated synthesis of ultrasmall photoluminescent carbon-nanodot arrays for optical applications. Angew Chem Int Ed Engl. 2017, 56(24), 6853–6858.
[84]   Zhou K, Zhang WJ, Luo YZ, Pan CY. Photoluminescent carbon dots based on a rare 3d inorganic-organic hybrid cadmium borate crystal. Dalton Trans. 2018, 47(22), 7407–7411.
[85]   Li Z, Liu W, Ni P, Zhang C, Wang B, Duan G, Chen C, Jiang Y, Lu Y. Carbon dots confined in n-doped carbon as peroxidase-like nanozyme for detection of gastric cancer relevant d-amino acids. Chem Eng J. 2022, 428, 131396.
[86]   Miao X, Qu D, Yang D, Nie B, Zhao Y, Fan H, Sun Z. Synthesis of carbon dots with multiple color emission by controlled graphitization and surface functionalization. Adv Mater. 2018, 30(1), 1704–1740.
[87]   Ye Y, Yang D, Chen H, Guo S, Yang Q, Chen L, Zhao H, Wang L. A highefficiency corrosion inhibitor of n-doped citric acid-based carbon dots for mild steel in hydrochloric acid environment. J Hazard Mater. 2020, 381, 121019.
[88]   Ma CA, Yin C, Fan Y, Yang X, Zhou X. Highly efficient synthesis of n-doped carbon dots with excellent stability through pyrolysis method. J Mater Sci. 2019, 54(13), 9372–9384.
[89]   Supchocksoonthorn P, Hanchaina R, Sinoy MCA, de Luna MDG, Kangsamaksin T, Paoprasert P. Novel solution- and paper-based sensors based on label-free fluorescent carbon dots for the selective detection of pyrimethanil. Appl Surf Sci. 2021, 564, 150372–150379.
[90]   Lin L, Rong M, Lu S, Song X, Zhong Y, Yan J, Wang Y, Chen X. A facile synthesis of highly luminescent nitrogen-doped graphene quantum dots for the detection of 2,4,6-trinitrophenol in aqueous solution. Nanoscale. 2015, 7(5), 1872–1878.
[91]   Choi Y, Kang B, Lee J, Kim S, Kim GT, Kang H, Lee BR, Kim H, Shim SH, Lee G, Kwon OH, Kim BS. Integrative approach toward uncovering the origin of photoluminescence in dual heteroatom-doped carbon nanodots. Chem Mater. 2016, 28(19), 6840–6847.
[92]   Wang C, Jiang K, Wu Q, Wu J, Zhang C. Green synthesis of red-emitting carbon nanodots as a novel "turn-on" nanothermometer in living cells. Chem Eur J. 2016, 22(41), 14475–14479.
[93]   Liu R. Facile synthesis of magneto-fluorescent carbon dots by one-step microwave-assisted pyrolysis. J Alloys Compd. 2021, 855, 157456.
[94]   Liu H, He Z, Jiang LP, Zhu JJ. Microwave-assisted synthesis of wavelength tunable photoluminescent carbon nanodots and their potential applications. ACS Appl Mat Interf. 2015, 7(8), 4913–4920.
[95]   Shao J, Zhu S, Liu H, Song Y, Tao S, Yang B. Full-color emission polymer carbon dots with quench-resistant solid-state fluorescence. Adv Sci. 2017, 4(12), 1700395.
[96]   Zheng J, Wang Y, Zhang F, Yang Y, Liu X, Guo K, Wang H, Xu B. Microwave-assisted hydrothermal synthesis of solid-state carbon dots with intensive emission for white light-emitting devices. J Mater Chem. 2017, C 5(32), 8105–8111.

[97]  Wang HJ, Hou WY, Kang J, Zhai XY, Chen HL, Hao YW, Wan GY. The facile preparation of solid-state fluorescent carbon dots with a high fluorescence quantum yield and their application in rapid latent fingerprint detection. Dalton Trans. 2021, 50, 12188–12196.

[98]  Li LL, Ji J, Fei R, Wang CZ, Lu Q,, Zhang JR, Jiang LP, Zhu JJ. A facile microwave avenue to electrochemiluminescent two-color graphene quantum dots. Adv Funct Mater. 2012, 22(14), 2971–2979.

[99]  Hai X, Mao QX, Wang WJ, Wang XF, Chen XW, Wang JH. An acid-free microwave approach to prepare highly luminescent boron-doped graphene quantum dots for cell imaging. J Mater Chem B. 2015, 3 (47), 9109–9114.

[100] Feng J, Wang WJ, Hai X, Yu YL, Wang JH. Green preparation of nitrogendoped carbon dots derived from silkworm chrysalis for cell imaging. J Mater Chem B. 2016, 4(3), 387–393.

[101] Wang J, Wang CF, Chen S. Amphiphilic egg-derived carbon dots: Rapid plasma fabrication, pyrolysis process, and multicolor printing patterns. Angew Chem Int Ed. 2012, 51(37), 9297–9301.

[102] Yan H, He C, Li X, Zhao T. A solvent-free gaseous detonation approach for converting benzoic acid into graphene quantum dots within milliseconds. Diam Relat Mater. 2018, 87, 233–241.

[103] Park SY, Lee HU, Park ES, Lee SC, Lee JW, Jeong SW, Kim CH, Lee YC, Huh YS, Lee J. Photoluminescent green carbon nanodots from food-wastederived sources: Large-scale synthesis, properties, and biomedical applications. ACS Appl Mat Interf. 2014, 6(5), 3365–3370.

[104] Aslandas AM, Balcı N, Arık M, Akiroglu HS, Onganer Y, Meral K. Liquid nitrogen-assisted synthesis of fluorescent carbon dots from blueberry and their performance in fe3þ detection. Appl Surf Sci. 2015, 356, 747–752.

[105] Zhu H, Liu A, Shan F, Yang W, Zhang W, Li D, Liu J. One-step synthesis of graphene quantum dots from defective cvd graphene and their application in igzo uv thin film phototransistor. Carbon. 2016, 100, 201–207.

[106] He V, Yan H, Li X, Wang X. In situ fabrication of carbon dots-based lubricants using a facile ultrasonic approach. Green Chem. 2019, 21(9), 2279–2285.

[107] Zhou Y, Mintz KJ, Oztan CY, Hettiarachchi SD, Peng Z, Seven ES, Liyanage PY, De La Torre S, Celik E, Leblanc RM. Embedding carbon dots in superabsorbent polymers for additive manufacturing. Polymers. 2018, 10(8), 921–928.

[108] Zhou X, Zhang Y, Wang C, Wu X, Yang Y, Zheng B, Wu H, Guo S, Zhang J. Photo-fenton reaction of graphene oxide: A new strategy to prepare graphene quantum dots for DNA cleavage. ACS Nano. 2012, 6(8), 6592–6599.

[109] Routh P, Das S, Shit A, Bairi P, Das P, Nandi AK. Graphene quantum dots from a facile sono-fenton reaction and its hybrid with a polythiophene graft copolymer toward photovoltaic application. ACS Appl Mat Interf. 2013, 5(23), 12672–12680.

[110] Zhu H, Liu A, Xu Y, Shan F, Li A, Wang J, Yang W, Barrow C, Liu J. Graphene quantum dots directly generated from graphite via magnetron sputtering and the applications in the thin film transistors. Carbon. 2015, 88, 225–232.

[111] Xia J, Zhuang YT, Yu YL, Wang JH. Highly fluorescent carbon polymer dots prepared at room temperature, and their application as a fluorescent probe for determination and intracellular imaging of ferric ion. Microchim Acta. 2017, 184(4), 1109–1116.

Abhinay Thakur, Ashish Kumar*

# Chapter 4
# Carbon dots in biosensing

**Abstract:** Carbonaceous nanoparticles specifically are at the frontline of the recent up-surge in using nanomaterials for biosensing activities. New carbon materials called carbon dots (CDs) have drawn a lot of recognition because of their intriguing characteristics, including their inherent nontoxicity, elevated solubility in a variety of solvents, outstanding electronic characteristics, strong chemical stability, massive specific surface region, plentiful angle spots for functionalization, exquisite biocompatibility, relatively inexpensive, and flexibility, and also their capacity to be altered with appealing surface precursors and other modifiers/nano mask. Because of these characteristics, they are very intriguing contenders for manufacturing a variety of high-performance biosensors. In this chapter, we outline the top-down and bottom-up synthesis pathways of CDs, emphasize their modification techniques, and discuss how they are used in several key biosensor applications. Additionally, the challenges and potential outcomes of the use of CDs for biosensors are also addressed.

**Keywords:** Carbon dots, biosensors, cancer, metal ion sensing, pollutants, real-time analysis

## 4.1 Introduction

Owing to the intriguing physiochemical characteristics of nano-sized materials (1–100 nm) and their linkages with other entities that function at the molecular scale, the cutting-edge area of nanotechnology has recently given rise to unique applications in a wide range of scientific fields [1–4]. The creation of nanodevices has shown that several substances perform better than their massive equivalents. Various physical (top-down) and chemical (bottom-up) manufacturing techniques could be used to create nanomaterials that could take place in the gaseous state, liquid stage, or solid stage. Laser pyrolysis, microwave irradiation, pulsed laser ablation in liquid, chemical vapour deposition, etching, electric explosion, physical vapour deposition, electric discharge, and high-energy ball milling are a few instances of physical synthesis methods. Hydrothermal synthesis, coprecipitation, solvothermal synthesis, sol–gel, chemical reduction, microemulsion, spray

*Corresponding author: Ashish Kumar, NCE, Department of Science and Technology, Government of Bihar, Bihar, India, e-mail: drashishchemlpu@gmail.com
Abhinay Thakur, Department of Chemistry, School of Chemical Engineering and Physical Sciences, Lovely Professional University, Phagwara, Punjab, India

https://doi.org/10.1515/9783110799958-004

pyrolysis, electrochemical synthesis, biological synthesis, and green synthesis are a few examples of processes used in chemical synthesis. The choice of an effective process for synthesizing nanomaterials relies on a number of variables including the equipment that is accessible, prices, efficiency, big volume, and environmental protection. Furthermore, each synthesis technique enables the production of certain physical chemistry and nanomaterial features. Optimizing the particulate sizes and forms could result in materials with these outstanding features (mechanical, surface area, optical, magnetic, quantum effect, high thermal resistance, and electrical conductivity). Nanomaterials are thus great prospects to address, enhance, or produce novel, exciting solutions in a variety of fields, including semiconductors, power storage, pharmacology, microbiology, pharmacological sensing, photocatalytic degradation, and water purification. The most utilized nanomaterial in biomedicine and biosensing has huge surface regions and photonic, magnetic, and thermo characteristics.

As their exploration via the electrophoretic detoxification of single-walled carbon nanotubes (CNTs) formed as by-products of the arc-discharge of soot, CDs have attracted consideration for their intriguing unexpected physiochemical characteristics including luminescent optical characteristics, elevated quantum efficiency, rigidity to photo-bleaching, outstanding photo-stability, outstanding water dissolution rate, biocompatibility, chemical stability, and nontoxicity [5–9]. Presently, it is understood that CDs differ from their analogues in that they are quasi-spherical nanoparticles having diameters under 10 nm that are made up of amorphous $sp^3$ carbon matrices around an $sp^2$ carbogenic core. To clarify the internal photoluminescence (PL) generating process of CDs, several empirical methods have been presented, accompanied by theoretical computational modellings including time-dependent density functional theory (DFT) computations and DFT. According to several studies, despite metallic quantum dots (QDs), the emitted colour of CDs is thought to be influenced by a variety of parameters, such as the structure of the initial synthetic precursor, energetic levels, particle emitter's traps, interface flaws, and optical mixing of excitation energy. The cause of the emissions, according to some researchers, is thought to be the creation of molecular fluorophores as by-products of synthesizing; nevertheless, the origin of the PL is still unclear. The significant concentration of carbon atoms and heteroatoms in CDs gives these chemical properties like strong solubility in water and, at the same moment, superior biocompatibility from a biological and chemical perspective. Specifically, scanning of numerous species objectives (cells, vesicles, microbes, fungus, and animals), tracking of diverse metallic ions and biomolecules in living cellular components, and imaging-guided therapies for malignancies (Figure 4.1).

In an experiment, Nguyen et al. [11] presented a straightforward, inexpensive, and environmentally benign approach for producing activated carbon and high-fluorescence carbon dots (CDs) in a single hydrothermal phase utilizing banana peels. Intense, bright-blue PL (mean diameter: 3–6 nm) was produced by the dispersal of CDs and may be exploited in a future study on biosensing, semiconductors, and catalysts. Additionally, activated carbon having an elevated specified surface region (294.6 $m^{-2}$ $g^{-1}$) and a substantially permeable morphology were found in the precipitation that

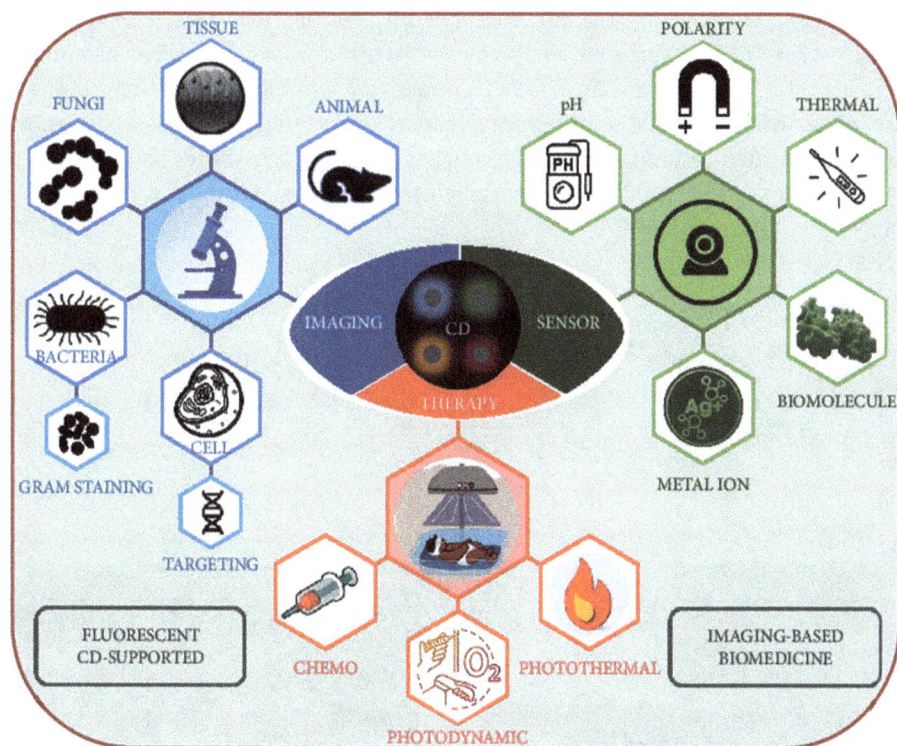

**Figure 4.1:** Fluorescent carbon dots are used in a variety of imaging-supported biomedical purposes, such as FL scanning of various biotargets, FL imaging-guided treatment, and FL imaging-supported sensors (adapted from Ref. [10] with permission from [Hindawi], [2022]. Distributed under the Creative Common Attribution-based License CCBY 4.0).

formed at the bottom of the hydrothermal technique. This material might be employed as an electrode for supercapacitors. With an elevated specific capacitance (199 F $g^{-1}$) in a soluble electrolyte (1 M KOH) and an elevated energy density of 54.15 Wh $kg^{-1}$ at a current density of 0.5 A $g^{-1}$, the three-electrode cell demonstrated outstanding capacity and consistency of activated carbon as the working electrode in numerous aqueous electrolytes. Banana peels, a plentiful bioresource, could thereby produce two outstanding items depending on the superior qualities of CDs and the potent electrochemical performance of activated carbon as an electrode component.

Similarly, when creating N-doped CDs, Lee et al. [12] added Mn(acetate)$_2$, and its impact on the enzymatic characteristics of Mn-induced N-CDs (Mn:N-CDs) was examined. X-ray photoelectron spectrometry and infrared spectroscopy were used to examine their chemical composition, and the findings demonstrate that Mn ions are responsible for the fluctuation in the density of complex formation in Mn:N-CDs but not in N-CD scaffolds. Mn, utilizing the hydrothermal technique, N-CDs were produced. The origins of carbon were citric acid and ethylenediamine, with a reaction mole fraction of 1:0.5.

Additionally, Mn(acetate)$_2$ was introduced, and its reaction with citric acid was 1:0.4 (Figure 4.2a). TEM imaging showed that Mn:N-CDs had a rounded form having a mean diameter of 6.7 nm (Figure 4.2b). Mn:N-CDs possessed a crystalline morphology having a lattice separation of 0.24 nm that complemented the (100) planar of graphite, as seen by the high-resolution (HR) TEM images (Figure 4.2c). Mn:N-CDs displayed a large signal at 19.14 in the XRD spectra, which corresponded to the graphite (002) plane (JCPDS card no. 26-1076).

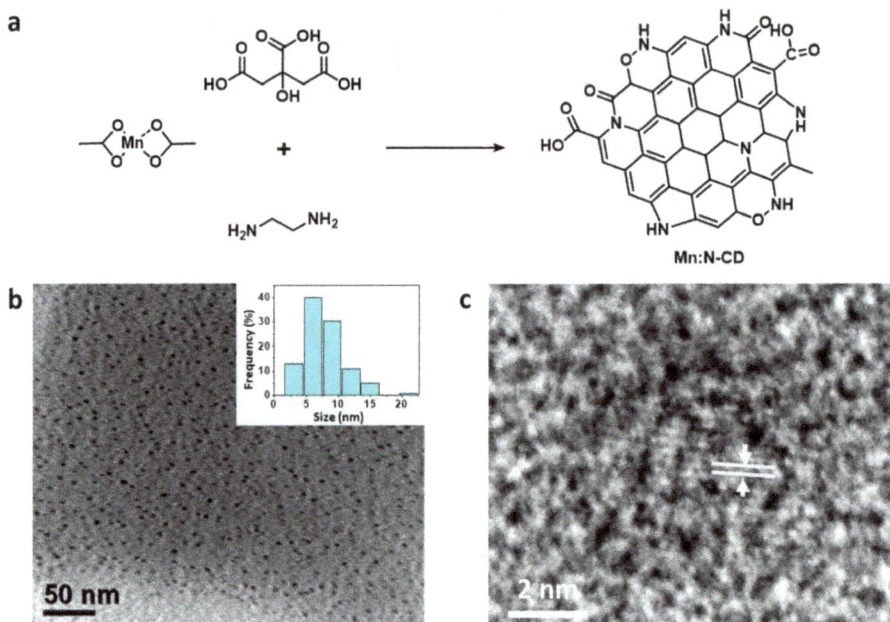

**Figure 4.2: (a)** The production of Mn-induced N-doped CDs is shown schematically (Mn:N-CDs); **(b)** pictures of the Mn:N-CDs took using low- and high-resolution transmission electron microscopy. Particle size dispersion inset in **(c)** (adapted from Ref. [12] with permission from [MDPI], [2021]. Distributed under the Creative Common Attribution-based License CCBY 4.0).

As the colour shift of a 3,3′,5,5′-tetramethylbenzidine/H$_2$O$_2$ mixture was studied in the addition of Mn:N-CDs and N-CDs, this architectural transition enhanced the enzymatic characteristics of Mn:N-CDs in comparison to those of N-CDs. A straightforward colorimetric method with Mn:N-CDs was utilized to identify γ-aminobutyric acid, a marker of brain-related disorder, depending on this improved enzymatic capability. As a result, they anticipate that Mn:N-CDs will make a fantastic enzymatic probing for the colorimetric sensor device.

Zhang et al. [13] created the CDs/GO nanocomposites in massive volumes using an easy and affordable method of stacking interface. The electrode interface of the CDs/GO nanocomposites was coated to increase electrochemical performance and serve as a bio-platform that concurrently engaged the targeted deoxyribonucleic acid probes. In comparison with pure GO on GCE or CDs on GCE, the CDs/GO/GCE was effectively manufactured and possessed superior high electrocatalytic behaviour, excellent biocompatibility, and great affinity against the targeted DNA sequence. With a LOD of 83 pM ($S/N$ = 3), the DNA biosensor exhibits outstanding sensing capability for identifying the essential pathogenic DNA of APL. They assume that the straightforward and affordable DNA biosensor offers the promise to be an efficient and robust technique for detecting pathogenic genes in medical diagnostics based on the findings of multiple experiments.

The thiol-functionalized CDs (S-CDs), according to Chen et al. [14], were used to create a robust and specific glucose biosensor. Simple hydrothermal carbonization and increased surface alteration were used to create S-CDs, which produced brilliant yellow-green light exhibiting outstanding pH/salt durability. S-CDs were converted into non-luminous S-CDs assemblies after being exposed to $H_2O_2$, depending on target-initiated catalytic oxidation of –SH into –S–S–. Because glucose oxidase may catalyse the breakdown of glucose into $H_2O_2$, S-CDs were also used to identify glucose having a minimal LOD (0.03 M) and great specificity for distinguishing glucose from other saccharides. In the end, S-CDs were successfully used to detect glucose in human serum.

Giang et al. [15] studied the use of visible light, alkaline phosphatase (ALP) and pyrophosphates (PPi), activity-responsive CDs to aim at cancer cells utilizing carbonized hyaluronic acid (CD(HA)), employing electrochemical impulses produced from surfaces covered with CD(HA), $TiO_2$, and $Cu^{2+}$. With an anti-fouling function to visualize the cellular adherence interaction with the surface and the identification of cancer cells, the stimuli-responsive layer offered an efficient way for wireless sensing to operate inside cells. A transition in the electrochemical transmitter in cell-cultured biosensors was caused by the strongly distinguishable intracellular PPi and ALP tiers of cancer cells likened to regular cells and CD44 receptor-mediated endocytosis by the CD(HA) chain into cancerous cells. These electrochemical methodologies included EIS and a wireless autonomous sensor framework to recognize this transformation. The anti-fouling capabilities of CD(HA)/$TiO_2$/$Cu^{2+}$ have used the detect malignancy. This was confirmed by the time-dependent cell dissociation happening on the layer formed in the vicinity of visible light. In comparison to usual cells (152 Ωk), the resistivity of CD(HA)/$TiO_2$/$Cu^{2+}$ biosensors grown in tumour cells significantly improved (248 Ωk), that was attributed to the internalization of CD (HA) by cancerous cells and the separation of $Cu^{2+}$ in the device by PPi and ALP in cancer cells. The CD(HA)/$TiO_2$/$Cu^{2+}$ wireless sensor system presents a potential method which enables malignancy prognosis with anti-fouling impacts to research the engagement among cells and modified substrates. It has elevated precision and specificity for malignancy identification by visible region, enzyme sensitivity with minimal LOD value (2.31 cells $mL^{-1}$) on the premise of electrochemical strategy, and 70.05 cells $mL^{-1}$ for electro-optic strategy.

## 4.2 Synthesis of CDs

Several synthetic methods have been put forth in recent years to produce CDs, and these methods may generally be divided into "top-down" and "bottom-up" techniques based on the size connection involving synthesis precursor and CDs. Concerning the "top-down" strategies, CDs are created by physically or chemically fragmenting greater carbon frameworks, including carbon rods, graphite oxide, and other large-size carbon precursor substances, into smaller pieces. Even though the granules have homogeneous size dispersion, the quantum yield (QY) is often poor and requires laborious post-surface passivation with stronger chemicals [9, 16–22]. Contrarily, "bottom-up" techniques generate CDs by dehydrating, polymerizing, cross-linking, and carbonizing tiny molecules to bigger molecules under specific circumstances. Due to the obvious abundance of resulting surface clusters, these methodologies became the mainstream for synthesizing high-performance CDs. Consequently, in most cases, specialized tools and drawn-out procedures are necessary. For the synthesis of several CDs, diverse techniques and the related equipment offer a variety of process conditions. Hence, selecting a suitable synthesis technique and particular approach will be the first factor to take into account. The most popular methods are solvothermal and hydrothermal, which effectively heat either aqueous media or an organic solvent that contains the required precursors. In order to produce CDs amid high-temperature and high-pressure circumstances, the combination of chosen precursors containing water or an organic solvent was typically added to a standard reactor and adjusted for reaction temperature, duration, and pressure. These two methods are suited for both "top-down" and "bottom-up" methods, and they are capable to achieve the reactions of the majority of precursors and solvent varieties. Comparing the two synthetic techniques, the "bottom-up" route produces CDs with more complicated geometries and a greater abundance of surface functional groups that are better suited for the targeted monitoring of toxins. Additionally, there is a lot of curiosity about the hydrothermal approach that uses microwaves as the source of heat. In order to enable their widespread use in multiple sensors, scientists have recently been extensively investigating the environmentally benign low-temperature hydrothermal process for synthesizing high-quality CDs having diverse morphologies, thickness, form, and surface functions. Figure 4.3 shows typical precursors for creating CD/polymer composites as well as the processes used to create them.

A key technique for expediting the production of CDs is microwave-assisted pyrolysis, which integrates microwave technologies alongside chemical synthesis to give effective and homogenous energy to the precursor liquid. To synthesize the many functionalities of CDs, variables including heating duration, microwave strength, and warming duration are typically modified. In consistently high situations, hydrosoluble or liposoluble CDs could be produced using this technique in under five minutes when using the "bottom-up" approach. The creation of CDs is somewhat constrained by the fact that water is typically utilized as the reaction media in a microwave oven setup and that the reaction temperature is unpredictable. It is advised to use this

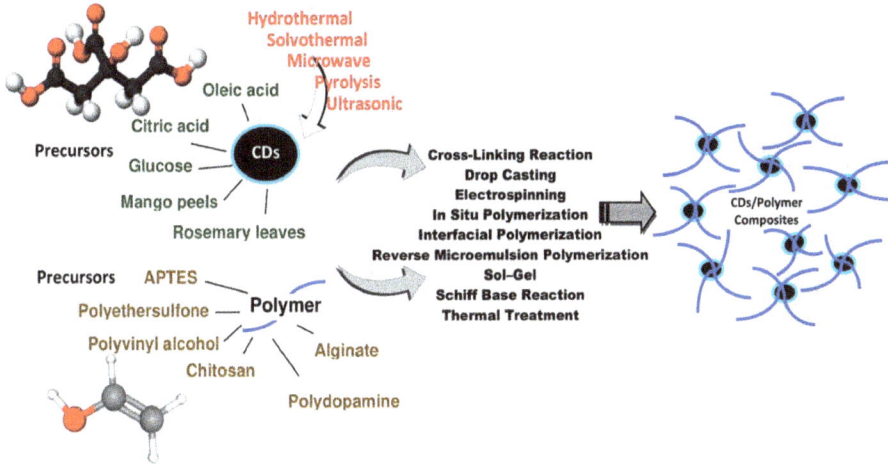

**Figure 4.3:** Polymer precursors, CD precursors, and techniques for generating composites of CD and polymer. Polyvinyl alcohol and citric acid, the two most common precursors for CD/polymer composites, are described in terms of their chemical compositions (adapted from Ref. [23] with permission from [MDPI], [2021]. Distributed under the Creative Common Attribution-based License CCBY 4.0).

technique for regular short-wavelength emissions CD production. A straightforward technique for creating CDs is chemical oxidation, which relies on the corrosive activity of potent acids and oxidants to remove huge carbon supplies through top to bottom or on redox processes to chemically create and processing of different CDs through the bottom to top. To create CDs, the precursors are combined using intense nitric acid or sulphuric acid for a few weeks in a corrosion-resistant vessel. The process is typically sped up using low-temperature warming, agitation, and ultrasonic techniques. Using the top-down approach, the amount of CD oxidation flaws is changed to alter the substance efficiency. Furthermore, this technique does not exfoliate CDs with much precision, and the molecular base composition is quickly broken, affecting their optical qualities. Therefore, to get around the aforementioned issues and enhance CD efficiency, chemical production through the bottom to top depending on redox processes could be used. The chemical oxidation process may quickly produce huge quantities of CDs without the requirement for complicated apparatus, but it has a small manufacturing capability owing to inconsistent manufacturing and requires costly oxidants, which pollute the atmosphere. A traditional technique is laser ablation, which involves utilizing a powerful laser beam to remove a part of a precursor substance from the substrate. It is frequently used to create different carbon nanomaterials including CDs, graphene, and CNTs because it enables a regulated quantity of radiation to be supplied to the precisely targeted region in terms of time and concentration. Therefore, the laser ablation process essentially simply creates a heterogeneous combination of moderate or strong PL solid carbon substances. Later, a highly efficient and accurate approach to producing a variety of CDs has been developed that employs carbon sedimentation in various solvents, whether by itself or those including organic compounds or polymers to effectively passivate the carbon NPs. The liquid-phase laser ablation technique created significantly shorter and much more homogeneous GQDs than the chemical oxidation technique did, making it a quicker and more hygienic one-step process for making CDs using fewer wastes and beginning chemicals. For environmentally friendly CD manufacturing, electrochemical oxidation has been extensively researched. In the electrochemical manufacture of CDs, a variety of electrode substances, including carbon fibre, CNTs, and graphite rode, could be used. The electrochemical production of a variety of CDs is appropriate for both the top-down and bottom-up approaches. By varying the electrode substance, electrode separation, power, and current intensity, the activation procedure is often performed in alkaline circumstances to get homogeneous CD sizes for realizing minimal ambient noise, little self-quenching, and excellent sensibility. Furthermore, it should be observed that the reaction raises the interface region among the electrode substance and the solution substrate that will have an impact on the manufactured CDs' performance. In order to solve this problem, Xu's group published an end-face electrochemical ripping method for reproducible graphene synthesis. This method offered a fresh concept for maintaining the interface current and enhancing the output of slicing substances and other carbon-based substances like CDs.

For detecting and biosensing purposes, Monday et al. [24] concentrated on creating CDs using palm kernel shells (PKS), a plentiful and benign waste product. In an autoclave batching processor, the solvothermal and hydrothermal procedures of one-pot synthesis were used to create L-ethylenediamine and phenylalanine-enriched CDs. Figure 4.4 illustrates that the primary components of PKS are hemicellulose, cellulose, and lignin. Particularly at a temperature over 400 °C for further over 2 h of carbonization period, the thermal decline of PKS in an ambient of inert nitrogen transforms the hemicellulose, cellulose, and lignin in PKS biomass into a black graphitic carbonaceous substance. In this study, CDs were created utilizing a hydrothermal autoclave procedure, which is straightforward and yields round CDs with roughly consistent diameters from the carbonized PKS. This research was a green production of CDs relying on the non-precursor utilized to make CDs with a significant quantum output of fluorescence and luminescence to be utilized in the construction of biosensors.

Assuming an excitation/emission frequency of 360/450 nm, the as-prepared N-CDs had good PL properties including QYs of 13.7% for ethylenediamine-doped N-CDs (CDs-EDA) and 8.6% for L-phenylalanine-doped N-CDs (CDs-LPh). The pictures from transmission electron microscopy (TEM) revealed that the mean size of the particles for both CDs in N-CDs was 2 nm. Results from UV–visible spectrophotometry indicated the C=C and C=O transitions. The N-CDs were found to have moiety including –OH, –C=O, and –$NH_2$ according to FTIR data, and the strong peaks in the X-ray diffraction (XRD) pattern indicated that the N-CDs were crystalline. This study showed that PKS biomass – often discarded as waste – manufacturing CDs featuring superior physico-chemical characteristics.

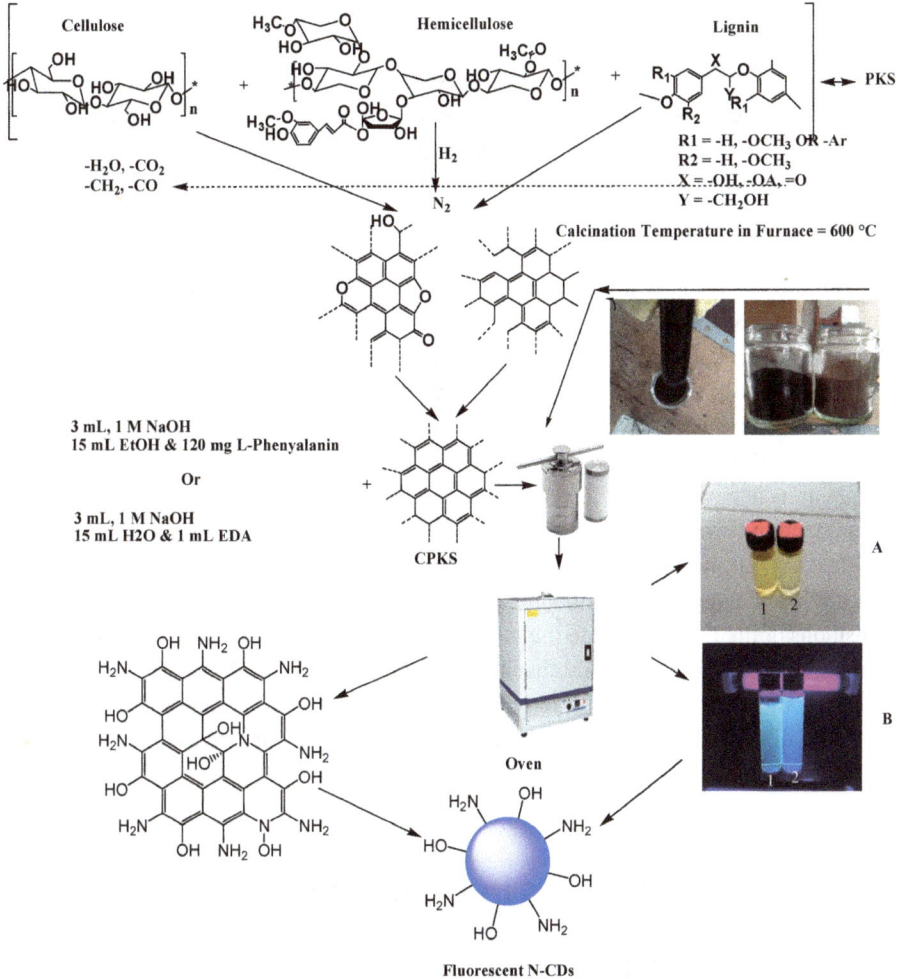

**Figure 4.4:** An example of the one-pot hydrothermal synthesis of CDs that have been produced with N-doping (adapted from Ref. [24] with permission from [MDPI], [2021]. Distributed under the Creative Common Attribution-based License CCBY 4.0).

# 4.3 Structural and photophysical characterization of CDs

The assessment of CDs is often carried out utilizing a variety of analytical approaches in order to thoroughly comprehend the intrinsic characteristics and distinctive characteristics displayed by these carbonaceous nanoparticles. XRD, HR transmission electron microscopy (HRTEM), Raman spectroscopy, X-ray photoelectron spectroscopy

(XPS), and Fourier-transform infrared spectroscopy could all be used to examine the morphological characteristics of CDs. Whereas XRD studies the crystalline phase of CDs and delivers confirmation on the unit cell diameters and crystal separation inside the crystalline carbon cores, HRTEM contributes important knowledge on the geometry, particulate sizes dispersion, and crystalline structure of the CDs. Another instrument for learning about the architectural characteristics of carbon atoms in CDs is the Raman spectrometer. The disorganized $sp^2$ carbons are responsible for the D band maximum in the Raman spectra of CDs, which is located at approximately 1,350 cm$^{-1}$, and the G band peak, which is located at about 1,600 cm$^{-1}$, is caused by the in-plane stretching vibration phase E2g of crystal graphite carbons. Information on the carbon structure, specifically the level of crystallinity and relative abundance of core carbon atoms as compared to surface atoms, could be gained from the proportion of strengths of these two distinctive Raman spectra. XPS and FTIR spectrum analyses are useful for illuminating the surface functionalities on the CDs. On the interface of CDs, individual atomic units are revealed by the XPS spectrum, and FTIR spectroscopy typically works in conjunction with XPS to disclose discrete details on functional groups. The absorption of CDs could be measured using a UV spectrophotometer, and the associated spectra typically show two prominent bands: one at 300–355 nm that corresponds to the $n$–$\pi^*$ transitions of the C=O group and the other at about 230–282 nm that represents the $\pi$–$\pi^*$ transition of the C=C group. Another notable characteristic of CDs that may be investigated via PL spectra is excitation-dependent fluorescence. Numerous applications make use of CDs' ability to tune their emission colour in response to changing excitation wavelengths.

## 4.4 Application of CDs in biosensing

Analytical sensing has a specialty called "biosensing" that pays particular emphasis to include molecular biometric components in the detection procedure. CDs are useful biosensing substances due to their superior photo-physical characteristics including PL and strong electrical conductance. Typically, the targeting moieties or interface functional moieties engage specifically with the analytes to affect the optical emission properties of CDs, which is the basis of the sensing mechanism of CDs. Theoretically, any optical variations, such as modifications in emission intensity, colorimetric frequency, or lifespan, could be used as quantifiable indicators to identify the associated analyte. This allows for the classification of CD-based biosensors into three groups: on–off, off–on, and fluorescence shift. In an experiment, focusing on "on–off–on" fluorescent nitrogen-doped CDs (NCDs), Ge et al. [25] created a simple fluorescent biosensor for Cu$^{2+}$ and S$^{2-}$ (N-CDs). Utilizing thermal processing and the precursor's ammonium citrate and hexamethylene-tetramine, N-CDs were initially created. The produced CDs had a bright green fluorescence, a 51.2% quantum output, and were very robust in aqueous immersion. For a 360 nm excitation, the highest fluorescence intensity spike was seen at 478 nm. Interestingly, $Cu^{2+}$ suppressed the fluorescence of N-CDs by a ratio of 6, using $Cu^{2+}$ experiments with LOD of 25 nM and ranges from 0.05 to 5 μM. It is noteworthy to note that additional $S^{2-}$ addition brought back the fluorescence of N-CD@$Cu^{2+}$ nanocomplex that was utilized to measure $S^{2-}$ over a wide spectrum from 0.05 to 10 M having an LOD of 32 nM. Additionally, it was shown that the N-CD and N-CD@$Cu^{2+}$ nanocomplexes were very selective for $S^{2-}$ and $Cu^{2+}$, correspondingly. Last but not the least, the sensor was effectively used to measure $Cu^{2+}$ and $S^{2-}$ in lake water, showing strong practical potential for analytic chemistry and eco-system monitoring.

For the quick and comfortable identification of micro (mi)RNA from CDs and GOs, Gao et al. [26] created a straightforward and extremely accurate DNA-based fluorescence biosensor. A single-stranded fuel DNA was effectively synthesized, coupled, and deposited onto the surface of GO through π–π layering, dimming fluorescence in the process. Figure 4.5 illustrates the fundamental concepts and procedures for the identification of the objective miRNA let-7a in a biological specimen (human serum). Recognition DNA (comprising grey, reddish, and yellow scenes) first found double-stranded (ds) DNA aiding HP DNA (blue series) across supplementary base pairing. An –HS group was added to the recognizing DNA's end and joined to the gold nanoparticles (AuNPs) via Au–S bonds to create dsDNA–AuNPs. The recognizing DNA strands have two distinctive toehold sections (grey and yellow patterns) at their two extremities. One of those toehold sections, the grey motif, linked to let-7a in the first reaction (TSDR1), whereas the yellow series coupled with fuel DNA in the second reaction (TSDR2). Without let-7a, single-stranded fuel DNA that had been labelled with CDs was adsorbable by GO via hydrophobic associations and π–π layering, which fully muted the fluorescence of the CDs by FRET. Let-7a activated TSDR1 when the targeted miRNA was present by attaching to the grey region of the recognizing DNA and taking the place of the HP DNA. Furthermore, the targeting was necessary for TSDR2 to occur since HP DNA produced throughout TSDR1 revealed the second toehold region. Toehold-mediated strands movement in TSDR2 caused fuel DNA-CDs to mix with the yellow region of the recognizing DNA and push let-7a out of the way. Let-7a was regenerated; as a result, the fuel DNA was removed from the GO interface, and CD fluorescence was recovered. Let-7a concentration and fluorescence restoration intensity were inversely correlated.

This fluorescence detecting approach could be used to monitor a range of targeted miRNAs and therefore may direct the construction of new biosensors with enhanced features due to its benefits of signal enhancement and excellent biocompatibility.

Tumour necrosis factor (TNF-a), a pro-inflammatory cytokine with important functions in apoptosis, division, survivability, multiplication, and migratory as well as immune system modulation, was the subject of Sri et al.'s [27] research. TNF-a is a perfect biomarker for diagnosing various diseases, especially cancer. TNF-a is less investigated for the detection of cancer than the other biomarkers of cancer. There are not many publications on the creation of biosensors that target TNF-a in specimens of human serum. CDs are also still being under-utilized in biosensor applications. In this respect, a

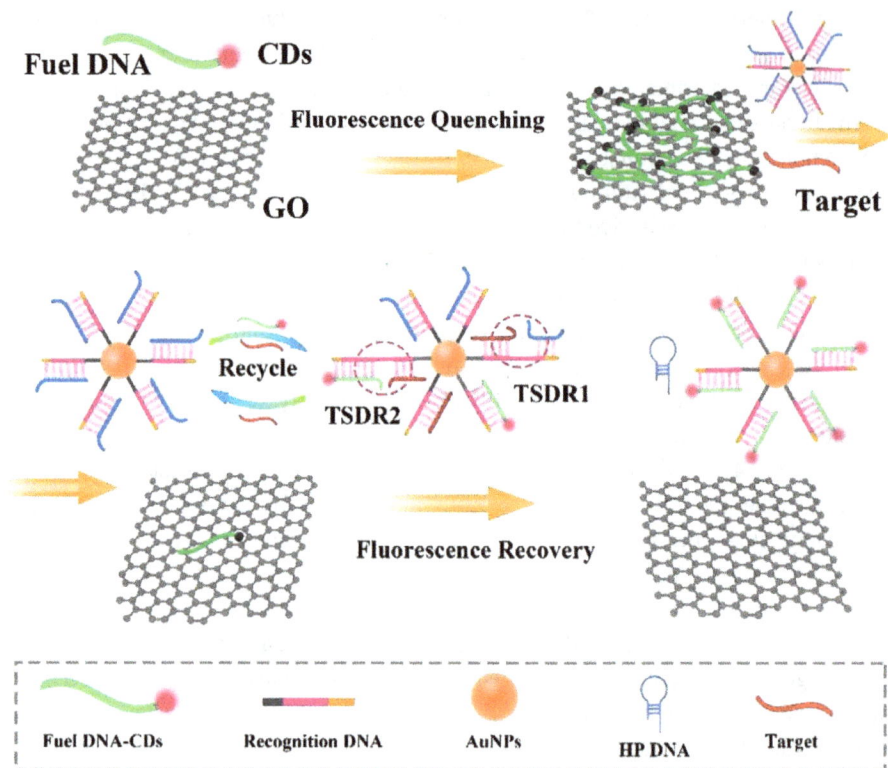

**Figure 4.5:** Depending on the CDs-labelled fluorescence probe and GO, a design for nonenzymatic miRNA recognition is shown (adapted from Ref. [26] with permission from [MDPI], [2021]. Distributed under the Creative Common Attribution-based License CCBY 4.0).

precise and affordable electrochemical biosensor centred on CDs has been created for the first moment. Microwave pyrolysis was used to easily yet simply create CDs. The matrices to store the CDs in the creation of the device were chosen as poly methyl methacrylate(PMMA). This brand-new CD-PMMA nanocomposite has a huge surface region, remarkable electrocatalytic conductance, and great cytocompatibility. To effectively bind TNF-a-specific antibodies and create an electrochemical immunoassay for the precise identification of TNF-a, CD-PMMA was used as the transducing component. The created biosensors had a dynamic spectrum of 0.05e160 pg mL$^{-1}$, a LOD of 0.05 pg mL$^{-1}$, and a sensibility of 5.56 pg mL$^{-1}$ cm$^{-2}$ for the identification of TNF-a. This immunosensor relying on CDs also maintains excellent levels of sensibility, specificity, and durability. For the earliest detection of malignancy in patient serum specimens, this immunosensor showed a strong association with the established method, enzyme-linked immunosorbent assay.

## 4.4.1 Cancer and malignancy

Cancer has been extensively studied earlier, especially clinically and academically, as it is among the most significant primary reasons for mortality in humans. Therefore, detecting cancer indicators quickly and very sensitively at an initial phase increases the likelihood of successful therapy and prolongs the lifespan of tumour patients, making accurate intervention of cancer crucial to lowering cancer-related fatalities. As a result, numerous analytical techniques have been created for the diagnosis of cancer and biomarker identification. Chemicals known as tumour indicators can be created by cancerous cells or by other bodily cells in reaction to the existence of a tumour. These compounds could be found in certain tumour patients' bloodstream, faeces, internal organs, or biological samples and could be raised by the existence of one or more tumour types. While proteins make up the majority of tumour indicators, DNA alterations and variations of gene transcription have also started to be utilized subsequently. Therefore, many tumour producers were discovered with diverse analytical suggestions for the detection and treatment of distinct types of malignancy. The suitable technique for creating nanomaterials with clearly delineated constructions, suitable physicochemical properties, and great bioactivity must be taken into consideration in order to increase the possibility of effective pre-clinical and clinical assessment of nanomaterial-based techniques for tumour markers sensing. Recently, a variety of nanomaterials have shown promise as instruments for biomedical applications. Among these nanomaterials are CDs.

Hepatocellular carcinoma (HCC), the second-leading source of cancer-related fatalities in China, was the subject of Li et al.'s [28] research. A particular antigen associated with HCC called glypican-3 (GPC3) is frequently employed in early diagnosis as a trustworthy biomarker of HCC. In this study, an extremely delicate homogeneous aptasensor for GPC3 sensing was developed using fluorescence resonance energy transfer (FRET), using magnetic graphene oxide ($Fe_3O_4$/GO) nanosheets as the acceptor and gold CDs that have been classified with the GPC3 aptamer (AuCDs-GPC3Apt) as the donor. To produce enough fluorescence, AuCDs were made using a single hydrothermal process. FRET occurs among AuCDs-GPC3Apt and $Fe_3O_4$/GO that reduces the overall system's fluorescence amplitude. The fluorescent AuCDs-GPC3Apt attaches to the targeted GPC3 and folds into a configuration that causes AuCDs-GPC3Apt to separate from $Fe_3O_4$/GO nanosheets whenever the targeted GPC3 was introduced to the FRET mechanism. Then, the $Fe_3O_4$/GO was magnetically detached, restoring the freely tagged AuCDs-fluorescence. GPC3Apt's LOD was 3.01 ng mL$^{-1}$ ($S/N$ = 3), and given ideal circumstances, the fluorescence recuperation frequency was significantly associated with the number of GPC3 (5–100 ng mL$^{-1}$). This technique offers a quick and efficient diagnostic technique for the measurement of GPC3 with tremendous possible applicability for earlier detection of HCC, with recoveries ranging from 98.76% to 101.29% in genuine human serum specimens.

Similarly, in order to create AuNPs and NPCD-AuNP composites using $Au^{3+}$, Le et al. [29] used the reductant capability of the phosphorus and nitrogen co-doped CDs (NPCDs) directly. A variety of spectroscopic and transmission electron methods, such as electrophoretic light scattering and XRD, were used to describe the composites. The fluorescence emission was quenched as a consequence of an efficient inner filter effect (IFE) in the composite substance caused by the overlapping of the fluorescence emission spectra of NPCDs and the absorption spectra of AuNPs. The fluorescence emission of the composite was regained in the addition of GSH, and it rose proportionately to raise the GSH content as shown in Figure 4.6. Additionally, with a LOD of 0.1 µM and reasonable outcomes, our GSH detecting technique demonstrated excellent selective and sensing capability in human serum.

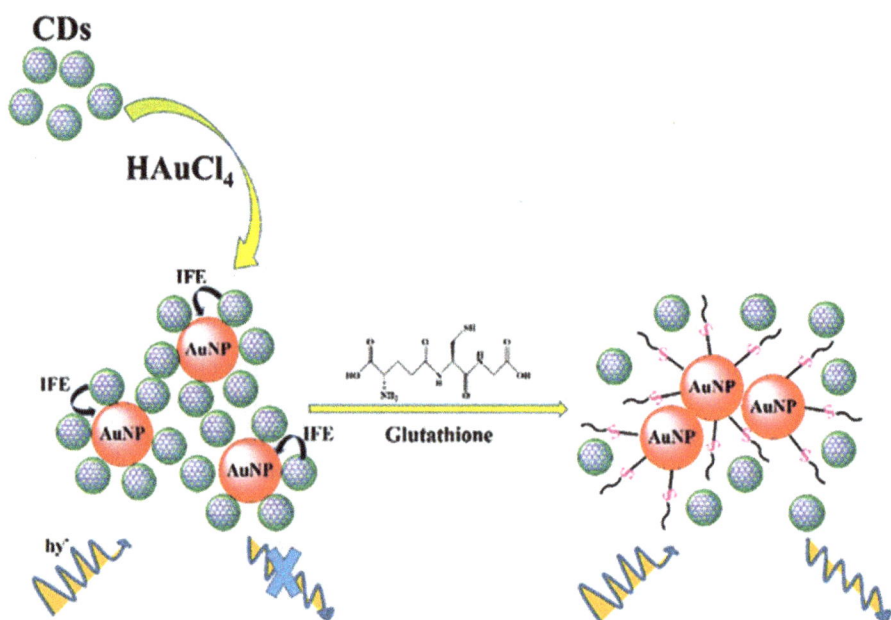

**Figure 4.6:** Glutathione detection demonstrated using the NPCD–AuNP composite substance's internal filter function (adapted from Ref. [29] with permission from [MDPI], [2022]. Distributed under the Creative Common Attribution-based License CCBY 4.0).

SEM, DLS, and TEM were used to examine the NPCD–AuNP composites' structure and shape. The produced AuNPs were mainly round and monodispersed underneath the action of NPCDs. They were small, ranging in size from 19.2 to 51.5 nm, with a mean range of 32.5 nm (Figure 4.7). A crystal of the AuNPs can be seen in a HRTEM view (inset of Figure 4.7D) that was subsequently characterized by XRD.

For the purpose of detecting genetically modified (GM) soybeans, Gao et al. [30] created a label-free electrochemical impedimetric biosensor relying on customized

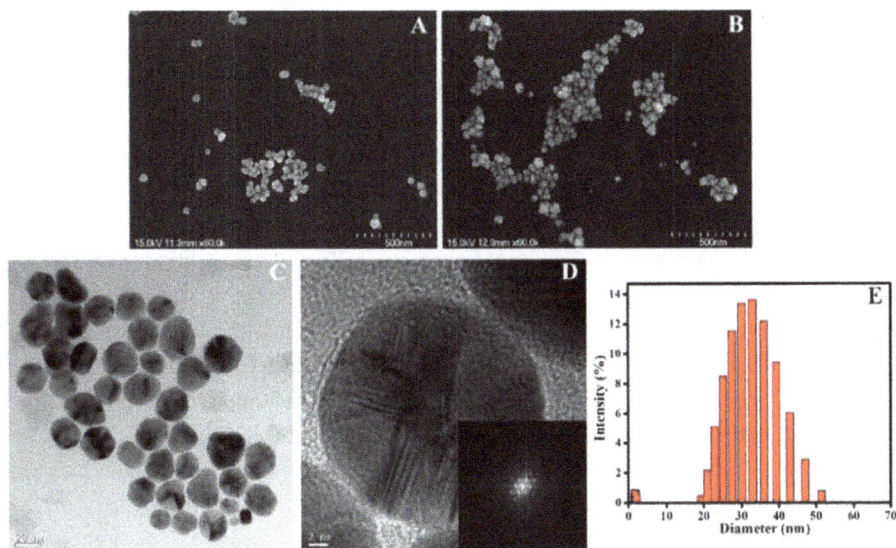

**Figure 4.7:** NPCD–AuNP composites as shown in SEM in the (A) lack and (B) addition of GSH. TEM picture (C). (D) A high-resolution TEM picture with an overlay of an FFT sequence. (E) NPCD–AuNP composites DLS evaluation (adapted from Ref. [29] with permission from [MDPI], [2022]. Distributed under the Creative Common Attribution-based License CCBY 4.0).

screen-printed carbon electrodes with gold CDs (GCDs) [30]. Investigations were done into GCDs' architecture and characteristics. The GCDs were used to increase electrical properties for the creation of DNA sensors and immediately attach to single-stranded DNA probes via the combination of thiol and Au. The shift in electron-transfer resistance (Ret) following DNA hybridization onto sensor surfaces served as a gauge for the measurement of the targeted DNA. With an LOD of $3.1 \times 10^{-14}$ M ($S/N = 3$), the Ret sensitivity (vs. Ag reference electrode) improved with the exponential of targeted DNA amounts under ideal parameters. Additionally, it was shown that the suggested DNA sensor had a significant sensitivity for differentiating between targeted DNA and discordant patterns. Additionally, the created biosensor was used to find SHZD32-1 in genuine specimens, and the outcomes were in good agreement with that of the gel electrophoresis technique. Owing to the unavailability of difficult DNA labelling, the label-free biosensor demonstrated a relatively simple substrate when contrasted to earlier studies for DNA recognition. As a result, the suggested technique demonstrated significant promise as a replacement for a straightforward, precise, targeted, and transportable DNA sensor.

On the basis of triplex DNA labelled using NCDs and AuNPs as acceptors and donors, respectively, in the FRET framework, Mahani et al. [31] formed an innovative turn-on fluorescent biosensor as a framework for the detecting of transcription key component NF p50. Various characterization methods were used to examine the synthesized NPs. When there was no targeted protein present, a labelled DNA molecule was created to create a triplex and to detect its creation via a shift in FRET efficacy. FRET among AuNP and NCD suppressed the fluorescence of CDs around 503 nm (excitation around 460 nm) when the triplex DNA was developing. Moreover, the existence of NF-kB p50 was accompanied by a significant increase in fluorescence strength spurred on by the emergence of single-stranded DNA that had been labelled with AuNPs from the triplex DNA framework, which was then utilized to sensitively identify the transcription component. For the measurement of NF-kB p50, this method demonstrated stability (R2 ¼ 0.9943) in the region of 20e150 pM having a LOD of 9 pM. Additionally, the sequence-specific triplex-based biosensor was highly sensitive in its ability to separate NF-kB p50 from other proteins. According to their findings, the biosensor offers a broadly applicable substrate for the quick identification of NF-kB p50 in synthesized media, offering promise for the prompt identification and treatment of malignancy.

Utilizing CDs, Mohammadi et al. [32] revealed a method for building dual-emission fluorescence sensors and verified its use for amperometric miRNA-21 detection and cell imaging of tumour cells in a micro system. With an excitation frequency of 360 nm, the composition of yellow CDs (Y-CDs) and blue CDs (B-CDs) exhibits dual-emission behaviour with peaks at 409 and 543 nm. The durable and precise attachment of DNA probe designed and synthesized B-CDs to complimentary miRNA-21 targeted caused disruptions in the probing framework and changed the fluorescence intensities in both wavelengths as the quantity of miRNA-21 increased. As a result, the turn-on signal to turn-off signal proportion was significantly changed. As-prepared BY-CDs was constructed as an effective substrate for ratiometric fluorescent miRNA-21 detection, having a broad linear region of 0.15 fM to 2.46 pM and a LOD of 50 aM, by measuring the intrinsic ratiometric fluorescence variance ($\Delta F540$ nm/$\Delta F410$ nm). The suggested test was also used to identify miRNA-21 in diluted human serum specimens exhibiting acceptable rehabilitation and in MCF-7 cell lines having a LOD of 3 cells in 10 µL, suggesting the assay's possibility for use in clinic treatment of the miRNA-associated disorder. More crucially, the photos demonstrated how MCF-7 cells that had been effectively labelled with BY-CDs might demonstrate the viability of the suggested microfluidic apparatus as a cell-capturing tool for bioimaging.

Kalkal et al. [33] proposed the theory that in the instance of COVID-19, CDs produced from AS (AS-CDs) normalized immunological abnormalities and suppressed the generation of pro-inflammatory cytokines. In the field of nanomedicine, CDs have historically been investigated as potential theranostic options for cell imaging and drug/gene distribution. Numerous researches have proven that AS has anti-oxidant and anti-fibrotic attributes. Allicin, the primary effective component of AS, was discovered to have powerful anti-oxidant and reactive oxygen species reducing properties. It explored how antibiotic, anti-microbial, and anti-virus properties could counteract inflammatory impacts and cytokine storms. A new tool in the therapeutic arsenal for the treatment of COVID-19 disease was made possible by the manufacture of theranostic CDs using AS, which also served as a COVID-19 diagnostic reagent. On the basis of the reasoning provided, it was proposed that AS-CDs aid in protection by preventing the spread of disease. The suggested theory might be used as a forward-looking strategy to pave the way for the creation of unique theranostic equipment which might be very effective in handling the continuing COVID-19 situation.

For the purpose of detecting miRNA-21 in MCF-7 cancerous cells, Mohammadi et al. [34] created CD-chitosan nanocomposite hydrogels by treating CDs made from different aldehyde precursors using chitosan. The Schiff base process, which forms imine bonds between the aldehyde groups on the surfaces of CDs and the amine in chitosan, was used to create three luminous hydrogels, which is more significant. Additionally, the hydrogel sheets, CDs, and CD-chitosan nanocomposite hydrogels were evaluated by SEM, TEM as well as UV–vis absorption. The DNA hydrogel bioassay method demonstrated excellent specificity and persistence for miRNA-21, with a LOD of 0.1–125 fM (0.03 fM). The biosensors for specimen assessment showed satisfactory predictability using MCF-7 malignancy cell counts of 1,000–25,000, 1,000–25,000, and 1,000–6,000 cells $mL^{-1}$ with LOD of 364,310 and 552 cells $mL^{-1}$ for nitrobezaldehyde, glutaraldehyde, and benzaldehyde-based nanocomposite hydrogels, correspondingly. Additionally, minimal probe toxicity was shown by cell vitality results; hence nanocomposite hydrogels were used for multi-colour illumination of MCF-7 malignancy cells.

For the quick and ultrasensitive analysis of norovirus (NoV), Achadu et al. [35] combined revolutionary surface-enhanced Raman scattering (SERS) and dual-modality fluorescence (FL) methods into a unitary probe (NoV). The recently created sulphur-doped CDs produced from agar serve as dual-signal improvements for the established FL–SER-based biosensor (S-agCDs). The anti-NoV antibody-conjugated S-agCDs and magnetic silver nanocubes [poly(dop)-MNPs-Ag NCs] produced a core-satellite immunocomplex as a consequence of the antigen-antibody immunoreaction. NoV-like particles (NoV-LPs) were identified using an immunomagnetic enriching technique and the SERS modalities on a single-layer graphene platform over a broad spectrum of 1 fg mL to 10 ng mL having an acceptable LOD of 0.1 fg mL. FL confocal scanning was used to track "hotspots" before SERS detected clinical NoV in faecal samples reached 10 RNA copies $mL^{-1}$, demonstrating the joint benefit of the biosensor's dual-signalling capabilities. The effectiveness of the suggested dual-modality biosensor raises the possibility of a quick and inexpensive NoV diagnosis and monitoring solution for the healthcare system.

## 4.4.2 Metal ion sensing

While certain heavy metals including copper (Cu), chromium (Cr), cobalt (Co), manganese (Mn), zinc (Zn), iron (Fe), and tin (Sn) are necessary for life processes in trace levels, they are poisonous and hazardous to people in higher quantities. Therefore, it is crucial to identify these metal ions, and numerous investigators have used CDs' fluorescence quenching phenomenon for this purpose [36]. For the cellular-level fluorescence turn-off sensing of $Fe^{2+}$, Phan et al. [36] created the NCDs. In this study, they presented inexpensive NCDs that could assess the existence of intracellular $Fe^{2+}$ ions and quickly distinguish $Fe^{2+}$ ions from $Fe^{3+}$ in aqueous media (Figure 4.8). Citric acid was used as a source of carbon and polyethylenimine with a molecular mass of 1800 (PEI1800) as a source of nitrogen in the synthesis of NCDs. These NCDs have minimal toxicity for cell imaging and a brilliant blue fluorescence which response to $Fe^{2+}$ ions. Owing to the coordination bond generation among $Fe^{2+}$ ions and O atoms on the interface of NCDs, the inclusion of $Fe^{2+}$ had a significant fluorescence emission impact, demonstrating that NCDs function as a fluorescent sensor for the durable and sensitive monitoring of $Fe^{2+}$ in living cells with no alterations.

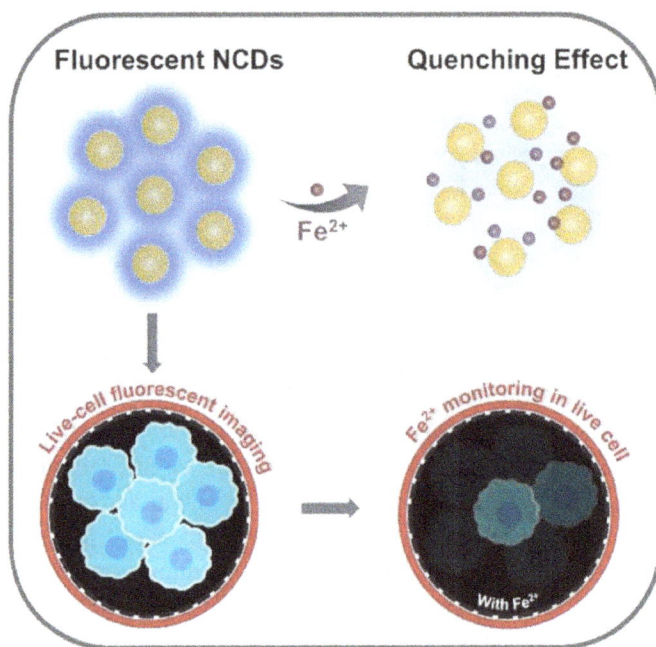

**Figure 4.8:** Fluorescence quenching caused by ferrous ions on CDs is represented, and its use for intracellular $Fe^{2+}$ ion detection is discussed (adapted from Ref. [36] with permission from [MDPI], [2022]. Distributed under the Creative Common Attribution-based License CCBY 4.0).

The fluorescence findings indicate that the fluorescence of NCDs was very responsive to the $Fe^{2+}$ levels in the solvent, which is necessary to systematically assess the detectability of NCDs to $Fe^{2+}$. Along a correlation coefficient ($R^2$) of 0.988, the $F_o/F$ value demonstrated a strong linear connection in the intensity region (0–50) µM as the amount of $Fe^{2+}$ ion raised. The lower LOD at 0.702 µM was estimated using the formula $3S/b$, wherein $S$ and $b$ were the calibrating curve's standard deviation of slope and intercept, correspondingly (Figure 4.9c). They evaluated the brightness variations of an equivalent number of NCDs in the context of various $Fe^{2+}$ ion levels. The fluorescence intensity of NCDs was measured at various periods following the introduction of $Fe^{2+}$ ion in establishing the ideal quench-ing period. By measuring the highest fluorescence emission levels at 445 nm under 365 nm excitation, the spectrums were used to calculate the magnitude of fluorescence strength. Figure 4.9a demonstrates that the fluorescence quenching proportion $F_o/F$ (in which $F_o$ is the fluorescence of the preliminary NCDs and $F$ is the fluorescence following the subjection of $Fe^{2+}$) was significantly enhanced (roughly 4.5-fold rise) all through the preliminary phase of reaction and upon 20 min, this quenching impact stayed un-changed. The following tests were therefore conducted for 2 min. When the amount of $Fe^{2+}$ raises from 0 to 50 µM underneath the ideal live-cell culture parameters (pH 7.4), the fluorescence outputs of NCDs continuously decrease. The fluorescence intensity of the as-prepared NCDs is virtually totally quenched in the addition of 50 µM $Fe^{2+}$ (Figure 4.9b).

Raman spectra validated the potential quenching strategy of $Fe^{2+}$. To gain a better understanding of the chemical characteristics of the as-prepared NCDs both prior and following their interactions with $Fe^{2+}$, the Raman investigation was conducted. Owing to the significant ambient fluorescence intensity in the testing environment, Figure 4.9d demonstrates that NCDs do not produce a Raman signal when $Fe^{2+}$ is absent. Owing to the production of coordination bonds among $Fe^{2+}$ ions and O atoms on the surface func-tional units of NCDs, the inclusion of $Fe^{2+}$ causes a shift in the surface chemistry of the NCDs through chelation. $FeSO_4$'s Raman spectra exhibit a distinct peak at 979 cm$^{-1}$, which is attributable to the existence of an S–O stretching bond. When contrasted to NCDs that were not subjected to $Fe^{2+}$ ions, the Raman spectra of the subjected NCDs show that the peaks at (1,322; 1,587; and 1,454) cm$^{-1}$, respectively, recognized as the disor-der (D) band, crystalline (G) band, and –CH$_2$ stretching vibration are more predominant. This suggests that the initial features of the treated NCDs arise upon the expulsion of the fluorescence context owing to the quenching fluorescent impact of $Fe^2$. As a result of the coordinate bond creation involving the $Fe^{2+}$ and O atoms on the surface of NCDs, a large spike at 609 cm$^{-1}$ observed in the $Fe^{2+}$-treated NCDs was associated with the production of the Fe–O coordinate bond, whereas a faint signal in the similar area was observed for $FeSO_4$ only. These findings supported the electron transmission to $Fe^{2+}$ ions that tran-spired as a consequence of the interactions among the plentiful functional clusters of NCDs and $Fe^{2+}$. In this process, the electrons from NCDs function as a donor of electron pairs, whereas the $Fe^{2+}$ acts as an acceptor that also affects fluorescence quenching as shown in Figure 4.9e. The fluorescence sensitivity of NCDs against $Fe^{2+}$ over factors ions has been studied to verify the specificity of the NCDs nanoprobe in sensing $Fe^{2+}$. In addi-tion to $Fe^{2+}$, NCDs at the similar dosage were also exposed with $Fe^{3+}$, $Ag^+$, $Na^+$, $K^+$, $Ca^{2+}$, $Pb^{2+}$, $Cu^{2+}$, $Co^{2+}$, and $Mg^{2+}$ metal ions. Figure 4.9f and g illustrates that amongst various metal ions, $Fe^{2+}$ treatment of NCDs results in the highest $F_o/F$ value, indicating a consid-erable quenching impact on NCD fluorescent intensity. Reduced fluorescence quenching impact was shown in NCDs exposed with other metal ions. The remarkable sensitivity of these NCDs to $Fe^{2+}$ over $Fe^{3+}$ is particularly apparent in the minimal quenching that was achieved using $Fe^{3+}$ ion. $Co^{2+}$ and $Cu^{2+}$ ions had a small quenching interference impact on NCDs, but this impact was significantly reduced when EDTA was added as a masking

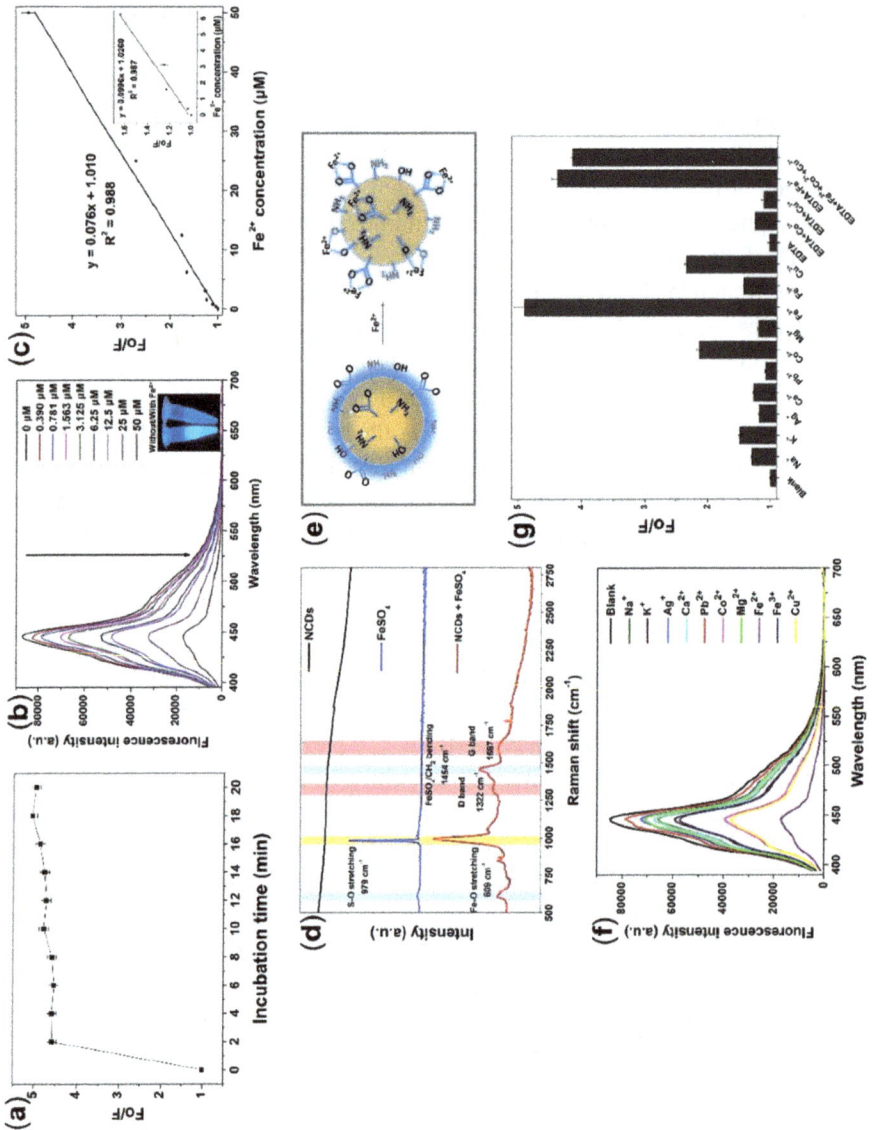

**Figure 4.9: (a)** The effectiveness of NCDs at quenching fluorescence when $Fe^{2+}$ is present and at various incubation times. **(b)** Fluorescence quenching spectrum at 365 nm under 365 nm excitation with $Fe^{2+}$ ion densities ranging from 0 to 50 M. Photos of NCDs solutions with and without ferrous ions taken at 365 nm UV light show the quenching action of the $Fe^{2+}$ ion in the inset. **(c)** Linear association among fluorescence quenching effectiveness ($F_0/F$) and $Fe^{2+}$ level of (0–50) M. $Fe^{2+}$ level of (0–6.25 M) and quenching efficacy are linearly related, exhibiting satisfactory linear calibration (inset). **(d)** The charge-transfer fluorescence quenching caused by chelation among NCDs and $Fe^{2+}$ is visible in the Raman spectrum of NCDs, $FeSO_4$, and NCDs with $Fe^{2+}$ ion. **(e)** Mechanism for chelating quenched fluorescence using electron transfer and $Fe^{2+}$ chelation. **(f)** NCD fluorescence spectrum (adapted from Ref. [36] with permission from [MDPI], [2022]. Distributed under the Creative Common Attribution-based License CCBY 4.0).

reagent. The inclusion of EDTA had no impact on the quenching impact produced by $Fe^{2+}$ because of the quick response. These findings show that the combination formed involving $Fe^{2+}$ and NCDs causes the NCDs fluorescence probe to be preferential for $Fe^{2+}$ over the other metal ions. The quenching impact of NCDs against $Fe^{2+}$ is not substantially influenced by the addition of other metal ions, demonstrating the specificity of NCDs against $Fe^{2+}$ in a complex ions specimen.

The interaction involving the fluorophore moieties on the interface of N-CDs and the OH formed by the catalysed process among copper(II) and cysteine permitted Chen et al. [37] to create a unique fluorescent biosensor for the quick monitoring of copper(II) (Cys). To be more precise, Cu(II) can catalyse the oxidation of Cys to generate cystine (Cys–Cys) and hydrogen peroxide ($H_2O_2$) as well as the breakdown of $H_2O_2$ to generate hydroxyl radicals (OH) through a process similar to Fenton. The surface morphology of N-CDs might be oxidized and destroyed by OH, which causes the N-CDs' fluorescence to be quenched. The LOD for Cu(II) is found to be 23 nmol $L^{-1}$, and the lLOQ was found to be 77 nmol $L^{-1}$ under ideal research circumstances. In addition, a few characterizations are offered to support the suggested idea. The technique has been used to effectively extract Cu(II) in ecological water and human serum with great specificity and improved precision.

Through a simple one-step hydrothermal production, Cai et al. [38] created fluorescent composite nanomaterials relying on an organic dye called 4-methylumbelliferone (4-MU) as modulator and D-arginine as carbon core, giving rise to CDs/4-MU hybrid nanomaterials (CDs-4-MU). Such nanoparticles may enhance the specificity and sensitivity of individual CDs against $Fe^{3+}$ ions in the distinct matrix. The electron transition involving CDs-4-MU and $Fe^{3+}$, which results in the quenching of the fluorescence, was the cause of the sensing method of CDs-4-MU against $Fe^{3+}$. In a Tris–HCl buffer media, the LOD and resulting linear spectrum are 0.68 µM and 2.29–200 µM, correspondingly. Additionally, using realistic specimens like tap water, mixed culture, and foetal bovine serum, this nanomaterial could monitor $Fe^{3+}$ ions. In addition, CDs-4-MU demonstrated excellent biocompatibility and was capable of absorption by MC3T3 cells, making it useful for the measurement of $Fe^{3+}$ ions at the molecular level and bioimaging. As a result, this research offered a flexible method for creating CDs-based mixed nanomaterials and opened a novel route for enhancing ion sensing in realistic specimens that is important for real-world applications.

Liang et al. [39] created a unique, very effective ECL emits by doping CDs and AuNPs into a zeolitic imidazolate framework (ZIF-8). Their research showed that ZIF-8 in the nanocomposites served as an ECL throttle to turn CDots into CDots•, which resulted in the increased ECL strength of CDots in $K_2S_2O_8$ solution in addition to serving as a transporter for ECL markings. To further improve the ECL signals, AuNPs were established in the cavities of ZIF-8, which possess high conductivity. In comparison to the pure CDots, the nanocomposites ultimately displayed an ECL intensity that was about seven times higher. The nanocomposites were perfect for creating biosensors because the AuNPs in them offered a framework for efficiently adhering the biomolecule. The produced CDots@ZIF-8/AuNPs were effectively used as a prototype for the quick and easy ECL monitoring of caspase-3 function throughout cell apoptosis. This inspires the development of new ECL luminous substances with potential uses in very sensitive bioanalysis.

By hydrothermally carbonizing the anaerobic bacterium *Fusobacterium nucleatum*, Liu et al. [40] created spectacular CDs that functioned as a biosensor for quick monitoring of intracellular $Fe^{3+}$. The produced *F. nucleatum*-CDs (Fn-CDs) had great biocompatibility, great durability, and bright fluorescence. The generated Fn-CDs had great capabilities for cell monitoring and biomedical labelling and were simple to internalize into both plant and human cells. Upon entering cells, Fn-CDs can continue to glow for up to additional 24 h. Additionally, the fluorescent Fn-CDs showed tremendous promise for use in in vivo sensing of $Fe^{3+}$ ions since they were very robust to the existence of $Fe^{3+}$ ions even in cells. The Fn-CDs were also safe for the mice, diffused quickly, and were quickly eliminated, offering them an ideal choice for in vivo cell imaging and biosensing.

### 4.4.3 Detection of organic molecules

Besides the detection of metal ions, the fluorescence feature of CDs was well utilized for the sensing of other organic species. For the label-free sensing of vitamin $B_{12}$ ($VB_{12}$), exogenous/endogenous peroxynitrite (ONOO) sensing, fluorescent pliable layer, and cell imaging fabrication, Meng et al. [41] created orange emitting luminous multi-functional CDs (O-CDs). Ethanol and safranine T were used as the precursors in a one-step hydrothermal method to create the O-CDs with excitation-independent properties. O-CDs' fluorescence was quenched by using $VB_{12}$ as a quencher because of the IFE. The LOD was determined to be 0.62 µM and the two-segment linear sensitivities were 1–65 µM and 70–140 µM. Additionally, ONOO, depending on static quenching, may diminish the fluorescence level of O-CDs (SQ). The LOD was 0.06 M, and the linearity limits are 0.3–9 µM and 9–48 µM. Due to their excellent bioactivity, minimal cytotoxicity, and good photostability, O-CDs were also used as a cellular imaging agent for intracellular $VB_{12}$ and exogenous/endogenous ONOO imaging. These findings suggest that O-CDs possess the ability to be employed as a highly selective and quick fluorescent probe for the quick monitoring of $VB_{12}$ and exogenous/endogenous ONOO in living cells. The O-CDs as suggested could also be used to create a fluorescent elastic layer by combining them with PVA. The aforementioned information demonstrates that O-CDs have a bright future in a variety of fields, including biosensing, biological labelling, therapeutic optical imaging, and fluorescent coatings.

Wang et al. [42] used citric acid as the source of carbon and ethylenediamine as the nitrogen supply to hydrothermally create water-soluble fluorescent CDs. The recurrent and scaled-up synthetic studies aimed to investigate the viability of CD manufacturing on a broad level. The combination of the CDs solutions and $Fe^{3+}$ solutions produced the

CDs/$Fe^{3+}$ hybrid. By using UV and fluorescence spectrometry, the optical characteristics, pH dependency, and durability behaviour of CDs or the CDs/$Fe^{3+}$ composite were investigated. The fluorescence emission of the CDs was evaluated at $\lambda_{ex}$ = 360 nm and $\lambda_{em}$ = 460 nm in accordance with the concepts of fluorescence quenching following the introduction of $Fe^{3+}$ and subsequently fluorescence restoration after the subjection of ascorbic acid. By doing a statistical examination of the fluctuating fluorescence strength, the ascorbic acid amount was determined. Figure 4.10 shows that ascorbic acid content rose, and CDs/$Fe^{3+}$ fluorescence intensity progressively improved.

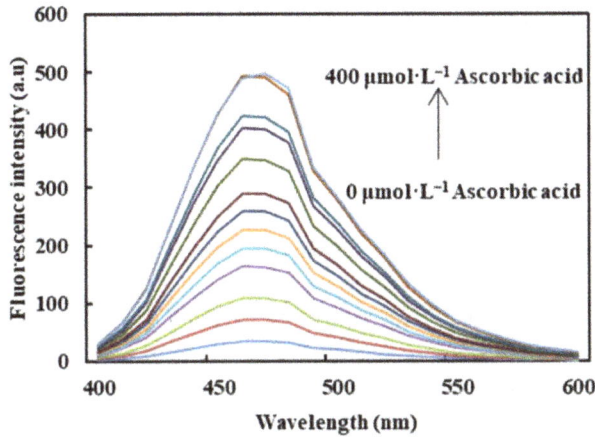

**Figure 4.10:** Impact of ascorbic acid quantity on the CDs/$Fe^{3+}$ composite's fluorescence strength (adapted from Ref. [42] with permission from [MDPI], [2021]. Distributed under the Creative Common Attribution-based License CCBY 4.0).

When the CDs/$Fe^{3+}$ composite was used to analyse various bioactive compounds, it was discovered that it specifically recognized ascorbic acid and displayed a remarkable linear association in the range of 5.0–350.0 μmol L$^{-1}$. The LOD was 3.11 μmol L$^{-1}$ as well. The hybrid mixture of CDs/$Fe^{3+}$ and potassium ferricyanide was yellow in the absence of ascorbic acid, as depicted in Figure 4.11. The response went from yellow to blue in the ascorbic acid-containing media. This event demonstrated that the ascorbic acid addition converted ferric to divalent iron ions in the CDs/$Fe^{3+}$ composite solution and established a combination with potassium ferricyanide. The outcomes demonstrate that the introduction of ascorbic acid changed the system's trivalent iron into bivalent iron and recovered the CDs/$Fe^{3+}$ composite solution's fluorescence.

For the "turn-on" assessment of pH and tetracycline antibiotics (TCs), such as oxytetracycline (OTC), TC, and chlortetracycline (CTC), Li et al. [43] created red-fluorescent CDs (R-CDs). Utilizing neutral red and thiourea as precursors, the CDs were synthesized utilizing a one-pot hydrothermal process. As-obtained CDs have minimal toxicity, excellent light absorption and bio-compatibility, and outstanding photobleaching resilience.

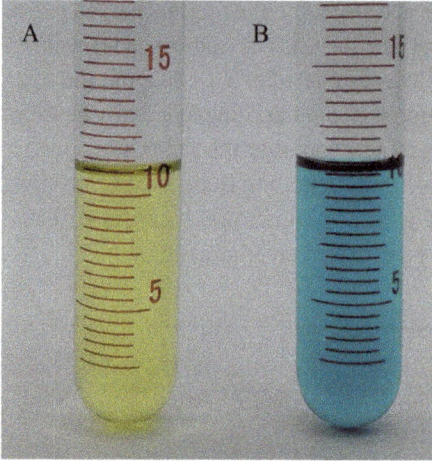

**Figure 4.11:** The colour modified upon the subjection of potassium ferricyanide to the CDs/Fe$^{3+}$ media excluding and including ascorbic acid (adapted from Ref. [42] with permission from [MDPI], [2021]. Distributed under the Creative Common Attribution-based License CCBY 4.0).

Significantly, following the insertion of TCs relying on the agglomeration-induced emission enhancement process, the CDs showed a considerable emission improvement. The LOD for TC, OTC, and CTC is 12, 23, and 25 nM, correspondingly, and the linear regression recognition limits were 3–35 M, 2–50 µM, and 5–50 µM, respectively. The suggested technique has been effectively used to identify TCs in milk products. The CDs also demonstrated a specific H$^{+}$-sensitive characteristic, which was utilized to create a fluorescent pH sensor having a linear region of 6.0–8.0 and a p$K_a$ of 7.08. Importantly, the nanosensor has been enhanced to track TCs and pH in cellular systems, illuminating the potential of the suggested approach for biosensing applications.

Utilizing a commercially available optical fibre spectrometer (OFS) and a customized colour sensor device, Choudhary et al. [44] showed a carbon quantum dot (CQD)-dependent fluorescence method for milk quality evaluation and monitoring of urea adulteration (CSD). A deviation from milk's ideal pH of 6.7 suggests spoiling. Phthalic acid (a carbon source) and triethylenediamine were used to create the CQDs (passive). The CQDs were discovered to have a quantum output of around 18.9% and to be selective in the pH region of 3–10. In terms of mean percentage yield, the reliability of spoilage detection utilizing CQDs was determined to be 99.2% ($R^2 = 0.97$) by the OFS and 99.59% ($R^2 = 0.98$) by the CSD. The mean percentage recovery for urea adulteration in milk determined by the OFS and CSD, respectively, was 103.0% ($R^2 = 0.98$) and 97.9% ($R^2 = 0.97$). The findings show that CQDs have outstanding possibilities for application as biosensors for detecting deterioration and adulteration.

CDs were created by Sangubotla and Kim [45] utilizing microwave irradiation, curcumin, and dimethylformamide. The resultant CDs were known as CDD–CDs. These were known as APT–CDs because they had been functionalized using the silicon precursor

3-(aminopropyl)-triethoxysilane. In order to create a unique bioprobe, laccase was additionally covalently attached to the APT-CDs. At 430, 380, and 360 nm excitation wavelengths, correspondingly, the CDD–CDs, APT–CDs, and bioprobe displayed orange ($\lambda_{em}$ = 586 nm), green ($\lambda_{em}$ = 533 nm), and blue ($\lambda_{em}$ = 476 nm)-coloured emissions. Quantum rates of 14.8% and 10.2% were seen in the CDD–CDs and bioprobe, correspondingly. In several solvents, the solvatochromism of the CDD–CDs was visible. With a LOD of 41.2 nM, the bioprobe revealed a considerable fluorescence quenching for dopamine in the linear region of 0–30 M. Leveraging ethyl cellulose and a straightforward dip-coating technique, a bioprobe was mounted on a tapered optical fibre and tested for use in multi-colour imaging techniques. In the dopamine range of concentrations of 0–10 M, the resultant tapered optical fibre reached a good LOD of 46.4 nM. The bioprobe showed outstanding multi-colour imaging capability, excellent biocompatibility, lengthy light absorption, and heat resistance. It also had adequate cytotoxicity against human neuroblastoma cells (SH-SY5Y). Human serum and cerebrospinal fluid were used to test the bioprobe's viability.

## 4.4.4 Pesticides

Pesticides have been used extensively to preserve agricultural crops against pests, weeds, and diseases for a long time, encompassing the earlier mineral kinds and the modern chemically manufactured varieties. Pesticide usage increased globally by 69.7% between 1990 and 2018, and it is even more than twice in large agricultural generating regions like Asia and North America. It has been a worldwide issue that the massive residue left behind from long-term misuse, overuse, and mismanagement of pesticides will seriously endanger the environment and population health. If a pesticide is not reduced over the period, it will persist in the vegetation, sediment, and groundwater. It will then be reallocated geographically via circulating water and develop over time. Consuming foodstuffs and related goods over the long term that have high pesticide residue levels will result in metabolic diseases, immune response issues, and other ailments. Including a significant number of deaths each year, the health threat for professional groups subjected to elevated pesticide concentration is probably certain. The build-up of pesticides has sped up some extinction of species and decreased the effectiveness of natural remedies. Then, numerous nations and worldwide organizations around the globe have implemented a number of metrics to enhance the influence of pesticides. These include, but are not restricted to, outlawing and expelling high-toxic pesticides from the industry promptly, limiting the utilization of high-risk pesticides and progressively substituting other low-toxic agrochemicals, establishing the highest residue concentrations, establishing innovative environmentally friendly and generally safe bioactive pesticides, and more. The reliable and sensitive detection of pesticide residues in various matrices is necessary for the execution of these procedures. In this respect, optical sensing technology has recently drawn a lot of interest in detecting trace pesticides due to their inherent ease, excellent sensibility, and compact apparatus and also simple, rapid, cost-effective on-site detection. For ensuring the great sensibility of the optical sensing systems for pesticides, it is crucial that the optical signals acquired on these detecting systems are properly originated from the luminous substances. While rare earth nanomaterials, organic fluorescent dyes, semiconductor QDs, and proteins are frequently utilized as conventional luminous probes due to their unique benefits, they have unavoidable drawbacks that limit their practical deployment. The last few generations have seen a rise in the exploration on CDs, functioning as novel substitutes due to their distinct chemical compositions and optical characteristics. This is due to the emergence of improved functional materials.

Beginning with 4-chloro-1,2-diaminobenzene and dopamine, Yang et al. [46] synthesized chlorine and nitrogen dually doped CDs (N,Cl-CDs) utilizing the hydrothermal process. Intense orange fluorescence was present in the N, and Cl-CDs, having excitation/emission peaks at 420/570 nm and relatively significant quantum efficiency (15%). Acetylcholinesterase (AChE) action and organophosphate pesticides (OPs), that are enzyme inhibitors, were both detected using the N,Cl-CDs. AChE catalyses the enzymatic breakdown of acetylthiocholine to create thiocholine that causes Ellman's reagents to degrade and yield a yellow-coloured result (2-nitro-5-thiobenzoate anion). The material has an IEF on the N,Cl-CDs' fluorescence. The LOD is 2 mU $L^{-1}$, and fluorescence declines exponentially in the 0.017–5.0 U $L^{-1}$ AChE action spectrum. IFE is less-produced and fluorescence recovery increases as a result of the function of AChE becoming further hindered in the presence of organophosphates. This was applied in order to quantify OPs. Performance is exponential in the OP range of concentrations of 0.3–1,000 µg $L^{-1}$, with a LOD of 30 ng $L^{-1}$.

Ingeniously, Huang et al. [47] created a very precise ratiometric fluorescent probe based on the IFE among nitrogen-doped CDs (N-CDs) and 2,3-diaminophenazine to measure OPs in tap water and food as well as to assess AChE activity (DAP). 1,2-Ethanediamine and pancreatin were used as precursors in a one-pot hydrothermal process to create N-CDs. On human cervical cancer HeLa cells and human embryonic kidney 293 T cells, N-CDs demonstrated outstanding fluorescence characteristics and minimal cytotoxicity, pointing to potential future medical uses. Acetylcholine was catalysed by the combination of AChE and choline oxidase to form choline, which was then further oxidised to generate $H_2O_2$. o-Phenylenediamine and $H_2O_2$ interacted in the context of horseradish peroxidase to create fluorescent DAP. So, among N-CDs having a 450 nm fluorescence signal and DAP with a 574 nm fluorescence signal, there was a ratiometric fluorescent probing substrate via IFE. The enzymatic action of AChE was permanently inhibited by OPs, which eventually caused the quantity of DAP to drop and the ratiometric fluorescent signals to change. Such fluorescent probe demonstrated comparatively reduced LOD of 0.38 U $L^{-1}$ for AChE, 3.2 ppb for dichlorvos, and 13 ppb for methyl-parathion in perfect circumstances. This ratiometric fluorescent probe's actual use in OP detection was subsequently confirmed in tap water and food specimens, exhibiting positive outcomes that were substantially consistent with the GC–MS findings.

Utilizing a new conjugated polymer called poly[1-(5-(4,8-bis(5-(2-ethylhexyl)thio-phen-2-yl)benzo[1,2-b:4,5-b′] dithiophen-2-yl]furan-2-yl] (PFTBDT) and CDs, Yasa et al. [48] created an appropriate and innovative framework to detect catechol. The op-toelectronic characteristics of PFTBDT were studied using spectroelectrochemical and electrochemical experiments after the synthesis of PFTBDT and CDs [48]. The lac-case enzyme was immobilized on the graphite electrode's produced layer substrate. The designed biosensor was discovered to possess a straight spectrum of 1.25–175 µM, a narrow LOD of 1.23 µM, and sensitivity of 737.44 µA mM$^{-1}$ cm$^{-2}$. Subsequently, a tap water specimen was used to assess the suitability of the suggested enzyme biosensor, and an acceptable recovery (96–104%) for catechol measurement was attained.

## 4.5 Mechanism of CDs

### 4.5.1 On–off mechanism

The on–off classification (also known as fluorescence quenching) is perhaps the most prevalent sensing layout for sensors using CDs between the disclosed sensing mecha-nisms. The method has been used to find different anions, cations, and tiny molecules. For example, electrons are transmitted from CDs to the vacant d orbitals of metallic ions as a result of the interaction of metallic ions by CDs, which causes fluorescence emission. The exterior functional units of CDs which could coordinate metal ions (in-cluding Cu2þ and Fe3þ) in biological processes have been the subject of numerous in-vestigations. For instance, CDs having a lot of phenolic hydroxyl groups had a high binding propensity for Fe$^{3+}$, which caused the Fe$^{3+}$ d orbitals to break during chela-tion. Fluorescence was produced as a consequence of a substantial electron transmis-sion from the photo-excited CDs to Fe3þ. Additionally, certain biologically active chemicals, like glucose, could preferentially attach to CDs made from phenylboronic acid, enabling "on-off" sensing of glucose once more. The cis-diols unit of glucose re-acts with the boronic acid units on the surfaces of the CDs in this procedure, suppress-ing the fluorescence via agglomeration phases.

### 4.5.2 Off–on mechanism

The "off–on" technique makes use of a rise in CDs' fluorescence intensity when an ana-lyte is present. By putting the repressed CDs back into the emission phase, this method allows for the measurement of the concentration of the analyte. The contact among the CDs and the quencher is broken down by a more favourable association or interaction with the analyte, allowing the CDs to restore their distinctive fluorescence. There are typically two methods for such "off–on" monitoring. A CD's surface must first have the quencher removed, and then it must be released from the quencher. Cation- or anion-mediated techniques are typically used in the first scenario. For example, following fluorescence quench, the luminescent characteristics of the CDs could be restored by the introduction of pyrophosphate (PPi), allowing for the identification of PPi as well as the "off–on" monitoring of PPi for fluorescence microscopy of intracellular PPi. Additionally, CDs treated with iodine could be utilized to assess the fluorescence turn-on of biological thiols as biological thiols and iodine may perform a substitution activity (Cys, GSH, and Hcy). For the alternative technique, metal nanoparticles (e.g. Au and Ag NPs) and multi-layer transition-metal oxides or disulphides (e.g. $MoS_2$ and $MnO_2$), GO, and other nanomaterials are typically used as quenchers. Accordingly, the presence of analytes causes CDs to be released from the nanomaterials, allowing for fluorescence restoration.

### 4.5.3 Fluorescence shift mechanism

The relative spectrum overlapping among donor emissions and receiver absorption enables the ratiometric identification of solutes, in contrast to the "on–off" detection method that quells the emissions of a particular fluorophore. By contrasting the strengths at two emission spectra, the quantity of the analyte could be expressed as a proportion. In comparison to assessments relying on a particular fluorescence maxima, the proportion shift offers a much more precise and robust output signals through the self-calibration of the two emission zones. Utilizing the IFE involving copper nano-clusters (CuNCs) and S-doped CDs and a hybrids matrix, the dual output ratiometric fluorescence probes was effectively used in the specific ratiometric detecting of di-notefuran (DNF). Presently, scientists are focusing mostly on monitoring specimens in real-time while attempting to minimize the laborious and time-consuming specimen preparatory procedures frequently connected with traditional analytical tools. Owing to their accessibility and convenience of usage, smartphones are now frequently utilized for real-time surveillance as analysers alongside a growing requirement for sensitive diagnostic rates and excellent sensitivities.

## 4.6 Challenges and future outlooks

People's concerns over living a healthier life have been made worse by the rising levels of environmental pollutants around the globe, which has sped up the production of equipment for detecting effluent leftovers. In recent times, optical sensing techniques which incorporate the intriguing benefits of CDs with the outstanding benefits of optical sensing have drawn increasing amounts of recognition in a variety of disciplines, offering strong precedents and a range of affordable and practical alternatives for environmental monitoring. As a type of multi-function nanomaterial, CDs could be used alone or in conjunction with other substances to generate optical sensors for the accurate detection of chemicals. CDs could also serve as luminophores, identifying components, transducers, and catalytic. By modifying the synthesis methods and circumstances, it is possible to modify the internal characteristics of CDs and create a variety of optical sensors. As a vast collection, CDs revealed a variety of intriguing +98 functions that can be used for optical chemical sensing research. There are various difficulties, including a shortage of coherence or an absence of association with agrochemical fluorescence sensing; however, the synthesis procedures, sensing processes, and operational relevance of CDs have all been extensively discussed in the literature. While CDs-based optical sensors have shown promising uses and wide-ranging potential in chemical assessment, development-related optical sensors still confronts several difficulties, including the lack of high-performance CDs and insufficiently effective methods for large-scale production and processing. The synthesized CDs' luminescence sensitivities are often weaker than those of traditional sensing elements as well as some CDs are just barely able to detect trace pollutants. The creation of pertinent sensors is hampered by the inadequate investigation of good CDs having long-wavelength emission, luminescent, and up-conversion luminosity. Due to the incompleteness of CD description approaches, they are sometimes used for blind detection. The production of new synthetic methods, including the magnetostrictive hyperthermia technique, and the enhancement of synthesis procedures via the introduction of some effective purification methods for large-scale fabrication, and the advancement of a thorough productivity assessment system, must be the primary goals of future studies. Inadequate purification procedures are another factor that has contributed to misunderstandings regarding the composition and features of CDs. Further research must focus more on the creation of diverse purification techniques to raise the sensibility of dictation, lower the price of analysis, and increase the purification, efficiency, and detection of the impact of CDs.

## 4.7 Conclusion

One-pot CD formulation is rapidly gaining popularity, mostly because of its ability to minimize waste and by-products. Natural C sources have been investigated as basic resources for the production of CDs from a variety of resources comprising food products, biomass production, organic wastes, and animal materials. Environmentally sustainable natural resources, in contrast to synthesized precursors, are rich in C and N supplies in the form of carbohydrates and also act as self-passivating reagents to produce surface-functionalized CDs. Since they are the repositories for organic compounds, green resources do not require an external layer. The utilization of green sources has many benefits because they are readily available, inexpensive, somewhat clean in their reactions, and non-toxic. As powerful and persistent biosources for the production of carbonaceous nanodots in aqueous solutions, parts of the plant like petals, berries, nuts, and stalks that include a variety of basic, acidic, and neutral bioactive components compounds are particularly intriguing. Despite numerous efforts to offer a reasonable solution, the process of CD production is still not fully recognized. The output, uniformity, and luminescence quality of CDs are all influenced by several variables including the ambient temperature, the solvent used, precursor amount, and pH. While CDs are thought to be prospective biosensing possibilities, some obstacles need to be removed, including their inability to a breakdown organic solvents other than water and their weak long-wavelength luminescence. Even though the cause of the PL in CDs remains up for debate, the existence of surface flaws, particulate density, and various functional units all have a significant role in the fluorescence qualities. Although nanotechnology has expanded into every field, it still has problems with large-scale production and simplicity of replication. The cost-effective usage of nanoparticles is constrained by the need for costly chemical compounds and advanced lab equipment; as a result, using natural sources reduces costs and provides an opportunity to mass manufacturing. Further research might use readily accessible natural resources to create CDs with the requisite high stability and outstanding fluorescence characteristics. Consequently, the need for more simple synthetic techniques and environmentally friendly sources to produce superior photoluminescent CDs on a massive scale remains immediate. We predict that this assessment will act as a springboard for additional study in this field to create intelligent and smart embedded sensors.

# References

[1]    Ravichandran R, Nanjundan S, Rajendran N. Effect of benzotriazole derivatives on the corrosion of brass in NaCl solutions. Appl Surf Sci. 2004, 236(1–4), 241–250.

[2]    Khaled KF, Amin MA. Dry and wet lab studies for some benzotriazole derivatives as possible corrosion inhibitors for copper in 1.0 M HNO3. Corros Sci [Internet]. 2009, 51(9), 2098–2106. Available from: http://dx.doi.org/10.1016/j.corsci.2009.05.038.

[3]    Khan PF, Shanthi V, Babu RK, Muralidharan S, Barik RC. Effect of benzotriazole on corrosion inhibition of copper under flow conditions. J Environ Chem Eng [Internet]. 2015, 3(1), 10–19. Available from: http://dx.doi.org/10.1016/j.jece.2014.11.005.

[4]    Abdullayev E, Price R, Shchukin D, Lvov Y. Halloysite tubes as nanocontainers for anticorrosion coating with benzotriazole. ACS Appl Mater Interfaces. 2009, 1(7), 1437–1443.

[5]    Thakur A, Kaya S, Abousalem AS, Kumar A. Experimental, DFT and MC simulation analysis of Vicia Sativa weed aerial extract as sustainable and eco-benign corrosion inhibitor for mild steel in acidic environment. Sustain Chem Pharm [Internet]. 2022, 29(July), 100785. Available from: https://doi.org/10.1016/j.scp.2022.100785.

[6]    Thakur A, Kumar A. Recent advances on rapid detection and remediation of environmental pollutants utilizing nanomaterials-based (bio)sensors. Sci Total Environ [Internet]. 2022, 834 (January), 155219. Available from: https://doi.org/10.1016/j.scitotenv.2022.155219.

[7]    Thakur A, Kumar A, Sharma S, Ganjoo R, Assad H. Materials Today: Proceedings Computational and experimental studies on the efficiency of Sonchus arvensis as green corrosion inhibitor for mild steel

in 0. 5 M HCl solution. Mater Today Proc [Internet]. 2022, xxxx. Available from: https://doi.org/10.1016/j.matpr.2022.06.479.

[8]   Ganjoo R, Sharma S, Thakur A, Kumar A. Thermodynamic study of corrosion inhibition of Dioctylsulfosuccinate Sodium Salt as corrosion inhibitor against mild steel in 1 M HCl. Mater Today Proc [Internet]. 2022, xxxx, 1–5. Available from: https://doi.org/10.1016/j.matpr.2022.05.594.

[9]   Thakur A, Kaya S, Abousalem AS, Sharma S, Ganjoo R, Assad H, et al. Computational and experimental studies on the corrosion inhibition performance of an aerial extract of Cnicus Benedictus weed on the acidic corrosion of mild steel. Process Saf Environ Prot [Internet]. 2022, 161, 801–818. Available from: https://doi.org/10.1016/j.psep.2022.03.082.

[10]  Minh L, Phan T. Fluorescent carbon dot-supported imaging-based biomedicine. Bioinorg Chem Appl. 2022, 2022, 1–32.

[11]  Nguyen TN, Le PA, Phung VBT. Facile green synthesis of carbon quantum dots and biomass-derived activated carbon from banana peels: Synthesis and investigation. Biomass Convers Bioref. 2022, 12(7), 2407–2416.

[12]  Lee A, Kang W, Choi JS. Highly enhanced enzymatic activity of mn-induced carbon dots and their application as colorimetric sensor probes. Nanomaterials. 2021, 11(11), 3046.

[13]  Zhang ZY, Huang LX, Xu ZW, Wang P, Lei Y, Liu AL. Efficient determination of pml/rara fusion gene by the electrochemical dna biosensor based on carbon dots/graphene oxide nanocomposites. Int J Nanomedicine. 2021, 16, 3497–3508.

[14]  Chen J, Yan J, Dou B, Feng Q, Miao X, Wang P. Aggregatable thiol-functionalized carbon dots-based fluorescence strategy for highly sensitive detection of glucose based on target-initiated catalytic oxidation. Sensors Actuators B Chem [Internet]. 2021, 330(August 2020), 129325. Available from: https://doi.org/10.1016/j.snb.2020.129325.

[15]  Giang NN, Won HJ, Lee G, Park SY. Cancer cells targeted visible light and alkaline phosphatase-responsive $TiO_2/Cu^{2+}$ carbon dots-coated wireless electrochemical biosensor. Chem Eng J [Internet]. 2021, 417(February), 129196. Available from: https://doi.org/10.1016/j.cej.2021.129196.

[16]  Thakur A, Kumar A. Sustainable inhibitors for corrosion mitigation in aggressive corrosive media: A comprehensive study. J Bio- Tribo-Corrosion [Internet]. 2021, 7(2), 1–48. Available from: https://doi.org/10.1007/s40735-021-00501-y.

[17]  Bashir S, Thakur A, Lgaz H, Chung I-M, Kumar A. Computational and experimental studies on Phenylephrine as anti-corrosion substance of mild steel in acidic medium. J Mol Liq. 2019, 293, 111539.

[18]  Bashir S, Lgaz H, Chung IM, Kumar A. Effective green corrosion inhibition of aluminium using analgin in acidic medium: An experimental and theoretical study. Chem Eng Commun [Internet]. 2020, 0(0), 1–10. Available from: https://doi.org/10.1080/00986445.2020.1752680.

[19]  Thakur A, Kumar A, Kaya S, Vo DVN, Sharma A. Suppressing inhibitory compounds by nanomaterials for highly efficient biofuel production: A review. Fuel [Internet]. 2022, 312(September 2021), 122934. Available from: https://doi.org/10.1016/j.fuel.2021.122934.

[20]  Bashir S, Thakur A, Lgaz H, Chung I-M, Kumar A. Corrosion inhibition performance of acarbose on mild steel corrosion in acidic medium: an experimental and computational study. Arab J Sci Eng [Internet]. 2020, 45(6), 4773–4783. Available from: https://doi.org/10.1007/s13369-020-04514-6.

[21]  Thakur A, Kumar A. Potential of natural product extract as green corrosion inhibitors for steel- a review. J Gujarat Res Soc. 2019, 21(8), 200–212.

[22]  Kumar A, Thakur A Encapsulated nanoparticles in organic polymers for corrosion inhibition [Internet]. Corrosion Protection at the Nanoscale. Elsevier Inc.; 2020. 345–362 p. Available from: http://dx.doi.org/10.1016/B978-0-12-819359-4.00018-0

[23]  Zulfajri M, Sudewi S, Ismulyati S, Rasool A, Adlim M, Huang GG. Carbon dot/polymer composites with various precursors and their sensing applications: A review. Coatings. 2021, 11(9), 1100.

[24] Monday YN, Abdullah J, Yusof NA, Rashid SA, Shueb RH. Facile hydrothermal and solvothermal synthesis and characterization of nitrogen-doped carbon dots from palm kernel shell precursor. Appl Sci. 2021, 11(4), 1–17.

[25] Ge J, Shen Y, Wang W, Li Y, Yang Y. N-doped carbon dots for highly sensitive and selective sensing of copper ion and sulfide anion in lake water. J Environ Chem Eng [Internet]. 2021, 9(2), 105081. Available from: https://doi.org/10.1016/j.jece.2021.105081.

[26] Gao Y, Yu H, Tian J, Xiao B. Nonenzymatic DNA-based fluorescence biosensor combining carbon dots and graphene oxide with target-induced DNA strand displacement for microRNA detection. Nanomaterials. 2021, 11(10), 2608.

[27] Sri S, Lakshmi GBVS, Gulati P, Chauhan D, Thakkar A, Solanki PR. Simple and facile carbon dots based electrochemical biosensor for TNF-α targeting in cancer patient's sample. Anal Chim Acta [Internet]. 2021, 1182, 338909. Available from: https://doi.org/10.1016/j.aca.2021.338909.

[28] Li G, Chen W, Mi D, Wang B, Li H, Wu G. et al. A highly sensitive strategy for glypican - 3 detection based on aptamer/gold carbon dots/magnetic graphene oxide nanosheets as fluorescent biosensor. Anal Bioanal Chem [Internet]. 2022, Available from: https://doi.org/10.1007/s00216-022-04201-5.

[29] Le TH, Kim JH, Park SJ. Turn on fluorescence sensor of glutathione based on inner filter effect of co-doped carbon dot/gold nanoparticle composites. Int J Mol Sci. 2022, 23(1), 190.

[30] Gao H, Cui D, Zhai S, Yang Y, Wu Y, Yan X. et al. A label-free electrochemical impedimetric DNA biosensor for genetically modified soybean detection based on gold carbon dots. Microchim Acta [Internet]. 2022, 189(6), 1–9. Available from: https://doi.org/10.1007/s00604-022-05223-7.

[31] Mahani M, Taheri M, Divsar F, Khakbaz F, Nomani A, Ju H. Label-free triplex DNA-based biosensing of transcription factor using fluorescence resonance energy transfer between N-doped carbon dot and gold nanoparticle. Anal Chim Acta [Internet]. 2021, 1181, 338919. Available from: https://doi.org/10.1016/j.aca.2021.338919.

[32] Mohammadi S, Salimi A, Hoseinkhani Z, Ghasemi F, Mansouri K. Carbon dots hybrid for dual fluorescent detection of microRNA-21 integrated bioimaging of MCF-7 using a microfluidic platform. J Nanobiotechnology [Internet]. 2022, 20(1), 1–15. Available from: https://doi.org/10.1186/s12951-022-01274-3.

[33] Kalkal A, Allawadhi P, Pradhan R, Khurana A, Bharani KK, Packirisamy G. Allium sativum derived carbon dots as a potential theranostic agent to combat the COVID-19 crisis. Sensors Int [Internet]. 2021, 2(May), 100102. Available from: https://doi.org/10.1016/j.sintl.2021.100102.

[34] Mohammadi S, Mohammadi S, Salimi A. A 3D hydrogel based on chitosan and carbon dots for sensitive fluorescence detection of microRNA-21 in breast cancer cells. Talanta [Internet]. 2021, 224, 121895. Available from: https://doi.org/10.1016/j.talanta.2020.121895.

[35] Achadu OJ, Abe F, Hossain F, Nasrin F, Yamazaki M, Suzuki T. et al. Sulfur-doped carbon dots@polydopamine-functionalized magnetic silver nanocubes for dual-modality detection of norovirus. Biosens Bioelectron [Internet]. 2021, 193(July), 113540. Available from: https://doi.org/10.1016/j.bios.2021.113540.

[36] Phan LMT, Hoang TX, Cho S. Fluorescent carbon dots for sensitive and rapid monitoring of intracellular ferrous ion. Biosensors. 2022, 12(1), 0–9.

[37] Chen S, Chen C, Wang J, Luo F, Guo L, Qiu B. et al. A bright nitrogen-doped-carbon-dots based fluorescent biosensor for selective detection of copper ions. J Anal Test [Internet]. 2021, 5(1), 84–92. Available from: https://doi.org/10.1007/s41664-021-00162-3.

[38] Cai H, Zhu Y, Xu H, Chu H, Zhang D, Li J. Fabrication of fluorescent hybrid nanomaterials based on carbon dots and its applications for improving the selective detection of Fe (III) in different matrices and cellular imaging. Spectrochim Acta – Part A Mol Biomol Spectrosc [Internet]. 2021, 246, 119033. Available from: https://doi.org/10.1016/j.saa.2020.119033.

[39] Liang GX, Zhao KR, He YS, Liu ZJ, Ye SY, Wang L. Carbon dots and gold nanoparticles doped metal-organic frameworks as high-efficiency ECL emitters for monitoring of cell apoptosis. Microchem J [Internet]. 2021, 171(June), 106787. Available from: https://doi.org/10.1016/j.microc.2021.106787.

[40] Liu L, Zhang S, Zheng X, Li H, Chen Q, Qin K. et al. Carbon dots derived from: Fusobacterium nucleatum for intracellular determination of $Fe^{3+}$ and bioimaging both in vitro and in vivo. Anal Methods. 2021, 13(9), 1121–1131.

[41] Meng Y, Jiao Y, Zhang Y, Lu W, Wang X, Shuang S. et al. Facile synthesis of orange fluorescence multifunctional carbon dots for label-free detection of vitamin B12 and endogenous/exogenous peroxynitrite. J Hazard Mater [Internet]. 2021, 408, 124422. Available from: https://doi.org/10.1016/j.jhazmat.2020.124422.

[42] Wang T, Luo H, Jing X, Yang J, Huo M, Wang Y. Synthesis of fluorescent carbon dots and their application in ascorbic acid detection. Molecules. 2021, 26(5), 1–11.

[43] Li L, Shi L, Jia J, Eltayeb O, Lu W, Tang Y. et al. Red fluorescent carbon dots for tetracycline antibiotics and pH discrimination from aggregation-induced emission mechanism. Sensors Actuators B Chem [Internet]. 2021, 332(November 2020), 129513. Available from: https://doi.org/10.1016/j.snb.2021.129513.

[44] Choudhary S, Joshi B, Joshi A. Translation of Carbon Dot Biosensors into an Embedded Optical Setup for Spoilage and Adulteration Detection. ACS Food Sci Technol. 2021, 1(6), 1068–1076.

[45] Sangubotla R, Kim J. Fiber-optic biosensor based on the laccase immobilization on silica-functionalized fluorescent carbon dots for the detection of dopamine and multi-color imaging applications in neuroblastoma cells. Mater Sci Eng C [Internet]. 2021, 122(December 2020), 111916. Available from: https://doi.org/10.1016/j.msec.2021.111916.

[46] Yang M, Liu M, Wu Z, He Y, Ge Y, Song G. et al. Carbon dots co-doped with nitrogen and chlorine for "off-on" fluorometric determination of the activity of acetylcholinesterase and for quantification of organophosphate pesticides. Microchim Acta. 2019, 186(8), 1–8.

[47] Huang S, Yao J, Chu X, Liu Y, Xiao Q, Zhang Y. One-step facile synthesis of nitrogen-doped carbon dots: A ratiometric fluorescent probe for evaluation of acetylcholinesterase activity and detection of organophosphorus pesticides in tap water and food. J Agric Food Chem. 2019, 67(40), 11244–11255.

[48] Yasa M, Deniz A, Forough M, Yildirim E, Persil Cetinkol O, Udum YA. et al. Construction of amperometric biosensor modified with conducting polymer/carbon dots for the analysis of catechol. J Polym Sci. 2020, 58(23), 3336–3348.

Palesa Seele*, Penny Mathumba

# Chapter 5
# Carbon dots in detection of biological molecules and metal ions

**Abstract:** Various types of nanomaterial have been exploited to further advance modern technology in the field of bio-applications, including theranostics, bioimaging, and biosensing. This has had an evolutionary impact on research and development as well as translating their use for modern-day medicine. Nanoparticles, in particular, carbon dots (CDs), are an exciting discovery in that they represent superior characteristics over other nanoparticles, including their excellent biocompatibility and low cytotoxicity, photostability, tuneable surface groups, and with unique fluorescence capabilities. Ample research still needs to be dedicated to investigate the biocompatibility and hemocompatibility of CDs, which are crucial aspects in determining the safety and risks accompanying their use for human and environmental consumption. Using CDs for detecting biological molecules can be appreciated in the recognition of biomarkers particularly for diagnostics, enhancing the functionality and stability of proteins, monitoring of biological processes, sub-cellular labelling of organelles, cell signalling, and transduction as well as in drug discovery and delivery.

CDs have been widely used in the colorimetric detection of biological material and metal ions, a method that exhibits ease of use and rapidity. Metal ions are inherent to normal physiological functions and homeostasis. Aberrant concentrations in cells and the environment are detrimental to animal health, thus requiring to be monitored and assessed per given time. CDs can be appreciated in applications that require colorimetric detection for rapid diagnosis purposes. Some of the examples that the research has been directed to include detection of HIV-p24 antigen for HIV diagnosis, analysis of *Enterobacter sakazakii* parasite in infant formulae samples, and detection of nucleic acid malaria biomarkers for all four *Plasmodium* species. Another potentially great innovative approach for TB testing that has been in the pipeline is the screening of volatile organic compounds from the breath of patients using CDs for photoluminescence detection of the biomarkers. The intrinsic properties of CDs are endowed by the carbon precursors used in synthesis, and post-synthetic modifications such as passivation, doping, and conjugation to other functional moieties. These are inherent in enabling interactions between the CDs and the targeted molecules, affecting the sensitivity and specificity with which biomolecules are detected.

*Corresponding author: Palesa Seele, MINTEK, South Africa
**Penny Mathumba**, MINTEK, Advanced Materials Division (AMD), 200 Malibongwe Drive, Randburg 2125, South Africa

The original version of this chapter was revised. Unfortunately, the co-author Penny Mathumba of this chapter was not mentioned in the original publication. This has been corrected. We apologize for the mistake.
https://doi.org/10.1515/9783110799958-005

**Keywords:** Carbon Dots, Nanoparticles, Biological molecules, Metal ions, Theranostics, Biocompatibility.

# 5.1 Introduction

The application of nanotechnology is gaining much-anticipated momentum in the field of theranostics with a positive impact on its evolution. Carbon dots (CD) represent a new and exciting type of nanotechnology that promises to outperform other nanoparticle types. This owes to their superior fluorescence (FL) properties, photostability, biocompatibility and low cytotoxicity, ease of synthesis, and its tuneable surface area which can be endowed with functional chemical groups that allow interactions with a broad spectrum of biological molecules.

The current chapter delves into the role of CDs, detailing their versatility in the detection of several biological entities and metal ions, by breaking it down into sections that are outlined in Figure 5.1. Briefly, the success of detecting biological molecules rests on the ability of CDs to invade the cells and induce their specified functions without much alteration in the overall function and integrity of the cell. This is defined as biocompatibility. Factors that describe biocompatibility such as cytotoxicity, immunotoxicity, hemocompatibility, the pathways and cellular uptake of CDs, induced apoptosis and/or necrosis, and clearance of the CDs are discussed in this chapter. These are instrumental in predetermining the safety of CDs in drug therapy and delivery, diagnostics, bioimaging, and biosensing. Delivering CDs to specific cellular organelles has been made possible with the exploitation of different preparation methods, followed by the addition of specific functional groups on the CD surface. This is an integral aspect of CDs that will be unpacked further in sections to follow, depicting the advantages of the fine-tuneable capabilities of CDs. The endocytic pathway with which the CDs are internalized by the cells depends on these methods. Upon the correct localization, biological processes including metabolic processes, cell signalling and transduction, synthesis, and degradation of biomaterial can thus be monitored, also in real time.

CDs have immense potential in detecting various biomaterials for diagnosis of infectious and non-infectious diseases and monitoring levels of metal ions in animals and the environment. In these cases, the colorimetric properties of CDs offer the advantage of rapidity and ease of use in analyte detection. Cancer is a quintessential disease where CDs have been immensely researched and developed for applications in theranostics. These studies on cancer also serve as models for other pathologies, hence the purpose of reviewing the role of CDs in diagnosis (see Section 5.5). Knowledge of the mechanism of interaction between CDs and relevant molecules is vital in understanding and predicting the cellular effects of CDs. The mechanism of interactions has been associated with the functional groups embedded on the CDs, including the zeta potential of the CDs and the respective physicochemical properties of the target molecule. In turn, these effectors are influenced by the microenvironment which is characterized by pH, salt concentration, and molecular crowding. Thus, it may not be surprising that molecular interactions including hydrophobic, van der Waals forces, ionic interactions, salt bridges, hydrogen bonds, covalent bonds, and π–π stacking are important determinants of the mechanism exploited by CDs [1, 2]. These bonds also qualify in playing a role in the stabilization and functionalization of proteins, especially enzymes when conjugated with CDs [3]. The types of CDs and conjugation with biological molecules are discussed in detail, and future perspectives are discussed in Section 5.3.

Despite the great potential that CDs display in biomedical applications, immerse research still needs to be directed to hemocompatibility and biocompatibility studies. CDs have been shown to have good distribution in mice [4]; therefore, more in vivo studies can be done to investigate hemocompatibility by varying the surface components on CDs and documenting their effect. To decrease cytotoxicity and toxic environmental outcomes the "bottom-up" approach was invented as an alternative synthesis method. Although an overall improvement has been achieved, this method does present various challenges such as broad size distribution, low yields, high energy costs, and poor size control [4, 5]. This ultimately impedes their bioapplications [7] since their size has a direct effect on their FL properties. Maintaining the long-term stability, especially of organically synthesized CDs, has also been reported to be difficult [7]. These challenges can be overcome by investigating the synthesis method on the stability of each CD component, conjugating CDs with molecules that preserve stability such as a cellulose matrix [8], and adding functional groups that increase biocompatibility such as glycosides in N-glycosylation.

In conclusion, future research and development should be directed to biocompatibility and hemocompatibility studies in vivo. This will further improve knowledge

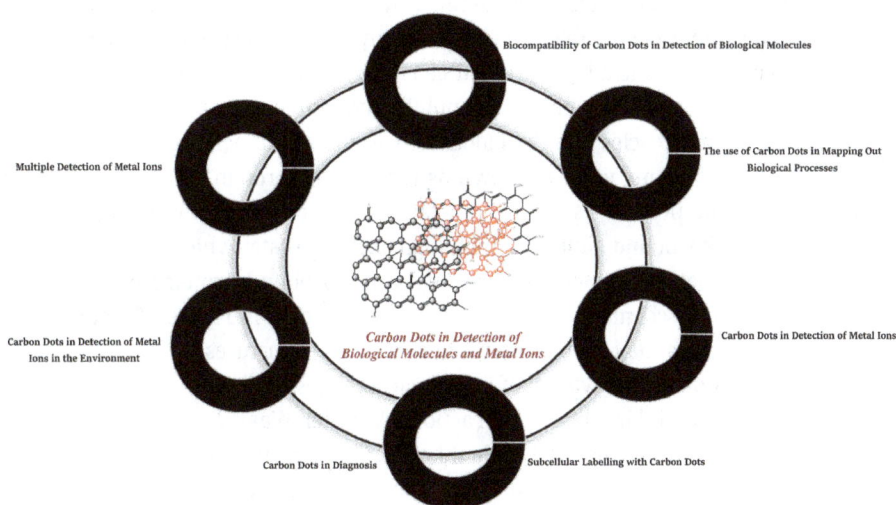

**Figure 5.1:** Overview of the chapter.

and understanding of the physicochemical properties of CDs that may be eliciting toxic effects. The potential of CDs in rapid diagnostic tests has not been explored enough. Refining the synthesis methods so that a more heterogeneous population of CDs is produced is one of the important factors in developing a successful rapid test. Therefore, outstanding evaluations and developments need to be done for CDs to accomplish globally standardized approvals and commercial use from organizations such as the Food and Drug Administration and World Health Organization.

## 5.2 Biocompatibility of carbon dots in detection of biological molecules

One of the chief characteristics that popularized the application of CDs in the biological field is their good biocompatibility. This has qualified them for various applications in the detection of biological molecules from bioimaging and biosensing as well as downstream applications in theranostics including drug and gene delivery, tissue engineering, diagnostics, and other pharmaceutical applications. Additionally, of high importance is hemocompatibility wherein the effects of contact of CDs with blood components are considered. In relating the structure of the nanomaterial and behaviour inside the cell, the physiological effects that it may have may be predictable.

### 5.2.1 Hemocompatibility of carbon dots

The composition of blood is intricately defined by liquid plasma and solid blood cells. Plasma contains numerous soluble proteins, electrolytes, and other biomolecules, whilst the cellular component includes red blood cells (RBCs), white blood cells (monocytes, macrophages, neutrophils, basophils, eosinophils, and lymphocytes), and platelets. The introduction of nanoparticles into the blood induces the engulfment of the particles by protein molecules forming what is known as the "corona protein" [3, 4]. The protein layer of the "corona protein" is "hard" and can sometimes be surrounded by a "soft" outer layer of weakly bound molecules with high exchange rates which is in contrast to the "hard" layer that is characterized by high affinity bound proteins with low exchange rates [3, 4]. A schematic representation of it is shown in Figure 5.2. Interaction with the nanoparticles may induce alterations in native structures of the proteins resulting in dysfunction and activation of pathological processes. Various interactions can be involved including hydrophobic interactions, Van der Waals forces, ionic interactions, salt bridges, hydrogen bonds, covalent bonds, and $\pi-\pi$ stacking [1, 2].

The effects of CDs on blood components have not been well-documented, only a few studies have looked at it. Fedel [11] reviewed the effects of CDs on coagulation, blood compatibility, haemolysis and activation of immune cells, platelets, leukocytes,

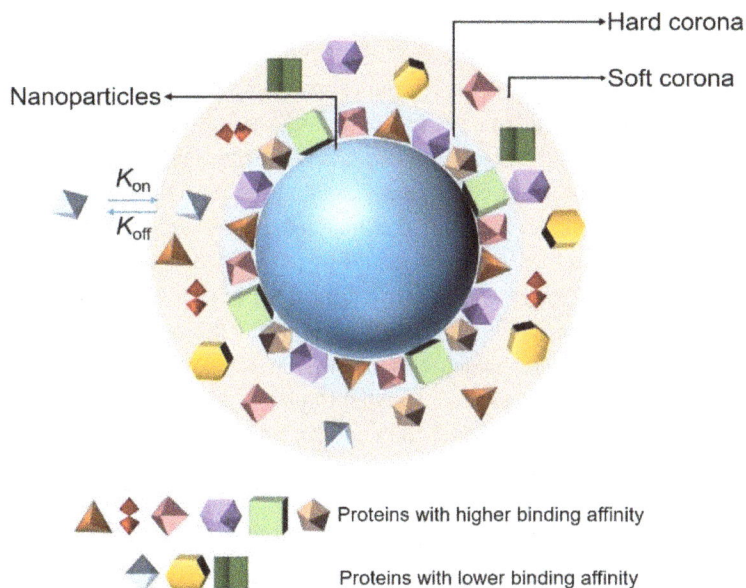

**Figure 5.2:** Schematic diagram illustrating the structure of the "corona protein" (the figure was reproduced with permission granted by Lee et al., 2021 [133]).

and complement cells. Herein they described how the biological system recognizes and reacts based on diverse properties of CDs including

– Size
– Physicochemical properties and geometrical structure
– Surface functionalized groups
– Degree of purity
– Solubility and dispersity in an aqueous medium
– Specific target cell

Structure and size are determined by the method of preparation and functionalization; however, CDs usually have a crystalline or amorphous core that is coated by a polar or nonpolar shell [11]. On the other hand, solubility and dispersity are a function of the reactive groups on the surface area which tend to be carboxylic, hydroxylic, and amide groups.

Li et al. [12] investigated the effects of α-cyclodextrin-prepared CDs on blood components. What the group observed was an overall adverse effect at concentrations higher than 0.1 mg mL$^{-1}$ and that the effects were concentration-dependent. High concentrations of CDs induced morphological changes and lysis of RBCs, dysfunction of the coagulation-related processes both in vivo and in vitro, and disruption in the fibrinogen microenvironment, and activation of the complement system and platelet. The authors postulated that hydrophobic interactions between CDs and RBCs, contributed by the hydrophobic groups on the surface area of CDs and lipid bilayer of the RBCs, drive the lytic effect. The hydrophobicity of CDs was also attributed to the hydrophobic aromatic rings residing in the inner core of CDs that were synthesized from saccharides [12]. Interestingly, the loss in the tertiary structure of fibrinogen was also accounted for by the hydrophobic interactions between the fibrinogen and CDs. Since the complement system forms part of the immune system which becomes activated in response to foreign materials such as bacteria, viruses, and others, it was also analysed. In this respect the surface hydroxyl and carboxyl groups and hydrophobic groups were reported as possible moieties activating the complement system. Platelets are critical for blood coagulation and clotting and contact with CD concentrations that were between 0.001 and 0.1 mg mL$^{-1}$ caused high activation of the platelets which was also driven by hydrophobic interactions [12]. Therefore, hydrophobic forces are suggested to play a major role in eliciting responses to contacts made by CDs with blood components. On the contrary, N,S-doped CDs that were synthesized from *m*-phenylenediamine and tobias acids were found to be of low cytotoxicity and good blood compatibility. The haemolysis rate was 0.5% at different concentrations of the CDs, additionally the morphology of the erythrocytes was unmodified and no aggregation was observed [13]. These studies suggest that the less hydrophobic the CD the better the chances of achieving good hemocompatibility.

## 5.2.2 Biocompatibility of carbon dots

The exploitation of nanomaterial for bioapplications has been impeded by cytotoxicity and the incompatible nature of the material with biological entities. In overcoming these challenges, refinement methods for CDs have been instrumental and this has been proved by both in vivo and in vitro studies. Preparation of the CDs by "green synthesis" followed by surface modification strategies such as doping can enhance the compatibility of CDs with biological molecules. Challenges with biocompatibility arise from CD insolubility, low cellular uptake, low distribution and dispersibility, morphological changes that affect the normal function of the biological molecules, and immunotoxicity. Some of the assays that have been used to analyse biocompatibility are cytotoxicity, apoptosis, relative gene expression, and immunotoxicity tests [8, 9].

A study conducted by Tian et al. [14] prepared carbon quantum dots (CQDs) by laser ablation in liquid and investigated their biocompatibility in cells, including Raw264.7 immune cells and the zebrafish. The authors postulated that since the CD surface is populated with carboxylic and carbonyl groups, these chemical moieties may react adversely with some cellular proteins [14]. Nevertheless, looking at the overall zeta potential of the CDs, which was about −33.7 mV, reaction with most proteins would be less probable. Their results show that there was minimum cytotoxicity and immunotoxicity induced by the CDs. This was seen by the lack of morphological changes observed in the zebrafish, meaning no lysis nor did major cytotoxicity occur. Additionally, the plasma membrane remained intact, and this was attributed to the small size of the CDs. Good dispersibility and ease in cellular uptake were also credited to the small-sized character of the CDs. To measure the immunotoxicity, which was barely evident, gene expression of immune proteins produced in the liver, levels of ROS, and expression cells of the adaptive and innate immune system (CD/cluster of differentiation markers) were recorded [14].

Although size has been cited as one of the major defining factors of biocompatibility, CDs that were cross-linked and embedded within a polyethyleneimine (PEI) matrix were 100 nm (which is about 10-fold larger than normal CDs) in diameter but still showed good biocompatibility [16]. Due to their unusually large-sized diameter, the quantum yield was 66% as well as high pH and photoluminescence stability. There was no evidence of pathophysiological damage such as necrosis, apoptosis, organ inflammation, and cytotoxicity. The clearance pathway of the CDs was also a vital aspect that was included in the analysis, and it was found that the preferred route of excretion was mainly hepatic rather than renal [16]. It is commonly known that the compounds or drugs that need to be metabolized in order to be eliminated are transported to the liver where the necessary modifications occur prior renal excretion. For example, hydrophobic compounds usually require modifications, and other intrinsic factors do dictate the clearance route including the pH, size, and polarity of the CDs.

The wide use of gold nanoparticles (GNPs) in diagnosis and pharmaceutical applications parallels that of CDs due to their unique optical properties, inert nature, ease of synthesis, and conjugation to biomolecules [11–13]. On the other hand, the biocompatibility of GNPs is still controversial and not yet well-established and documented. In vitro and In vivo studies conducted by Kadhim et al. [17] have reported no GNP-induced cytotoxic effect against the reference cell line and no morphological changes, also, mice which were intraperitoneally injected with the GNPs had no weight alterations and no histopathological alterations [17]. On the downside, Fan et al. [18] discovered that although human cells exhibited 80% survival rate when treated with 15 and 30 nm GNPs, cell viability decreased to less than 60% after 5 days whilst cell necrosis increased with treatment with 5 and 15 nm GNPs, respectively. Intriguingly, green synthesis and post-synthesis modifications to GNPs are suggested to improve biocompatibility [12, 13]. All of these studies suggest that the intrinsic properties of nanomaterials and the post-modifications determine not only their function but also dictate their biological fate.

## 5.3 Types of carbon dots and bioconjugation strategies

According to their various formation mechanisms, micro-/nanostructures, and properties, CDs are currently primarily categorized as graphene quantum dots (GQDs) [21], CQDs [22], and carbonized polymer dots (CPDs) [23] (Figure 5.3). Associations between these three types of CDs can be created by varying the graphene layer and carbonization level.

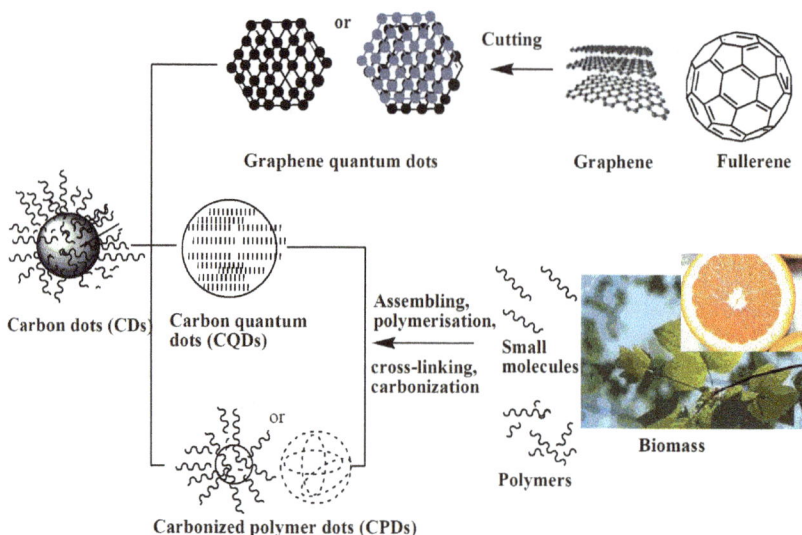

**Figure 5.3:** The various types of carbon nanomaterial.

GQDs have connected chemical groups on the surface/edge or within the interlayer defect that are arranged in one- or multiple-layer graphite structures [18, 19]. They are often produced by "top-down" preparation methods, which include cutting larger graphitized carbon materials, such as graphite powder, carbon rods, carbon fibres, carbon nanotubes, carbon black, or graphene oxide, into smaller pieces [9, 20–22]. These materials feature clear graphene lattices. Their optical characteristics are primarily controlled by the surface architecture and size of conjugated domains. GQDs are anisotropic and have lateral dimensions that are bigger than their height [29], whereas CQDs and CPDs are typically spherical and are frequently made from tiny molecules, polymers, or biomass using "bottom-up" procedures such as assembly, polymerization, cross-linking, and carbonization (e.g. combustion and thermal treatment) [30]. Their PL mechanism is intrinsic state luminescence and the quantum confinement effect of size.

Notably, CPDs were initially proposed in 2018 [31] due to the formation process, architectures, and PL mechanism, revealing aggregated/cross-linked and carbonized polymer hybrid nanostructures [32]. They have unique "core–shell" nanostructures, which are composed of carbon cores with a diameter of less than 20 nm, highly dehydrated cross-linking polymer frames, or slight graphitization, and shells with a profusion of functional groups and polymer chains. These nanostructures give CPDs greater stability, compatibility, ease of modification and functionalization, and a wider range of applications [32, 33]. Particularly, unlike the PL characteristic and mechanism of GQDs and CQDs, the optical properties of CPDs primarily result from the molecular state and cross-link enhanced emission effect, which makes the relationship between structure and performance of CPDs more controlled [23].

The diversity of CDs is brought about by a variety of synthesis methods and abundant carbon sources. The synthesis, PL mechanism, and uses of CDs have been succinctly described in some review publications. The excellent optical properties of CDs are discussed in this review, which seeks to take a unique stance, along with recent significant advancements in applications in a variety of fields, including optical (sensor, information encryption), energy (catalysis, light-emitting diodes, photovoltaics, rechargeable batteries), and promising biomedical applications (bioimaging, phototherapy, drug/gene delivery, and nanomedicine) [29–33]. Finally, we briefly go over the current issues and future prospects for this field in terms of planning, PL implementation, and CD use. In the near future, we anticipate that this review will provide critical analysis and comparative perspectives that will spark new and exciting work on CDs for energy, medicinal, and environmental applications.

Many different applications, including catalysis, imaging, drug administration, and biosensing, have made use of various nanoparticles and quantum dots (QDs). However, their toxicity and other negative health consequences, including pulmonary toxicity, translocation to extrapulmonary regions, and escaping phagocytosis, have restricted their long-term use. On the other hand, because of their unique catalytic properties and selectivity, enzymes are increasingly used in the food, pharmaceutical, and energy industries [40]. However, enzymes have a poor reusability and limited operating stability [35, 36]. They are sensitive to pH and temperature, which is a downside. The stability and reusability of enzymes have been maintained by using a variety of immobilization approaches. In this situation, using both inorganic and organic supports can adequately alter the chemical and physical characteristics of the enzyme, hence reducing its activity [37–40]. Additionally, the enzyme has a high likelihood of being easily leached out when the pH of the medium is changed when it is physically adsorbed through weak hydrogen bonds, van der Waals interactions, and electrostatic forces [41–43].

Although it has become relatively popular to use polymers, carbon nanotubes, various metal nanoparticles, and CDs in biosensing, nothing has been done to incorporate CDs onto the enzymes. In order to obtain special features, CDs (non-toxic nanoparticles) were just recently integrated onto protein molecules. For instance, Liu and colleagues [50] have demonstrated that the toxin ricin's immunomodulatory effect was boosted when CDs were conjugated with it. Hydrogen peroxide sensing has made use of immobilizing horseradish peroxidase using CDs/CoFe-layered double hydroxides [51]. The inhibition of protein aggregation by iron oxide nanoparticles coupled with glutamine- and proline-based osmolytes has also been demonstrated [52]. The surface properties of the QDs at the protein-QD interface were greatly impacted by the interaction of QDs with proteins, as demonstrated by Hosseinzadeh et al [53]. The capability of naturally receptor-conjugated conducting nanoparticles for biosensing was recently reviewed. They used enzyme-galactosidase to conjugate non-toxic gel-like CDs that were made in their lab and surprisingly discovered that the enzyme's stability had significantly improved. Likewise, even when the enzyme was kept at room temperature for more than two months it did not lose activity.

The exact mechanism explaining the behaviour of the employed nanoparticles, the improvement of the signal-to-noise ratio, the transduction, and the amplification of signals are all constrained by a number of difficulties in the construction of a highly sensitive and advanced biosensor. Although metal nanoparticles have been utilized extensively to lower detection limits, they were mostly poisonous and had large analyte detection fluctuations during batch measurements due to the tiny differences in the density of the metal nanoparticles [54]. Several methods, including cross-linking, physical adsorption, and covalent entrapment, have been used in the past to immobilize-galactosidase in nanomaterial matrices for biosensors; however, they are prone to substantial loss of enzyme during the sensing process [49–51]. In order to increase the biosensor's overall sensing capacity, they developed a method for conjugating-galactosidase with CDs in the presence of the cross-linking agent 1-ethyl-3-(3-dimethylaminopropyl) carbodiimide. CD-based nanozymes are also an exciting application of CDs, wherein the shortfalls of enzymes such as instability, limited tuneability, and high costs are compensated. Thus the future of CDs in conjugation and enhancing their functionality is promising [3].

## 5.4 The use of carbon dots in mapping out biological processes

Understanding how, where, and when biological processes occur is fundamental to preventative, therapeutic, and diagnostic research as well as development. By virtue of their high biocompatibility, CDs have been widely used in mapping out biological processes mainly through bioimaging and biosensing. These applications have birthed the field of theranostics and with employment of CDs there has been tremendous growth in this field [5–15]. CDs have been used to study biological molecules ranging from whole cells, sub-cellular organelles, proteins, and nucleic acids to single molecules such as glucose and metal ions. To further understand biological processes it is important to consider that metal ions form an integral role in cell signalling and transduction, catalytic reactions, metabolic processes, and in regulating osmosis [59]. Deficiency and excessive exposure can lead to severe health problems; hence CDs have been exploited to monitor their levels both in vivo and in vitro.

The metabolic processes of microorganisms, plants, and humans, including their metabolome, can be exploited as biomarkers for disease detection, for monitoring of the level of metabolites, and in plants and bacteria as the carbon skeleton for the green synthesis of CDs [16–19]. The latter, which is also known as the bottom-up approach has been highly researched and has gained much interest because the raw material from plants is renewable, the method is simple to execute and importantly, it can enhance specificity and dictate the localization of the CDs.

## 5.4.1 Carbon dots in monitoring of metal ions

Impairment in the homeostasis of metal ions poses danger to the proper functioning of biological systems and processes such as catalysis, cell signalling and transduction, metabolism, and osmotic regulation. For example, the lysosome operates optimally on the basis that the metal ions $Ca^{2+}$, $Fe^{2+}$, $Zn^{2+}$, and $Cl^-$ are in homeostasis since they regulate its function. Neurodegenerative diseases and metabolic disorders are borne from dysregulation of these ions [16, 21]. Mostly, metal ions have been monitored through FL probes, such as CDs, which have proved to be beneficial in lowering the limit of detection (LOD) and enhancing selectivity. The detection of heavy metal ions including $Pb^{2+}$, $Hg^{2+}$, $Cd^{2+}$, $Cr^{3+}$, and $Ag^+$ has been achieved successfully through the chelation of functional groups that are embedded on the CDs. This would result in either quenching or increase in FL [65]. The carbon source from which the nanomaterial is prepared from [66] has mostly been cited for providing specificity to the targeted metal ion. A summary of targeted analytes has been summarized in Table 5.1 [21, 23]. The metal ion, $Fe^{3+}$, is an essential co-factor to various proteins and enzymes which are involved in oxygen transport, production of energy, neurotransmission, nucleic acid synthesis, cell replication, and growth [59]. It is one example that demonstrates that indeed the different preparation methods of CDs can alter the LOD (see Table 5.1).

Doping is a chemical process that is used to endow nanoparticles with particular chemical entities in order to fulfil specific objectives such as enhancing their optical, electrical, and biological activities of the material [68]. N-doping is the most common because of its similar-sized small atom to carbon; also, it can bind to many other carbons because of its five valency electrons. This type of doping is mostly reported to improve FL intensities of CDs [69]. In a study, the quantum yield of CD-monitored $Fe^{3+}$ was increased from 6.8% to 20.1% post N-doping with 1,2-ethylenediamine [67]. On the contrary, doping with phosphorous has been associated with higher wavelength emissions. A combination of these dopants, that is, N,P-doped CDs have been efficiently used for the detection of the ions, $MnO_4$ and $Fe^{3+}$ in water [70]. Similarly, the quantum yield of CDs was enhanced by N,P-doping in the detection of Mn(VII) with high selectivity [69]. In these cases, the selectivity of the N,P-doped CDs can be attributed to their respective carbon sources, where the first study used guanosine 5'-monophosphate as a carbon source whilst the latter study exploited 1,2-ethylenediamine. Another type of dopant that has been used is fluorine (F) in the intracellular monitoring of $Ag^+$ ions in HEK293 normal cells and B16F10 cancer cells. F-doping increased the solubility of CDs in the aqueous environment [71] which is significant for increasing the biocompatibility of the CDs.

Both quenching and FL have been used successfully to detect molecules. Aluminium ($Al^{3+}$) is one of the most abundant and useful elements in the environment; however in high concentrations it is associated with debilitating health outcomes including the development of Parkinson's disease, Alzheimer's disease, and dialysis encephalopathy, which affect the nervous system [72]. Hydrophobic CDs were made amphiphilic by modifications with cetyltrimethyl ammonium bromide. Through electrostatic interactions between the amino groups of CDs and the hydroxyl group of morin, the FL of CDs was quenched. Addition of $Al^{3+}$ meant that the CD was displaced resulting in formation of the $Al^{3+}$-morin complex which caused the recovery of FL [72]. The mechanism is summarized in Figure 5.4.

Designing assays for metal ion detection is not limited to a single ion but CD FL probes can be used for multiple ion monitoring. Multiple ion detection is discussed as a separate topic (5.7 Multiple detection of metal ions).

**Table 5.1:** Comparisons of synthesis methods on LODs of $Fe^{3+}$, $Hg^{2+}$, and $Pb^{2+}$ ions.

| Carbon source | Fluorescence probe | LOD (nM) | Linear range (nM) |
|---|---|---|---|
| **$Fe^{3+}$** | | | |
| Blueberry | N,S-CDs | $4 \times 10^3$ | $25–500 \times 10^3$ |
| Garlic | CDs | 200 | $0–500 \times 10^3$ |
| Cotton | CDs | 27 | $50–10^4$ |
| *Fructus lycii* | CDs | 21 | $0–30 \times 10^3$ |
| **$Hg^{2+}$** | | | |
| Folic acid | CDs | 230 | $0–25 \times 10^3$ |
| Citric acid | CDs | 100 | $1–18 \times 10^3$ |
| **$Pb^{2+}$** | | | |
| *Ocimum sanctum* leaves | CDs | 0.59 | 10–1,000 |
| Sodium citrate and polyacrylamide | CDs | 4.6 | 16.7–1,000 |
| Chocolate | CDs | 12.7 | 33–1670 |
| Calcein dye | CDs | 0.011 | 0.032–0.78 |
| *Ginkgo biloba* leaves | CDs | 0.055 | 0.1–20 |
| Glycerol and ethylenediamine | CDs | 15 | $6 \times 10^6$ |
| Kerosene soot | CDs | 2 | 500–1,100 |
| Chitosan hydrogel | CDs | 0.5 | NA |
| Potato-dextrose agar | CDs | 0.11 | $0–20 \times 10^3$ |
| Bovine serum albumin | CDs | $5.05 \times 10^3$ | $0–6 \times 10^6$ |

This table was reproduced from references [64, 72] with permission from Elsevier.

## 5.4.2 Sub-cellular labelling with CDs

In sub-cellular labelling, the specificity of CDs to an organelle is determined by the chemistry of the surrounding membrane as well as the modifications embedded on the surface of the CD and the cellular internalization pathway of the nanomaterial. The choice of CD synthesis method and design is guided by the charge, hydrophobicity, hydrophilicity, and isoelectric properties of the molecular make-up of the targeted organelle. A new phenomena known as "structure-inherent targeting" (SIT), wherein the CDs' intrinsic properties are inherited from the parent carbon source, is now a popular way

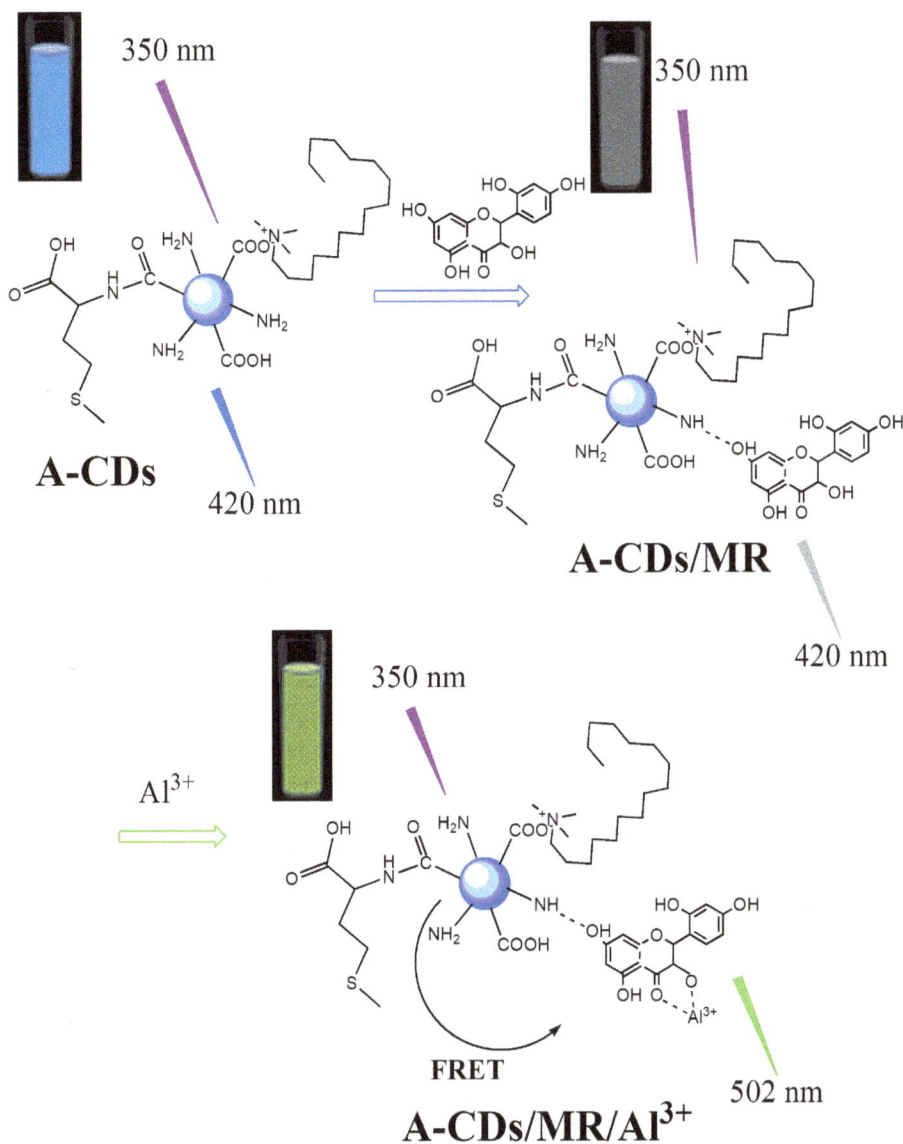

**Figure 5.4:** Diagram depicting the mechanism of Al³⁺ detection with carbon dots (the figure was reproduced from Kong et al. [72], with permission granted by Elsevier).

of basing the synthesis method on the physicochemical properties of the targeted analyte [74]. CDs have been known to accumulate in the cytoplasm but with the correct
physicochemical characteristics different organelles have been labelled exclusively including the lysosome, mitochondrion, endoplasmic reticulum (ER), nucleus, and the nucleolus [1]. The detection mechanism of the nanomaterial probe relies mainly on two

aspects: the moiety that recognizes and binds to the organelle and the fluorescent moiety. The strategies that are used to target the different organelles are discussed below.

### 5.4.2.1 Lysosome

The lysosome performs several vital functions of the cell, with almost 50 known hydrolases. Its main purpose is to degrade cellular molecules such proteins, lipids, polysaccharides, and nucleic acids. Dysfunctions that result in changes to its pH, morphology, and polarity may indicate pathology. CDs, more especially functionalized ones, have been used to monitor these properties especially the pH which requires a much simpler experimental design to monitor. Lysosomes are amphipathic membrane-enclosed vesicles with a size range of 0.5–1.2 µm and are known to have an acidic (pH 5/5.5) environment [64]. Accumulation of CDs has been mainly in the cytoplasm which has a pH higher than the lysosome [75]. Therefore, to specifically target the lysosome the CD probe has to contain a moiety that has a pH/$pK_a$ lower than that of the cytoplasm [2]. Acidotropic molecules such as acridine orange and neutral red are cationic dyes which diffuse through the membrane and accumulate in the lysosome due to their weakly basic nature [32–34]. These dyes are however not as specific to the lysosome and have been cited as unstable hence the development of LysoTrackers. The LysoTrackers also exploit $pK_a$ for targeting the lysosome with an improved specificity; however with longer incubation periods they cause FL quenching [79]. Chen et al. [79] report a Superior LysoProbe which comprises various N-linked glycans to the parent fluorophore displaying "superior" specificity to the lysosome compared to the LysoTracker. Their strategy was based on the fact that the lysosomal membrane is greatly N-linked glysosylated, additionally, increasing the hydrophobicity of the linker between the fluorescent probe and the lysosome-targeting moieties enhanced cellular uptake [79]. In addition to doping, glycosylation presents another modification strategy that could be introduced to the surface of CDs to enhance their specificity.

By bearing weakly positively charged amino acids CDs have been successfully localized to the lysosome; however heterogeneously charged CDs can also target the lysosome [1, 28, 36]. CDs that have been synthesized from p-benzoquinone and ethanediamine (EDA) are intrinsically functionalized with lipophilic amines thereby facilitating specific localization to the lysosome [1, 2]. It is believed that the lipophilic amine groups confer a weakly basic property which facilitates their accumulation in the lysosome. Contrastingly, Shuang et al. reported that the addition of the lipophilic amine, laurylamine, to amine-functionalized CDs resulted in localization in the ER [81]. The ability of the CDs to differentiate between the two organelles is attributable not only to their surface chemistry but also to their mechanism of internalization used. It is postulated that the latter is internalized via lipid rafts while internalization of amino functionalized CDs is mediated by clathrin [81].

Recently, morpholine-conjugated CDs have been widely accepted and successfully used to target lysosomes; this is due to the presence of amine groups [64]. Morpholines are lysosomotropic molecules typically with $pK_a$'s of 5 to 6 which allows them to reside inside the acidic lysosome. A review by Choi and colleagues [64] discusses several studies that have attached morpholine to their FL probes and have been effectively detected in the lysosome. Additionally, morpholine-functionalized CDs with PEI passivation (CD-PEI-ML) have been shown to increase the FL quantum yield and photostability [1, 2]. Due to its small-sized diameter of 3.15 nm, CD–PEI–ML could easily cross the membrane which is great for biocompatibility [1, 2]. The role that the state and extend of hydrophobicity and hydrophilicity in defining the localization of CD-based probes is inconclusive.

Alkyl-functionalized CDs were observed to be rapidly uptaken by cells requiring less energy and showing passive diffusion, also, mainly localizing to the lysosome. The mechanism of internalization may be different with hydrophobic CDs [1, 2]. Incorporating metal ions such as manganese (Mn) and ruthenium (Ru) in CDs can facilitate lysosome targeting. The MnPCNDs were prepared by carbonization of [M=Zn(II) or Mn(III)] of tetra-(meso-aminophenyl)porphyrin in the presence of citric acid (CA) [82]. To target cancer cells, Ru(II)complex-functionalized CDs were found to localize in the lysosome [83]. This means that the incorporation of certain metal ions, including their mode of internalization and synthesis methods, primes the CD for targeting the lysosome.

### 5.4.2.2 Mitochondrion

Given that the mitochondrion is a double-membraned power house of the cell that generates ATP molecules during cellular respiration, this aspect of the mitochondrion is exploited in designing probes that specifically target it. FL probes that contain a positively charged lipophilic moiety are most likely to be attracted to the negatively charged membrane potential of the mitochondria because they possess a higher positive zeta value. This justifies the usage of lipophilic cations such as triphenylphosphonium (TPP), methylpyridinium, and indolium for specifically targeting the mitochondria, with most studies using TPP [2]. TPP has been widely used as a mitochondriotropic moiety for carrying antioxidants. Using urea and CA as the carbon source of CD, it coats them with $-NH_2$, $-COOH$, and $-OH$ groups which enable conjugation with TPP. Ultimately, the increase in zeta potential occurring upon TPP conjugation drives localization of the CDs to the mitochondrion. This is also true for CDs conjugated with a coumarin-3-carboxylic acid moiety which resulted in a 5.9-fold increased zeta potential [2]. It has been well established in this chapter that the different methods of synthesis determine the CDs specificity this being also true for CDs localizing in the mitochondrion. Preparation of CDs from precursors such as mercaptosuccinic acid, ethylenediamine, and chitosan via a one-pot hydrothermal treatment; and preparation from m-aminophenol and CA endows CDs with chemical groups that enable them to be in the mitochondrion [2].

### 5.4.2.3 Nucleus

Also known as the control centre of the cell, the nucleus harbours nucleic acids and proteins enclosed by a membrane. The resident nucleic acids give this organelle an overall negative charge; thus surface charge of CDs plays a vital role in determining localization in the nucleus. It is difficult for molecules to penetrate the nucleus due to the phospholipid double membrane and the small nuclear pores. However, positively charged nanoparticles exploit the "proton sponge" effect to enter the cytosol by disturbing the endosomes and locating to the nucleus where they eventually bind to DNA [84]. In retrospect, positively charged CDs are able to evade the lysosome making it a complicated design process to exclusively select the nucleus. Fortunately, positively charged CDs that are conjugated to nuclear localization signals (NLSs) peptides have a much better chance of being exported to the nucleus where they interact with the nuclear protein, importin [2]. In some studies, CDs without the NLS peptide resulted in population of the CDs in the cytoplasm.

Another study showed that positively charged CDs that were coated with quaternary ammonium groups located in the nucleus of fibroblasts. With an average size of 7 nm, the CDs were reported to have high nucleus uptake [84]. CDs which were not functionalized but were able to accumulate in the nucleus had intrinsic properties that allowed binding to nucleic acids via interactions such as $\pi-\pi$ stacking, electrostatic interactions, and hydrogen bonding. CA as a carbon precursor has been successfully used to stain the nucleus. For example, when the ratio of CA to EDA was decreased so that the zeta potential decreased from $-17.9$ to $-2.84$ mV, the nucleus was abundantly stained. Another example is where Gd(III)-doped CDs that were prepared from CA and GdCl$_3$ specifically targeted the nucleus [2]. Also in an effort to develop a theranostic-based CD that delivers doxorubicin to the nucleus of cancer cells, Jung et al. [85] synthesized CDs using CA as a carbon source and β-alanine as a passivating agent. Thereafter, they ligated the CDs to a zwitterionic molecule with both negatively and positively charged groups which gave it an overall neutral charge. Their approach was based on two facts: that cells preferentially uptake negatively charged molecules and that once in the cell, the CDs would localize to the nucleus if the pH is 0.3–0.5 higher than that of the cytoplasm [85]. FL was successfully captured in the nucleus of the cells. When the doxorubicin-CDs (Dox-CDs) were intravenously injected and tested in nude mice expressing 4T1-luc2 breast cancer xenografts, the Dox-CDs were 13% more effective in inhibiting the growth of cancer cells compared to Dox delivered on its own. Thus, CA is an essential carbon precursor for CDs that target the nucleus.

### 5.4.2.4 Endoplasmic reticulum

The ER is composed of both smooth and rough ER which forms continuous folds that connect through the outer layer. The rough ER is embedded with ribosomes and is responsible for protein synthesis and secretion [86]. It is these ER-associated proteins that are strategically targeted for synthesis of probes [1]. Some of the proteins and the mechanisms with which they are targeted are discussed in detail in the section, receptor/membrane protein targeting. Like with targeting the lysosome, CDs that are amine-functionalized can also localize in the ER. For example, CDs synthesized using urea and CA thereafter functionalized with laurylamine confer hydrophobicity to CDs, and this is thought to cause localization to the ER. Another strategic approach is synthesizing the CDs using $p$-phenylenediamine as a precursor followed by doping with trace metals, $O$- and $N$-, abbreviated to MNOCNP. Conjugation of MNOCP with poly(ethylene glycol) caused a change in zeta potential of the nanoparticles and enhanced physiological stability and accumulation in the ER [1, 2]. Table 5.2 gives a summary of the different moieties that target the organelles.

**Table 5.2:** Probe moieties attached for targeting specific organelles and their analytes.

| Organelle | Targeting moiety | Targeted analyte |
|---|---|---|
| Lysosome | Morpholine | $H_2S$, CyS/Hcy, GSH, HClO, $SO_2$ |
| Lysosome | Monothio-bishydrazide | HClO |
| Lysosome | Benzothiazolium | pH |
| Lysosome | Hemicyanine | pH |
| Nucleus | Rhodium complex | Mismatched DNA |
| Nucleus | Benzothiazole | DNA G-quadruplex |
| Nucleus | Hoechst | AT-rich region in DNA |
| Nucleus | Py-Im polyamide | Telomeres |
| Nucleus | *NLS | Integrin and CD13 |
| Nucleus | *NLS | $H_2O_2$ |
| Mitochondria | Quinoline | $H_2O_2$ |
| Mitochondria | N-Methylpyridine | $H_2O_2$ |
| Mitochondria | Hemicyanine | $ClO^-$ |
| Mitochondria | Nile Blue | $ClO^-$ |
| Mitochondria | *TPP | Mitochondrial ultrastructure and membrane |
| Mitochondria | DQAsome | ATP |
| Mitochondria | Mitochondria-penetrating peptide | Mitochondrial polarity and viscosity |
| Mitochondria | Mitochondria-targeting peptoid | Mitochondrial serine protease activity |
| Endoplasmic reticulum | $p$-Toluenesulphonamide | NO |
| Golgi | Phenylsulphonamide | $H_2S$ |
| Golgi | Sphingosine | pH |
| Plasma membrane | Cholesterol | Membrane structure, $Zn^{2+}$ |
| Plasma membrane | Alkyl chain (C = 8–18) | $Zn^{2+}$ |
| Plasma membrane | Bis-sulphonic acids | Membrane structure |
| Plasma membrane | Alkyl chain (C = 17) | Membrane structure |
| Plasma membrane | Nile Red | Lipid composition |

*NLS, nuclear localization signal; TPP, triphenylphosphonium.

### 5.4.2.5 Receptor/membrane protein targeting

Receptors are biological molecules that are inherent in cell signalling and transduction, ion transportation, and immune responses, and they have been implicated in the pathophysiology of various infectious and non-infectious diseases. They are well-established and successful pharmaceutically drug targets, especially for cancer. Cancer cells are one of the first cells to be imaged using CDs. The nanomaterial was highly specific to folate receptors which are known to be abundantly expressed on cancer cells; sofolic acid was a carbon source from which the CDs were prepared [24, 43, 44]. In one interesting study, multi-CDs were used in combination with DNA aptamers for the detection of protein tyrosine kinase 7 (PTK 7), a receptor that is overexpressed in cancer cells. The aptamers specifically recognize and bind PTK 7, so that in its presence it reacts with the yellow FL CDs (yCDs)-modified aptamers which antagonizes quenching by $Fe_3O_4$ magnetic nanoparticles (MNPs). The blue FL CDs were used as an internal reference. In this case, CDs were indirect probes which did not bind directly to PTK 7. Additionally, with this method a loop amplification method was incorporated to further amplify the signal and improve the LOD [89].

CDs are an increasingly exploited new technology that is not only used to study molecular interactions and organization, but they have also been used to detect molecules at a single molecular scale. A distinction between receptor clustering and oligomerization can be a complex exercise without the correct tools. He and colleagues [90] were able to distinguish these by using CDs, wherein a G-coupled receptor, CRXCR4 was investigated. Spontaneous blinking of CDs conjugated to the Fab region of the anti-monoclonal CRCR4 antibody was used for detection. The CD and Fab antibody molecules assembled in a 1:1 ratio; thus, it was assumed that each localization (observed grey dot) represented one molecule. This enabled the researchers to conclude that the oligomerization state of the receptor is time-dependent and that clusters can form within the membrane [90].

These developments present great potential for using CDs as fluorescent probes not only due to their superior sensing capabilities but also due to their anti-microbial properties. CDs exhibit anti-microbial activities by direct interactions with the bacterial membrane and disrupting it, with simultaneous release of ROS which damage the cytoplasm. Heteroatoms such as nitrogen and sulphur that are embedded on CDs are implicated in the anti-microbial activities albeit with different mechanisms of activity [47, 48].

### 5.4.2.6 Targeting of nucleic acid

The targeting of nucleic acids with CDs may not be mutually exclusive with subcellular targeting of the nucleus organelle. The delivery of genetic material for therapeutic purposes and cell imaging has been advanced by the use of vehicles such as nanoparticles. Genetic material is unstable and is prone to degradation by nucleases; hence some sort of protective material has to be used to deliver it [93]. Based on the

SIT-based strategy, which considers the structure of nucleic acids, especially double-stranded DNA which comprises minor and major grooves, a negatively charged skeleton provided by the phospholipid backbone; quinoline derivatives have been used strategically to target nucleic acids. This is so because quinoline can bind efficiently to the groves of nucleic acid via electrostatic and π–π interactions [74]. Lebeau and his group [93] synthesized cationic CDs which successfully delivered siRNA and DNA plasmid to different cell types, including the delivery of DNA to experimental mice through airway transfection. The group characterized the CDs including their ability to form stable complexes with the targeted DNA in different cells [93]. The low cytotoxicity and good cellular uptake of the nanoparticles can also be associated with the ability of the DNA to condense with the cationic CDs, thus resulting in compacting and small size of the complex.

An interesting study where CDs and GNPs were used simultaneously to quantify DNA, CDs were assigned FL enhancers whilst GNPs were quenchers. The two nanomaterials have been previously paired since they are excellent FL fluorophores and quenchers, respectively. Peptide nuclei acid (PNA) was exploited as a fluorescence resonance energy transfer (FRET) regulator between the nanoparticles [94]. PNAs are interesting DNA analogues in that the sugar–phosphodiester backbone is replaced by a charge-neutral polyamide backbone. The interaction between the PNA and target DNA is via methylene carbonyl linkages between the purine and pyrimidine bases and the polyamide backbone [95], and this interaction leads to increased stability of the DNA. In the absence of DNA, PNA induces aggregation of GNPs inhibiting quenching while simultaneously causing FL of CDs. In contrast, the PNA–DNA complex causes dispersion of GNPs which bind the CDs and inhibiting their FL [94]. Thus, different approaches can be exploited to increase specificity to target nucleic acids.

## 5.5 Carbon dots in diagnostics

The significance of a timeous diagnosis for both infectious and non-infectious diseases such as cancer amongst several other diseases cannot be emphasized enough. Administering proper diagnosis requires that appropriate tools be used, this being the foremost link to prompt therapeutic intervention, curbing transmissibility and ultimately the morbidity and mortality associated with the disease. Recognizing biomarkers as valuable biological molecules that are dysregulated during the lifecycle of a disease is an important aspect in the design of modern diagnosis tools, and of equal importance is developing a robust detection system. The performance of traditional methods of diagnosis has been in the spotlight amidst the recent COVID-19 pandemic, once again proving their inadequacies in meeting the required high technical demands due to the obviously high volumes required for testing. The lack of accessibility to poor remote communities due to expensive equipment and infrastructure needed for analysis a compounding issue. Developing rapid diagnostic tools has been a highlight in research towards point of care diagnosis that is inexpensive, portable, easy to use, and accessible to all. Exploiting nanotechnology in the devices' detection system is an underpinning characteristic that confers rapidity since the results can be observed visually and even immediately with some devices such as the lateral flow immunoassays (LFIs). CDs have been widely used in the colorimetric detection of biological material. Their use can be appreciated in LFI detection systems where they have been investigated in the detection of the HIV-p24 antigen [96], analysis of *Enterobacter sakazakii*parasitein infant formulae samples [97], and in the detection of nucleic acid malaria biomarkers for all four *Plasmodium* species [98]. Another potentially great innovative approach for TB testing that has been in the pipeline is the screening of volatile organic compounds (VOCs) from the breath of patients using CDs for photoluminescence detection of the biomarkers [99].

HIV-p24 antigen is widely researched as a biomarker for HIV testing, though leaning on the use of GNPs [94–96]. CDs have been shown to be compatible with microwell plates and microchip-based sandwich assays for the detection of this antigen [96]. The use of microwell plates offers the much-needed advantage of high-throughput screening; however to translate the research into a POC device, the authors also designed a microfluidic chip. Detection of the HIV-p24 antigen was achieved via a sandwich design by exploiting the amine groups on the CDs and conjugating them to streptavidin. This is a common strategy that is used even for GNP colorimetric detections. In this case a lateral flow assay is possible to develop for POC. A similar study where the neutravidin–biotin detection system is used has been documented for the detection of *Enterobacter sakazakii* parasite in the analysis of infant formulae [97]. In this case, nucleic acids were the targeted analyte, and visualized by a black/grey line exhibited by the fluorescent CDs that were conjugated to neutravidin. Protein-induced FL enhancement is used for monitoring interactions between proteins and nucleic acids which has been widely used to study the dynamic interactions between DNA and DNA polymerases [98]. This was employed in the detection of the malaria biomarker *Plasmodium falciparum* glutamate dehydrogenase (GDH). Singh and colleagues [98] designed and produced aptamers specific to the GDH which were conjugated to the CDs, where upon interaction with the GDH antigen the FL is increased. The carbon precursors of the CDs were various amino acids which probably endowed the CDs with functional groups that facilitated interactions with the aptamer. The CD-aptamer complex is reported to bind in the hydrophobic pocket of GDH [98]. Colorimetric detection of biological molecules has also been applied to VOCs and CDs have been realized as a convenient photoluminescent probe. Tuberculosis is an ancient pulmonary disease but still without efficient diagnostic tools with sputum collection as the remaining standard sample for testing. VOCs present a convenient and non-invasive way of testing, and CDs have been used for colorimetric detection [99].

# 5.6 Carbon dots in detection of metal ions in the environment

Although some heavy metal ions, including copper, iron, aluminium, and chromium (III), are nutritionally necessary and needed by some species, larger doses can be hazardous [97, 98]. The most prevalent non-biodegradable and hazardous contaminants in industrial effluents are heavy metal ions, even if they are present in trace concentrations. Examples include chromium(VI), lead(II), arsenic(III), cadmium(II), and mercury(II). To reduce water pollution and stop negative effects from occurring, sensors for detecting heavy metal ions must be both environmentally friendly and logistically practicable [99, 100]. Researchers are still looking into the best ways to create sensors that can detect pollutants in the environment and living things in real time. Fluorescent nanocarbons are replacing metal-based nanoparticles and organic dyes as the preferred method of contamination detection [107].

The surface oxygen moieties on many green CDs are what cause the metal ions to coordinate with them and induce PL quenching, which makes them useful for sensitive and selective colorimetric and fluorometric metal ion sensing. The primary factors for PL quenching are the energy transfers between metal ions and nanocarbons through selective interactions caused by functional groups and surface traps [102, 103]. Selectivity is affected by all of a nanocarbon's properties including its size, shape, and surface functions. In addition, by doping CDs with nitrogen, boron, sulphur, and phosphorus, the optoelectronic characteristics, stability, and applications can be managed [104–106].

The primary factors for PL quenching are the energy transfers between nanocarbons and metal ions through selective interactions caused by functional groups and surface traps [113, 114]. Selectivity is affected by all of a nanocarbon's properties, including surface functions, edge structure, size, and shape. Additionally, doping CDs with nitrogen, boron, sulphur, and phosphorus allows for control over their optoelectronic characteristics, stability, and applications [109–111].

Different detection mechanisms, which are based on either attenuation or enhancement of FL, have been used to design FL sensors, including Förster/FRET, photoinduced electron transfer (PET), inner filter effect (IFE), coordination-induced aggregation, and FL quenching [112–117]. By turning absorption data into FL readout, Stokes made the initial discovery that IFE is a crucial process to increase the detection sensitivity of fluorescent sensors. IFE requires a spectrum overlap between the emission of the donor molecules and the absorption of the acceptor molecules. Furthermore, it avoids the critical covalent contact between contaminants and sustainable CDs. When CDs are present, the FL lifespan measurement of the receptors changes less, indicating the presence of the FRET effect [111, 118]. An energized electron is transported from CDs to the receptor in a redox reaction known as PET. While the static and dynamic quenching processes both entail the transfer of an electron from the donor to the acceptor, the

**Table 5.3:** Oxidation and reduction peaks for GCE electrodes modified with N-CQD-Bi + Nafion.

| | Oxidation peak (V) 20 ppm | Reduction peak (V) 20 ppm | Oxidation peak (V) 30 ppm | Reduction peak (V) 30 ppm | Oxidation peak (V) 40 ppm | Reduction peak (V) 40 ppm |
|---|---|---|---|---|---|---|
| GCE | −0.6063 | −0.4342 | −0.609 | −0.0119 | −0.5871 | −0.4770 |
| GCE + Nafion | −0.6312 | −0.4899 | −0.5853 | −0.4071 | −0.6251 | −0.4819 |
| GCE + N-CQD | −0.4492 | 0.40001 | −0.5822 | −0.4051 | −0.4476 | −0.3491 |
| GCE + N-CQD + Nafion | −0.5995 | −0.4836 | −0.5892 | −0.4100 | −0.4615 | −0.3860 |
| GCE + N-CQD-Bi | −0.6014 | −0.5343 | −0.6094 | −0.4087 | −0.6007 | −0.5002 |
| GCE + N-CQD-Bi + Nafion | −0.4816 | −0.5927 | −0.4576 | −0.6748 | −0.4395 | −0.5802 |

former involves FL quenching between the acceptor and donor [107, 108]. The oxidation and reduction peaks for GCE electrodes modified with N-CQD-Bi and Nafion are depicted in Table 5.3.

Due to their capacity to produce a wide range of nanomaterials, including carbon nanotubes, graphene, fullerene, CDs, carbon nanowires, and nano-diamonds, carbon-based materials are now playing a crucial role in nanotechnology [125]. CQDs are commonly utilized in potable detectors due to their large abundance, low toxicity, high stability, and affordable nature. The surface of CDs may be easily modified, making water-soluble CDs possible. High FL quantum yields are the consequences. The physiochemical properties of fluorophores, in particular, undergo a significant alteration as a result of the binding interaction between CDs and heavy metals. A few of these are lifetime, FL intensity, anisotropy, and so on.

Metal oxides are quite helpful for interacting with target molecules. By modifying the surface, these metal oxides' sensing capabilities can be increased. The electrical properties of metal oxides are substantially dissimilar from those of metals, semi-conductors, magnets, and insulators. Semi-conductor metal oxide-based sensors are crucial for environmental and human health safety [125, 126]. These metal oxide-based sensors work on the incredibly straightforward premise that changes in electrical conductivity result from the interaction of oxygen between grain boundaries and ambient gases. $ZnO$, $Nb_2O_5$, $NiO$, and other metal oxide sensors have received the most research.

In an electrochemical investigation, several materials have recently been used to modify the electrode. Bismuth has been extensively researched in a variety of sectors due to their great biocompatibility. In truth, bismuth is a special element with a wide range of uses, including the detection of heavy metals, electrocatalysis, and photocatalysis. Bismuth nitrate ($Bi(NO_3)_3$) and bismuth oxide ($Bi_2O_3$)-coated electrodes have been employed as an alternative to hazardous mercury-based electrodes for the electrochemical stripping analysis of heavy metals. The non-toxicity to the environment and operational biosafety of modified electrodes made from bismuth-based materials are their major benefits. Because bismuth is more easily oxidized than mercury is, the bismuth-based electrode frequently performs rather well in a relatively small potential window, especially when it is in a more negative anodic range. Bismuth nanoparticles are one of versatile electrode modifiers because of their wide electrochemical window and low toxicity.

## 5.7 Multiple detection of metal ions

One of the focuses of new technological advances is developing high-throughput screening assays that are designed to make multiplex monitoring possible. Like so, multiple metal ion monitoring is feasible with CDs and has been done by building multidimensional sensors [67]. In this instance, the selectivity, specificity, photostability, cytotoxicity, and biocompatibility of the CD may be a concern but fortunately, there are studies that have proved otherwise.

Moorinta and colleagues [127] have demonstrated the multi-functionality of CDs, wherein, CDs that were prepared from yoghurt were used for the detection of formic acid vapour and could also be used for the detection of metal ions. These CDs were able to discriminate between different concentrations of folic acid as well as detect various metal ions including $Co^{2+}$, $Mg^{2+}$, $Sn^{2+}$, $Fe^{2+}$, $Cu^{2+}$, $Pb^{2+}$, $Zn^{2+}$, $Na^+$, $Fe^{3+}$, $Ni^{2+}$, $Ag^+$, $Ca^{2+}$, and $K^+$. The ions were observed to bind at various sites of the CDs, followed by differentiation with principal component analysis [127]. In the study, biocompatibility and cytotoxicity were not the limiting factors. Multiple ion detection and cell imaging have been attained by exploitation of mPD-CDs which were prepared from $m$-phenylenediamine, a substrate that enables staining of animal, bacterial, and fungal cells. The mPD-CDs employed FL quenching for metal ion detection of $Cd^{2+}$ and $Zn^{2+}$ and a FL increase for $Cu^{2+}$, $Hg^{2+}$, $Gd^{3+}$, $Fe^{2+}$, $Fe^{3+}$, or $I^-$. The authors were able to show the efficiency of their technique in zebrafish and living cells [128]. Thus, a single type of CD can be sufficient for use in multiple ion detection.

Another great example demonstrating the versatility of CD application was depicted in a study where $Fe^{3+}$ and pH monitoring were successfully used in imaging MCF-7 cells. Blue fluorescent CDs were applied not only in $Fe^{3+}$ detection but also for the pH monitoring of river water [129]. The green fluorescent CDs on the other hand showed an increase in FL upon increment in pH between 1.98 and 8.95. Therefore, using two types of CDs in a single experiment is possible and less time-consuming in terms of preparation of separate experiments. Other strategies that have been used for the multidetection of metal ions is the use of a substrate with multiple binding sites. Glutathione (GSH) is a tripeptide that is endogenously expressed in cells, functioning as an antioxidant. It is composed of glycine, glutamine, and cysteine residues. CDs conjugated with GSH (GSH-CD) were exploited as colorimetric probes for the qualitative detection of $Zn^{2+}$, $Cu^{2+}$, and $Mn^{2+}$. The GSH-CDs exhibited dual FL emissions at 460 and 683 nm which in combination resulted in cyan colour under UV light. Fascinatingly, the metal ions could be monitored individually or simultaneously [130]. It is probable that the multi-detector ability of the CDs is due to the tripeptide which offers multiple binding sites so that there is less competition for binding on the sites. The authors of the study postulated that while $Zn^{2+}$ bound to the HS-site the $Cu^{2+}$ and $Mn^{2+}$ bound to the $-NH_2$ termini site of the GSH–CD. $Zn^{2+}$ induced a colour change from cyan to pink with a blue shift in wavelength from 683 to 652 nm and from 460 to 520 nm. In contrast, the $Cu^{2+}$ or $Mn^{2+}$ resulted in a blue colour change of the GSD–CD cyan; however, they were distinguishable by the lower FL peak of $Mn^{2+}$ with a slight red shift when compared with $Cu^{2+}$. The presence of all three metals displayed a nattier blue colour conversion from the cyan and with no change in the wavelength peak at 460 nm. Similar findings were observed with real water samples from the lake versus ultra-pure water, showing that colorimetric and FL properties can be employed in differentiating the metal ions [130].

Ratiometric FL sensing is a popular strategy exploited in the quantitative analysis of metal ions relying on both quenching and enhanced FL [65]. The main components of the assay would include either a mixture of CDs, dye-modified CDs, or other fluorescent material. The concentrations of the carcinogenic and very toxic metal ions and $Pb^{2+}$ and $Hg^{2+}$ in ultrapure water and river water were analysed with this approach; interestingly, the two types of CDs used were synthesized from the same biomass. The ratiometric method has also been explored for developing point-of-care (POC) diagnostic tools. Making use of the aggregation versus the monomeric state of a probe has been explored over assay probes that exhibit a shift in wavelength and/or FL based on the concentration of analyte [131]. The ratiometric technique is advantageous over single luminescence emission for the two following reasons:
– Photobleaching of the probe during detection leads to unreliable results.
– Variations in the excitation source can be a problem.

The rapid detection levels of human serum albumin microfluidic paper-based analytical device (μPAD), a protein that acts as a biomarker for various diseased states, were made possible with the ratiometric analysis [131]. Standard assays such as the ELISA have a long turnaround time, hence the need to explore a more rapid, cost-effective, and easy method such as the μPAD. μPAD was paired with 2'-hydroxychalcone derivatives as fluorescent probes so that changes can be visualized with the naked eye. A combination of green fluorescent CDs embedded within the μPAD and the smartphone technology has been developed in detecting anthrax, which is an infectious disease borne from bacillus anthracis species. Dipicolinic acid (DPA) is an anthrax biomarker which forms 5–15% of the spore. Europium ($Eu^{3+}$), which has a good binding affinity for DPA, was coordinated via the amine and carboxyl groups on the CDs playing the capturing role of the assay [132]. In the paper-based assay, emission at 530 nm remains unchanged acting as a reference whilst an increase in FL is observed at 630 nm in the presence of DPA, which then is served as a response signal.

# 5.8 Conclusion

The detection of biological molecules and metal ions in animals and the environment are crucial, particularly, their rapid detection is important for timeous diagnosis that will facilitate administration of both therapeutic and preventative interventions. This has been realized by the bio-application of nanoparticles in theranostics. CDs are a discovery that has had a dynamic effect on theranostics mainly due to their biocompatibility, photostability, tuneable surface area, good FL properties, and ease of synthesis. Their intrinsic chemical properties are provided by the carbon source used in the synthesis process which has a direct effect on their sensitivity and specificity. Fortunately, CDs can be modified to endow sensitivity and specificity by either doping or conjugation methods. Immerse research still needs to be directed to hemocompatibility, biocompatibility, and diagnostic studies.

# References

[1]    Unnikrishnan B, Wu RS, Wei SC, Huang CC, Chang HT. Fluorescent carbon dots for selective labeling of subcellular organelles. ACS Omega. 2020, 5(20), 11248–11261. doi: 10.1021/acsomega.9b04301.

[2]    Liu H. et al. Lighting up individual organelles with fluorescent carbon dots. Front Chem. 2021, 9(November), 1–16. doi: 10.3389/fchem.2021.784851.

[3]    Jin J, Li L, Zhang L, Luan Z, Xin S, Song K. Progress in the application of carbon dots-based nanozymes. Front Chem. 2021, 9(September), 1–8. doi: 10.3389/fchem.2021.748044.

[4]    Bao X. et al. In vivo theranostics with near-infrared-emitting carbon dots – Highly efficient photothermal therapy based on passive targeting after intravenous administration. Light Sci Appl. 2018, 7(1), 1–11. doi: 10.1038/s41377-018-0090-1.

[5]    Biswal MR, Bhatia S. Carbon dot nanoparticles: Exploring the potential use for gene delivery in ophthalmic diseases. Nanomaterials. 2021, 11(4), 1–12. doi: 10.3390/nano11040935.

[6]    Ross S, Wu RS, Wei SC, Ross GM, Chang HT. The analytical and biomedical applications of carbon dots and their future theranostic potential: A review. J Food Drug Anal. 2020, 28(4), 677–695. doi: 10.38212/2224-6614.1154.

[7]    Mishra V, Patil A, Thakur S, Kesharwani P. Carbon dots: Emerging theranostic nanoarchitectures. Drug Discov Today. 2018, 23(6), 1219–1232. doi: 10.1016/j.drudis.2018.01.006.

[8]    Chen Y. et al. Enhanced fluorescence and environmental stability of red-emissive carbon dots via chemical bonding with cellulose films. ACS Omega. 2022, 7(8), 6834–6842. doi: 10.1021/acsomega.1c06426.

[9]    Cedervall T. et al. Understanding the nanoparticle-protein corona using methods to quntify exchange rates and affinities of proteins for nanoparticles. Proc Natl Acad Sci U S A. 2007, 104(7), 2050–2055. doi: 10.1073/pnas.0608582104.

[10]   Lundqvist M. et al. The nanoparticle protein corona formed in human blood or human blood fractions. PLoS One. 2017, 12(4), 1–15. doi: 10.1371/journal.pone.0175871.

[11]   Fedel M. Hemocompatibility of carbon nanostructures. 2020, (Figure 1), 1–36.

[12]   Li S, Guo Z, Zhang Y, Xue W, Liu Z. Blood compatibility evaluations of fluorescent carbon dots. ACS Appl Mat Interf. 2015, 7(34), 19153–19162. doi: 10.1021/acsami.5b04866.

[13]  Zhong J. et al. Blood compatible heteratom-doped carbon dots for bio-imaging of human umbilical vein endothelial cells. Chinese Chem Lett. 2020, 31(3), 769–773. doi: 10.1016/j.cclet.2020.01.007.

[14]  Tian X. et al. Carbon quantum dots: In vitro and in vivo studies on biocompatibility and biointeractions for optical imaging. Int J Nanomedicine. 2020, 15, 6519–6529. doi: 10.2147/IJN. S257645.

[15]  Iravani S, Varma RS. Green synthesis, biomedical and biotechnological applications of carbon and graphene quantum dots. A review. Environ Chem Lett. 2020, 0123456789. doi: 10.1007/s10311-020-00984-0.

[16]  Liao J, Yao Y, Lee CH, Wu Y, Li P. In vivo biodistribution, clearance, and biocompatibility of multiple carbon dots containing nanoparticles for biomedical application. Pharmaceutics. 2021, 13(11). doi: 10.3390/pharmaceutics13111872.

[17]  Kadhim RJ, Karsh EH, Taqi ZJ, Jabir MS. Biocompatibility of gold nanoparticles: In-vitro and In-vivo study. Mater Today Proc. 2021, 42, 3041–3045. doi: 10.1016/j.matpr.2020.12.826.

[18]  Fan JH, Hung WI, Li WT, Yeh JM. Biocompatibility study of gold nanoparticles to human cells. IFMBE Proc. 2009, 23, 870–873. doi: 10.1007/978-3-540-92841-6_214.

[19]  Gurunathan S, Han JW, Park JH, Kim JH. A green chemistry approach for synthesizing biocompatible gold nanoparticles. Nanoscale Res Lett. 2014, 9(1), 1–11. doi: 10.1186/1556-276X-9-248.

[20]  Parboosing R, Govender T, Maguire GEM, Kruger HG. Synthesis, characterization and biocompatibility of a multifunctional gold nanoparticle system for the delivery of single-stranded RNA to lymphocytes. South African J Chem. 2018, 71, 1–14. doi: 10.17159/0379-4350/2018/v71a1.

[21]  Facure MHM, Schneider R, Mercante LA, Correa DS. A review on graphene quantum dots and their nanocomposites: From laboratory synthesis towards agricultural and environmental applications. Environ Sci. 2020, 7, 3710–3734.

[22]  Wang Y, Hu A. Carbon quantum dots: Synthesis, properties and applications. J Mater Chem C. 2014, 2, 6921–6939.

[23]  Tao S. et al. Confined-domain crosslink-enhanced emission effect in carbonized polymer dots. Light Sci Appl. 2022, 11, 56.

[24]  Sharma V, Kagdada HL, Jha PK. Four-fold enhancement in the thermoelectric power factor of germanium selenide monolayer by adsorption of graphene quantum dot. Energy. Apr 2020, 196, 117104. doi: 10.1016/J.ENERGY.2020.117104.

[25]  Chini MK, Kumar V, Javed A, Satapathi S. Graphene quantum dots and carbon nano dots for the FRET based detection of heavy metal ions. Nano Struct Nano Obj. Jul 2019, 19, 100347. doi: 10.1016/J. NANOSO.2019.100347.

[26]  El-Shabasy RM, Elsadek MF, Ahmed BM, Farahat MF, Mosleh KM, Taher MM. Recent developments in carbon quantum dots: Properties, fabrication techniques, and bio-applications. Processes. 2021, 9(2). doi: 10.3390/pr9020388.

[27]  Liu H. et al. Direct carbonization of organic solvents toward graphene quantum dots. Nanoscale. 2020, 12(20). doi: 10.1039/d0nr01903h.

[28]  Xu C. et al. Fabrication of centimeter-scale light-emitting diode with improved performance based on graphene quantum dots. Appl Phys Express. 2017, 10(3). doi: 10.7567/APEX.10.032102.

[29]  Mansuriya BD, Altintas Z. Graphene quantum dot-based electrochemical immunosensors for biomedical applications. Materials Basel. 2020, 13, 96.

[30]  Zeng Q, Feng T, Tao S, Zhu S, Yang B. Precursor-dependent structural diversity in luminescent carbonized polymer dots (CPDs): The nomenclature. Light Sci Appl. 2021, 10, 142.

[31]  Dimos K. Tuning carbon dots' optoelectronic properties with polymers. Polymers (Basel). 2018, 10, 1312.

[32]  Tao S, Zhu S, Feng T, Zheng C, Yang B. Crosslink-enhanced emission effect on luminescence in polymers: Advances and perspectives. Angew Chemie. 2020, 132, 9910–9924.

[33] Feng Z, Adolfsson KH, Xu Y, Fang H, Hakkarainen M, Wu M. Carbon dot/polymer nanocomposites: From green synthesis to energy, environmental and biomedical applications. Sustainable Mater Technol. 2021, 29. doi: 10.1016/j.susmat.2021.e00304.

[34] Zulfajri M, Sudewi S, Ismulyati S, Rasool A, Adlim M, Huang GG. Carbon dot/polymer composites with various precursors and their sensing applications: A review. Coatings. 2021, 11(9). doi: 10.3390/coatings11091100.

[35] Molaei MJ. Principles, mechanisms, and application of carbon quantum dots in sensors: A review. Anal Methods. 2020, 12(10). doi: 10.1039/c9ay02696g.

[36] Nazri NAA, Azeman NH, Luo Y, Bakar AAA. Carbon quantum dots for optical sensor applications: A review. Opt Laser Technol. 2021, 139. doi: 10.1016/j.optlastec.2021.106928.

[37] Cao L. et al. Carbon dots for energy conversion applications. J Appl Phys. 2019, 125(22). doi: 10.1063/1.5094032.

[38] Azam N, Najabat Ali M, Javaid Khan T. Carbon quantum dots for biomedical applications: review and analysis. Front Mater. 2021, 8. doi: 10.3389/fmats.2021.700403.

[39] Alaghmandfard A. et al. Recent advances in the modification of carbon-based quantum dots for biomedical applications. Mater Sci Eng C. 2021, 120. doi: 10.1016/j.msec.2020.111756.

[40] Robinson PK. Enzymes: Principles and biotechnological applications. Essays Biochem. 2015, 59, 1–41.

[41] Liang J, Liang K. Multi-enzyme cascade reactions in metal-organic frameworks. Chem Record. 2020, 20(10). doi: 10.1002/tcr.202000067.

[42] Drout RJ, Robison L, Farha OK. Catalytic applications of enzymes encapsulated in metal–organic frameworks. Coord Chem Rev. 2019, 381. doi: 10.1016/j.ccr.2018.11.009.

[43] Zdarta J, Meyer AS, Jesionowski T, Pinelo M. A general overview of support materials for enzyme immobilization: Characteristics, properties, practical utility. Catalysts. 2018, 8(2). doi: 10.3390/catal8020092.

[44] Zahirinejad S. et al. Nano-organic supports for enzyme immobilization: Scopes and perspectives. Coll Surf Biointerf. 2021, 204. doi: 10.1016/j.colsurfb.2021.111774.

[45] Sepahvand H, Heravi MM, Saber M, Hooshmand SE. Techniques and support materials for enzyme immobilization using Ugi multicomponent reaction: An overview. J Iran Chem Soc. 2022, 19(6). doi: 10.1007/s13738-021-02449-9.

[46] Melo ADQ, Silva FFM, Dos Santos JCS, Fernández-Lafuente R, Lemos TLG, Dias Filho FA. Synthesis of benzyl acetate catalyzed by lipase immobilized in nontoxic chitosan-polyphosphate beads. Molecules. 2017, 22(12). doi: 10.3390/molecules22122165.

[47] Pan Y. et al. How do enzymes orient when trapped on metal-organic framework (MOF) surfaces? J Am Chem Soc. 2018, 140(47). doi: 10.1021/jacs.8b09257.

[48] Grollmisch A, Kragl U, Großeheilmann J. Enzyme immobilization in polymerized ionic liquids-based hydrogels for active and reusable biocatalysts. Syn Open. 2018, 2(2). doi: 10.1055/s-0037-1610144.

[49] Song J. et al. Construction of multiple enzyme metal–organic frameworks biocatalyst via DNA scaffold: A promising strategy for enzyme encapsulation. Chem Eng J. 2019, 363. doi: 10.1016/j.cej.2019.01.138.

[50] Li Y. et al. Hybrids of carbon dots with subunit B of ricin toxin for enhanced immunomodulatory activity. J Colloid Interface Sci. 2018, 523, 226–233.

[51] Wang Y, Wang Z, Rui Y, Li M. Horseradish peroxidase immobilization on carbon nanodots/CoFe layered double hydroxides: Direct electrochemistry and hydrogen peroxide sensing. Biosens Bioelectron. 2015, 64, 57–62.

[52] Pradhan NRJN, Jana NR. Inhibition of protein aggregation by iron oxide nanoparticles ocnjugated with glutamine and proline based Osmosis. ACS Appl Nano Mater. 2018, 1, 1098.

[53] Hosseinzadeh G, Maghari A, Farniya SMF, Keihan AH, Moosavi-Movahedi AA. "Interaction of insulin with colloidal ZnS quantum dots functionalized by various surface capping agents,". Mater Sci Eng C. 2017, 77. doi: 10.1016/j.msec.2017.04.018.

[54] Kulpa-Koterwa A, Ossowski T, Niedziałkowski P. Functionalized $Fe_3O_4$ nanoparticles as glassy carbon electrode modifiers for heavy metal ions detection – A mini review. Materials Basel. 2021, 14, 7725.

[55] Vaghari H. et al. Application of magnetic nanoparticles in smart enzyme immobilization. Biotechnol Lett. 2016, 38, 223–233.

[56] Leo E, Vandelli MA, Cameroni R, Forni F. Doxorubicin-loaded gelatin nanoparticles stabilized by glutaraldehyde: Involvement of the drug in the cross-linking process. Int J Pharm. 1997, 155, 75–82.

[57] Gupta K, Jana AK, Kumar S, Maiti M. Immobilization of amyloglucosidase from SSF of Aspergillus niger by crosslinked enzyme aggregate onto magnetic nanoparticles using minimum amount of carrier and characterizations. J Mol Catal B Enzym. 2013, 98, 30–36.

[58] Panwar N. et al. Nanocarbons for biology and medicine: sensing, imaging, and drug delivery. Chem Rev. 2019, doi: 10.1021/acs.chemrev.9b00099.

[59] Zheng X, Cheng W, Ji C, Zhang J, Yin M, Access O. "Detection of metal ions in biological systems: A review,". 2020, 231–246.

[60] Wang Z. et al. The green synthesis of carbon quantum dots and applications for sulcotrione detection and anti-pathogen activities. J Saudi Chem Soc. 2021, 25(12), 101373. doi: 10.1016/j.jscs.2021.101373.

[61] Chahal S, Macairan JR, Yousefi N, Tufenkji N, Naccache R. Green synthesis of carbon dots and their applications. RSC Adv. 2021, 11(41), 25354–25363. doi: 10.1039/d1ra04718c.

[62] Li F. et al. Highly fluorescent chiral N-S-doped carbon dots from cysteine: Affecting cellular energy metabolism. Angew Chemie – Int Ed. 2018, 57(9), 2377–2382. doi: 10.1002/anie.201712453.

[63] Lin F, Li C, Chen Z. Bacteria-derived carbon dots inhibit biofilm formation of Escherichia coli without affecting cell growth. Front Microbiol. 2018, 9(FEB), 1–9. doi: 10.3389/fmicb.2018.00259.

[64] L. J-H, Na-Eun Choi JL, Lee J-Y, Park E-C. Recent advances in organelle-targeted fluorescent probes. Chinese J Org Chem. 2021, 41(2), 611–623. doi: 10.6023/cjoc202006046.

[65] Liu Z. et al. Ratiometric fluorescent sensing of $Pb^{2+}$ and $Hg^{2+}$ with two types of carbon dot nanohybrids synthesized from the same biomass. Sensors Actuators B Chem. 2019, 296, 126698. doi: 10.1016/j.snb.2019.126698.

[66] Darinel S. et al. iScience Heavy metal ion detection using green precursor derived carbon dots. doi: 10.1016/j.isci.

[67] Koutsogiannis P, Thomou E, Stamatis H, Gournis D, Rudolf P. Advances in fluorescent carbon dots for biomedical applications. Adv Phys X. 2020, 5(1). doi: 10.1080/23746149.2020.1758592.

[68] Sharma N, Jandaik S, Singh TG, Kumar S. Nanoparticles: Boon to Mankind and Bane to Pathogens. 2016, Elsevier Inc.

[69] Du F. et al. Facile, rapid synthesis of N,P-dual-doped carbon dots as a label-free multifunctional nanosensor for Mn(VII) detection, temperature sensing and cellular imaging. Sensors Actuators B Chem. 2018, 277(March), 492–501. doi: 10.1016/j.snb.2018.09.027.

[70] Wang M. et al. Highly luminescent nucleoside-based N, P-doped carbon dots for sensitive detection of ions and bioimaging. Front Chem. 2022, 10(June), 1–11. doi: 10.3389/fchem.2022.906806.

[71] Minh L, Phan T. Fluorescent carbon dot-supported imaging-based biomedicine: a comprehensive review. 2022, 2022.

[72] Kong D, Yan F, Luo Y, Ye Q, Zhou S, Chen L. Amphiphilic carbon dots for sensitive detection, intracellular imaging of Al3+. Anal Chim Acta. 2017, 953, 63–70. doi: 10.1016/j.aca.2016.11.049.

[73] Sun X. et al. Green synthesis of carbon dots originated from Lycii Fructus for effective fluorescent sensing of ferric ion and multicolor cell imaging. J Photochem Photobiol B Biol. 2017, 175(August), 219–225. doi: 10.1016/j.jphotobiol.2017.08.035.

[74] Xu N. et al. Carbon dots inspired by structure-inherent targeting for nucleic acid imaging and localized photodynamic therapy. Sensors Actuators B Chem. 2021, 344(June), 130322. doi: 10.1016/j.snb.2021.130322.

[75] Esteves da Silva JCG, Gonçalves HMR. Analytical and bioanalytical applications of carbon dots. TrAC – Trends Anal Chem. 2011, 30(8), 1327–1336. doi: 10.1016/j.trac.2011.04.009.

[76] Vermes I, Haanen C. Apoptosis and programmed cell death in health and disease. Adv Clin Chem. 1994, 31(C), 177–246. doi: 10.1016/S0065-2423(08)60336-4.

[77] Ates G, Vanhaecke T, Rogiers V, Rodrigues RM. Assaying cellular viability using the neutral red uptake assay. Methods Mol Biol. 2017, 1601, 19–26. doi: 10.1007/978-1-4939-6960-9_2.

[78] Manente S, De Pieri S, Iero A, Rigo C, Bragadin M. A comparison between the responses of neutral red and acridine orange: Acridine orange should be preferential and alternative to neutral red as a dye for the monitoring of contaminants by means of biological sensors. Anal Biochem. 2008, 383(2), 316–319. doi: 10.1016/j.ab.2008.09.015.

[79] Chen X. et al. Lysosomal targeting with stable and sensitive fluorescent probes (Superior LysoProbes): applications for lysosome labeling and tracking during apoptosis. Sci Rep. 2015, 5, 1–10. doi: 10.1038/srep09004.

[80] Zou G, Chen S, Liu N, Yu Y. A ratiometric fluorescent probe based on carbon dots assembly for intracellular lysosomal polarity imaging with wide range response. Chinese Chem Lett. 2022, 33(2), 778–782. doi: 10.1016/j.cclet.2021.08.076.

[81] Shuang E, Mao QX, Yuan XL, Kong XL, Chen XW, Wang JH. Targeted imaging of the lysosome and endoplasmic reticulum and their pH monitoring with surface regulated carbon dots. Nanoscale. 2018, 10(26), 12788–12796. doi: 10.1039/c8nr03453b.

[82] Wu F. et al. Red/near-infrared emissive metalloporphyrin-based nanodots for magnetic resonance imaging-guided photodynamic therapy in Vivo. Part Part Syst Charact. 2018, 35(9), 1–7. doi: 10.1002/ppsc.201800208.

[83] Zhang DY. et al. Ruthenium complex-modified carbon nanodots for lysosome-targeted one- and two-photon imaging and photodynamic therapy. Nanoscale. 2017, 9(47), 18966–18976. doi: 10.1039/c7nr05349e.

[84] Havrdová M, Urbančič I, Bartoň Tománková K, Malina L, Štrancar J, Bourlinos AB. Self-targeting of carbon dots into the cell nucleus: Diverse mechanisms of toxicity in NIH/3T3 and l929 cells. Int J Mol Sci. 2021, 22(11). doi: 10.3390/ijms22115608.

[85] Jung YK, Shin E, Kim BS. Cell nucleus-targeting zwitterionic carbon dots. Sci Rep. 2015, 5. doi: 10.1038/srep18807.

[86] Schwarz DS, Blower MD. The endoplasmic reticulum: Structure, function and response to cellular signaling. Cell Mol Life Sci. 2016, 73(1), 79–94. doi: 10.1007/s00018-015-2052-6.

[87] Chung KN. et al. Stable transfectants of human MCF-7 breast cancer cells with increased levels of the human folate receptor exhibit an increased sensitivity to antifolates. J Clin Invest. 1993, 91(4), 1289–1294. doi: 10.1172/JCI116327.

[88] Bhunia SK, Maity AR, Nandi S, Stepensky D, Jelinek R. Imaging cancer cells expressing the folate receptor with carbon dots produced from folic acid. ChemBioChem. 2016, 17(7), 614–619. doi: 10.1002/cbic.201500694.

[89] Ma Y, Wang Y, Liu Y, Shi L, Yang D. Multi-carbon dots and aptamer based signal amplification ratiometric fluorescence probe for protein tyrosine kinase 7 detection. J Nanobiotechnol. 2021, 19(1), 1–11. doi: 10.1186/s12951-021-00787-7.

[90] Nau WM. et al. Carbon dot blinking enables accurate molecular counting at nanoscale resolution. Anal Chem. 2021, 93(8), 3968–3975. doi: 10.1021/acs.analchem.0c04885.

[91] Ghirardello M. et al. Carbon dot-based fluorescent antibody nanoprobes as brain tumour glioblastoma diagnostics. Nanoscale Adv. 2022, 4(7), 1770–1778. doi: 10.1039/d2na00060a.

[92] Devkota A, Pandey A, Yadegari Z, Dumenyo K, Taheri A. Amine-coated carbon dots ($NH_2$-FCDs) as novel antimicrobial agent for Gram-negative bacteria. Front Nanotechnol. 2021, 3(November), 1–10. doi: 10.3389/fnano.2021.768487.

[93] Pierrat P. et al. Efficient in vitro and in vivo pulmonary delivery of nucleic acid by carbon dot-based nanocarriers. Biomaterials. 2015, 51, 290–302. doi: 10.1016/j.biomaterials.2015.02.017.

[94] Gao T. et al. A peptide nucleic acid–regulated fluorescence resonance energy transfer DNA assay based on the use of carbon dots and gold nanoparticles. Microchim Acta. 2020, 187(7). doi: 10.1007/s00604-020-04357-w.

[95] Chinmoy Mishra LS. Peptide nucleic acid. Prog Biochem Biophys. 1996, 23(3), 212–213. doi: 10.5455/aim.2011.19.118-123.

[96] Chunduri LAA. et al. Development of carbon dot based microplate and microfluidic chip immunoassay for rapid and sensitive detection of HIV-1 p24 antigen. Microfluid Nanofluidics. 2016, 20(12), 1–10. doi: 10.1007/s10404-016-1825-z.

[97] Blažková M, Javůrková B, Fukal L, Rauch P. Immunochromatographic strip test for detection of genus Cronobacter. Biosens Bioelectron. 2011, 26(6), 2828–2834. doi: 10.1016/j.bios.2010.10.001.

[98] Singh NK, Chakma B, Jain P, Goswami P. Protein-induced fluorescence enhancement based detection of plasmodium falciparum glutamate dehydrogenase using carbon dot coupled specific aptamer. ACS Comb Sci. 2018, 20(6), 350–357. doi: 10.1021/acscombsci.8b00021.

[99] Bhattacharyya D, Sarswat PK, Free ML. Quantum dots and carbon dots based fluorescent sensors for TB biomarkers detection. Vacuum. 2017, 146, 606–613. doi: 10.1016/j.vacuum.2017.02.003.

[100] He Q, Zhu Z, Jin L, Peng L, Guo W, Hu S. Detection of HIV-1 p24 antigen using streptavidin-biotin and gold nanoparticles based immunoassay by inductively coupled plasma mass spectrometry. J Anal At Spectrom. 2014, 29(8), 1477–1482. doi: 10.1039/c4ja00026a.

[101] Kosaka PM, Pini V, Calleja M, Tamayo J. Ultrasensitive detection of HIV-1 p24 antigen by a hybrid nanomechanical-optoplasmonic platform with potential for detecting HIV-1 at first week after infection. PLoS ONE. 2017, 12(2). doi: 10.1371/journal.pone.0171899.

[102] Gulati S, Singh P, Diwan A, Mongia A, Kumar S. Functionalized gold nanoparticles: Promising and efficient diagnostic and therapeutic tools for HIV/AIDS. RSC Med Chem. 2020, 11(11), 1252–1266. doi: 10.1039/d0md00298d.

[103] Briffa J, Sinagra E, Blundell R. Heavy metal pollution in the environment and their toxicological effects on humans. Heliyon. 2020, 6, 691.

[104] Balali-Mood M, Naseri K, Tahergorabi Z, Khazdair MR, Sadeghi M. Toxic mechanisms of five heavy metals: Mercury, lead, chromium, cadmium, and arsenic. Front Pharmacol. 2021, 12, 1–19.

[105] Pujol L, Evrard D, Groenen-Serrano K, Freyssinier M, Ruffien-Cizsak A, Gros P. Electrochemical sensors and devices for heavy metals assay in water: The French groups' contribution. Front Chem. 2014, 2, 1–24.

[106] Molaei MJ. A review on nanostructured carbon quantum dots and their applications in biotechnology, sensors, and chemiluminescence. Talanta. 2019, 196. doi: 10.1016/j.talanta.2018.12.042.

[107] Damera DP, Manimaran R, Krishna Venuganti VV, Nag A. Green Synthesis of Full-Color Fluorescent Carbon Nanoparticles from Eucalyptus Twigs for Sensing the Synthetic Food Colorant and Bioimaging. ACS Omega. 2020, 5, 19905–19918.

[108] Xu D, Lei F, Chen H, Yin L, Shi Y, Xie J. One-step hydrothermal synthesis and optical properties of self-quenching-resistant carbon dots towards fluorescent ink and as nanosensors for Fe3+ detection. RSC Adv. 2019, 9, 8290–8299.

[109] Wang J, Yang Y, Liu X. Solid-state fluorescent carbon dots: Quenching resistance strategies, high quantum efficiency control, multicolor tuning, and applications. Mater Adv. 2020, 1(9). doi: 10.1039/d0ma00632g.

[110] Ma X, Li S, Hessel V, Lin L, Meskers S, Gallucci F. "Synthesis of N-doped carbon dots via a microplasma process". Chem Eng Sci. 2020, 220. doi: 10.1016/j.ces.2020.115648.

[111] Ji H, Zhou F, Gu J, Shu C, Xi K, Jia X. Nitrogen-doped carbon dots as a new substrate for sensitive glucose determination. Sensors (Switzerland). 2016, 16(5). doi: 10.3390/s16050630.

[112] Bonet-San-Emeterio M, Algarra M, Petković M, Del Valle M. Modification of electrodes with N-and S-doped carbon dots. Evaluation of the electrochemical response. Talanta. 2020, 212. doi: 10.1016/j.talanta.2020.120806.

[113] Gharat PM, Pal H, Dutta Choudhury S. Photophysics and luminescence quenching of carbon dots derived from lemon juice and glycerol. Spectrochim Acta – Part A Mol Biomol Spectrosc. 2019, 209. doi: 10.1016/j.saa.2018.10.029.

[114] Wei J. et al. A novel fluorescent sensor for water in organic solvents based on dynamic quenching of carbon quantum dots. New J Chem. 2018, 42(23). doi: 10.1039/c8nj04365e.

[115] Wang L, Zhou HS. Green synthesis of luminescent nitrogen-doped carbon dots from milk and its imaging application. Anal Chem. Sep 2014, 86(18), 8902–8905. doi: 10.1021/ac502646x.

[116] Dong Y. et al. Carbon-based dots co-doped with nitrogen and sulfur for high quantum yield and excitation-independent emission. Angew Chemie Int Ed. Jul 2013, 52(30), 7800–7804. doi: 10.1002/anie.201301114.

[117] Dong W, Wang R, Gong X, Dong C. An efficient turn-on fluorescence biosensor for the detection of glutathione based on FRET between N,S dual-doped carbon dots and gold nanoparticles. Anal Bioanal Chem. 2019, 411, 6687–6695.

[118] Tang X. et al. Nitrogen-doped fluorescence carbon dots as multi-mechanism detection for iodide and curcumin in biological and food samples. Bioact Mater. 2021, 6(6). doi: 10.1016/j.bioactmat.2020.11.006.

[119] Yang L, Wen J, Li K, Liu L, Wang W. Carbon quantum dots: Comprehensively understanding of the internal quenching mechanism and application for catechol detection. Sensors Actuators B Chem. 2021, 333. doi: 10.1016/j.snb.2021.129557.

[120] Abdolmohammad-Zadeh H, Azari Z, Pourbasheer E. Fluorescence resonance energy transfer between carbon quantum dots and silver nanoparticles: Application to mercuric ion sensing. Spectrochim Acta – Part A Mol Biomol Spectrosc. 2021, 245. doi: 10.1016/j.saa.2020.118924.

[121] Mintz KJ, Guerrero B, Leblanc RM. Photoinduced electron transfer in carbon dots with long-wavelength photoluminescence. J Phys Chem C. 2018, 122(51). doi: 10.1021/acs.jpcc.8b06868.

[122] Dong B. et al. Fluorescence immunoassay based on the Inner-filter effect of carbon dots for highly sensitive amantadine detection in foodstuffs. Food Chem. 2019, 294. doi: 10.1016/j.foodchem.2019.05.082.

[123] Zhang J, Dong L, Yu SH. A selective sensor for cyanide ion (CN−) based on the inner filter effect of metal nanoparticles with photoluminescent carbon dots as the fluorophore. Sci Bull. 2015, 60(8). doi: 10.1007/s11434-015-0764-5.

[124] Miao S, Liang K, Kong B. Förster resonance energy transfer (FRET) paired carbon dot-based complex nanoprobes: Versatile platforms for sensing and imaging applications. Mater Chem Front. 2020, 4, 128–139.

[125] Khayal A. et al. Advances in the methods for the synthesis of carbon dots and their emerging applications. Polymers. 2021, 13(18). doi: 10.3390/polym13183190.

[126] Fine GF, Cavanagh LM, Afonja A, Binions R. Metal oxide semi-conductor gas sensors in environmental monitoring. Sensors. 2010, 10, 5469–5502.

[127] Moonrinta S, Kwon B, In I, Kladsomboon S, Sajomsang W, Paoprasert P. Highly biocompatible yogurt-derived carbon dots as multipurpose sensors for detection of formic acid vapor and metal ions. Opt Mater (Amst). 2018, 81(April), 93–101. doi: 10.1016/j.optmat.2018.05.021.

[128] Durrani S. et al. Carbon dots for multicolor cell imaging and ultra-sensitive detection of multiple ions in living cells: One stone for multiple birds. Environ Res. 2022, 212(PC), 113260. doi: 10.1016/j.envres.2022.113260.

[129] Diao H. et al. Facile and green synthesis of fluorescent carbon dots with tunable emission for sensors and cells imaging. Spectrochim Acta – Part A Mol Biomol Spectrosc. 2018, 200, 226–234. doi: 10.1016/j.saa.2018.04.029.

[130] Yu L. et al. Effective determination of Zn2+, Mn2+, and Cu2+ simultaneously by using dual-emissive carbon dots as colorimetric fluorescent probe. Eur J Inorg Chem. 2018, 2018(29), 3418–3426. doi: 10.1002/ejic.201800474.

[131] Luo Z. et al. Paper-based ratiometric fluorescence analytical devices towards point-of-care testing of human serum albumin. Angew Chemie. 2020, 132(8), 3155–3160. doi: 10.1002/ange.201915046.

[132] Wang J. et al. An europium functionalized carbon dot-based fluorescence test paper for visual and quantitative point-of-care testing of anthrax biomarker. Talanta. 2020, 220(June), 121377. doi: 10.1016/j.talanta.2020.121377.

[133] Li Y, Lee JS. Insights into characterization methods and biomedical applications of nanoparticle-protein corona. Materials Basel. 2020, 13(14). doi: 10.3390/ma13143093.

Manoj Kumar Banjare*, Kamalakanta Behera*, Ramesh Kumar Banjare
and Siddharth Pandey*

# Chapter 6
# Carbon dots in nanozymes

**Abstract:** The new artificial enzymes called nanozymes, which are built on carbon
dots (CDs), were created to address several inherent problems with natural enzymes,
such as their costing of storeroom, structural volatility, and chemical sympathy.
Owing to their high catalytic activity, biocompatibility, and ease of surface fictionali-
zations, CDs have attracted a lot of attention in the recent years and are now being
considered potential replacements in the biomedical, bio-sensing, detection, and
green areas. CDs nanozymes have spurred an increase in the study because of their
superior catalytic properties, biocompatibility, and environmental friendliness. Although significant advancements have been made, no book chapter devoted to CDs
nanozymes has yet been published. In this chapter of the book, we examine the procedures and building blocks utilized to create CDs with enzyme-like behaviour. Additionally, the crucial problems and difficulties of studying nanozymes are covered. To
offer future research possibilities, current obstacles to the thriving conversion of CDs
to possible relevance are addressed.

**Keywords:** Carbon dots, nanozyme, chemosensing, catalytic activity, bioimaging, properties and applications of CD nanozyme

**Acknowledgement:** The authors are grateful to HOD MATS School of Sciences, MATS University, Raipur,
Chhattisgarh

**Authors' contribution:** The manuscript was written through the contributions of all authors. All authors
have approved the final version of the manuscript.

**Note:** All authors declare no competing financial interest.

*Corresponding authors: Manoj Kumar Banjare, MATS School of Sciences, MATS University, Pagaria
Complex, Pandri, Raipur, Chhattisgarh 492004, India, e-mails: manojbanjare7@gmail.com,
manojbanjarechem111@gmail.com
*Corresponding authors: Kamalakanta Behera, Department of Chemistry, University
of Allahabad, Prayagraj, Uttar Pradesh 211002, India, e-mail: kamala.iitd@gmail.com;
*Corresponding authors: Siddharth Pandey, Department of Chemistry, Indian Institute of Technology
Delhi, Hauz Khas, New Delhi 110016, India, e-mail: sipandey@chemistry.iitd.ac.in
Ramesh Kumar Banjare, MATS College, MATS University, Aarang, Chhattisgarh 492004, India

https://doi.org/10.1515/9783110799958-006

## 6.1 Introduction

Proteins make up the majority of enzymes, which are potent biocatalysts, with a few catalytic RNA molecules tossed in for good measure [1, 2]. The primary function of enzymes is to catalyse the modification of bio-molecules, and these reactions are frequently conceded out in calm conditions. The nature of industrial catalysts depends on the high temperature and pressure, organic solvents, and settings with severe pH levels [3, 4]. Due to their high catalytic activity and substrate selectivity, natural enzymes have found extensive usage in the industrialized, medicinal, and biological domains [7–12]. The main work of enzymes is to catalyse the alteration of bio-molecules, and these processes typically take place under quiet circumstances [5, 6]. Unlike typical chemical or industrial catalysts, which are frequently used in difficult circumstances including high temperature, high pressure, organic solvents, and environments with intense pH stage [3, 4], their potential uses in the realms of food processing, biosensing, environmental protection, biomedicine, and other areas are all constrained by these disadvantages. To get around these problems, scientists have spent a lot of time investigating synthetic enzyme imitators [15–17]. In previous investigations, it has been demonstrated that many compounds, including fullerenes, cyclodextrins, polymers, dendrimers, porphyrins, metal complexes, and numerous bio-molecules, can act as artificial enzymes [18–24].

While the identification of iron oxide ($Fe_3O_4$) NPs as peroxides mimic in year 2007, numerous studies on synthetic enzymes based on non-material (referred to as "nanozymes") have been carried out [25–30]. Nanozymes, which have nanoscale dimensions of 1–100 nm and its enzyme catalytic properties, are non-materials [41, 42]. To catalyse the same bio-catalytic events as enzymes, non-materials have inherent enzymatic catalytic capabilities [43]. The advantages of both traditional chemical catalysts and biocatalysts are successfully combined by nanozymes. The discovery of non-material having inherent catalytic abilities has received the majority of attention recently since realizing the potential applications of non-material is an intriguing field. Nanozymes are superior to natural enzymes in that they are more affordable, highly stable, and long-lasting [26–28, 43] (Table 6.1). The ability of some nanozymes to catalyse synthetic bioprocesses like bio-orthogonal catalysis is even more exciting [44–46].

Based on these unique features, nanozymes have been applied in bio-sensing [47–49], remediation of the environment [50, 51], identification and treatment of disease [52–54], the use of antibacterial drugs [55–58], and cyto-protection against cellular bio-molecules [59–61] (Figure 6.1). Table 6.1 displays the examples of research on carbon dots (CDs) with enzyme-mimicking properties that were produced using each method.

**Table 6.1:** Doped CDs, pristine CDs, and CD-based hybrid nanozymes with enzyme-like activities.

| CDs nanozyme | Substrates | Carbon precursor | Synthesis methods | Enzyme-like activity | References |
|---|---|---|---|---|---|
| Doped N-, Fe-CDs | $H_2O_2$ and TMB | β-Cyclodextrin | Hydrothermal | Peroxidase | [31] |
| Se-CDs | $H_2O_2$ | Selenocystine | Hydrothermal | Superoxide dismutase (SOD) | [32] |
| GQDs/Fe$_3$O$_4$ | $H_2O_2$, TMB | Graphene oxide | Precipitation procedure | Peroxidase | [33] |
| CDs | $H_2O_2$, TMB | β-Cyclodextrin | Chemical oxidation | Peroxidase | [34] |
| Au, N-CDs | $KO_2$ | Citric acid | Hydrothermal | Peroxidase | [35] |
| N, S-CDs | $H_2O_2$, TMB | Citric acid | Hydrothermal | Peroxidase | [36] |
| CD-based hybrid nanozymes, N-CDs, and manganese oxide/ferric oxide hybrids | $H_2O_2$, TMB | CH$_3$COONa | Hydrothermal | Peroxidase | [15] |
| GQDs | $KO_2$ | Anthracite and bituminous coal | Chemical oxidation | Superoxide dismutase (SOD) | [10] |
| Pristine CDs CDs | $H_2O_2$, ABTS | Citric acid, L-glutamine, succinic acid, L-glutamic acid, and glycine | Pyrolysis | Peroxidase | [7] |
| Cu(II)/Cu$_2$O/N-GQDs | $H_2O_2$, ABTS | Citric acid | Pyrolysis | Peroxidase | [37] |

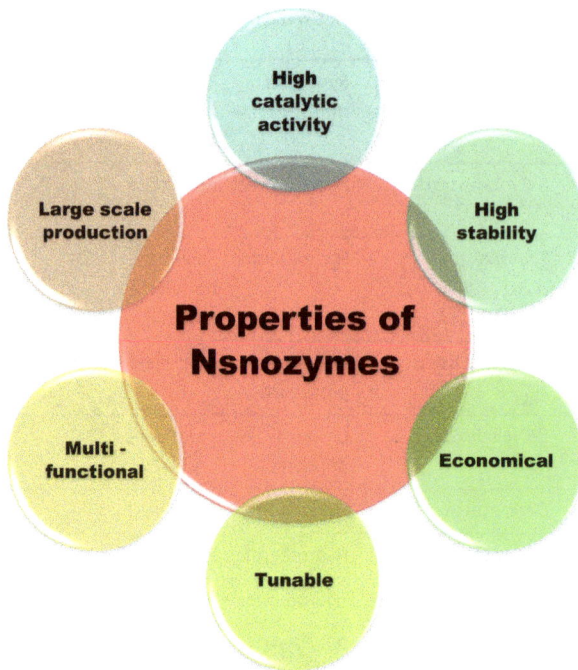

**Scheme 6.1:** Properties of nanozymes.

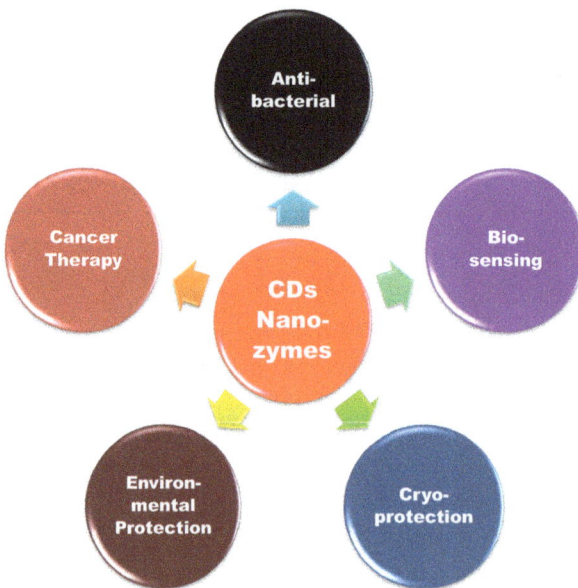

**Figure 6.1:** Recent applications of CD nanozymes.

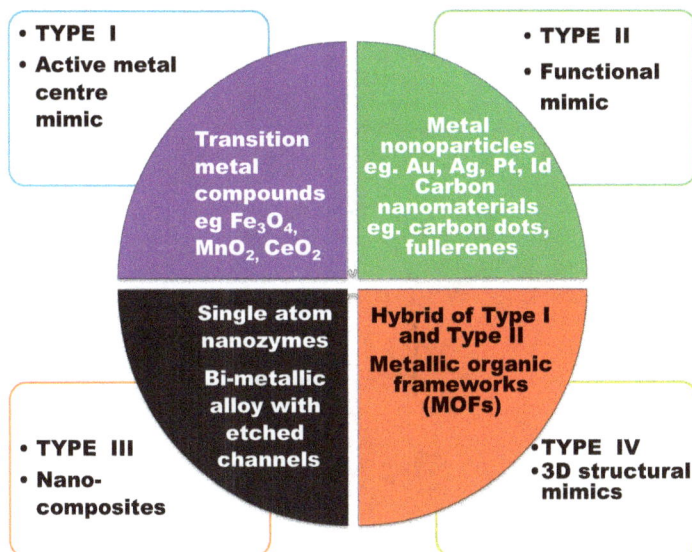

Figure 6.2: Types of nanozymes based on their mode of natural enzyme-mimicking behaviour.

## 6.2 Varieties of CD nanozymes

Over the past few decades, numerous nanomaterials have been shown to own basic enzyme-like activity [13, 36–40]. The ability to catalyse a specific chemical reaction is inherent to natural enzymes, usually at a single active site. For this analysis, we have categorized nanozymes into four groups based on how closely they mimic the activity of normal enzymes (Figure 6.2). One technique imitates the metal catalytic active site found in metalloenzymes by using artificial metal sites (type I nanozymes), that is, metal oxides (MOx) or metal sulphides [74]. $Fe_3O_4$ nanoparticle were used to carrier of nanozymes.

The oxidation of 2,2′-azino-bis(3-ethylbenzothiazoline-6-sulphonic acid) (ABTS), $o$-phenylenediamine dihydrochloride, and chromogenic 3,3′,5,5′-tetramethylbenzidine (TMB) by hydrogen peroxide ($H_2O_2$) peroxidases was examined by Gao et al. [25]. The highest catalytic activity was seen for the smallest $Fe_3O_4$ nanoparticles (30, 150, and 300 nm), indicating that the peroxidase-like activity was induced by the $Fe_3O_4$ nanoparticles' vast surface area [62–64]. According to Gao et al. [25], the ferrous and ferric ions in the nanoparticles mimicked the iron-heme-binding site in HRP. $CeO_2$ nanoparticles use their metal as a nanozyme because the metal has structural and biological properties that are similar to those of the iron ion, particularly in terms of protein interaction [75]. $CeO_2$ nanoparticles are multi-functional catalysts that exhibit, in addition to peroxidase-like activity, catalase- and dismutation of $O_2$ into $O_2$

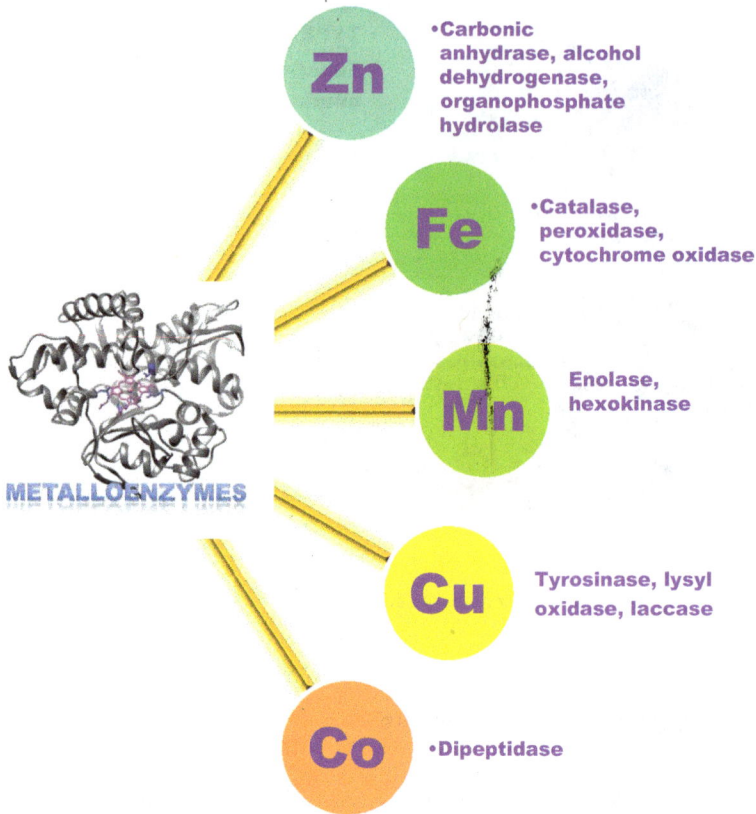

**Figure 6.3:** Instances of metalloenzymes.

and $H_2O_2$. Figure 6.3 shows copper, cobalt, and manganese ions, three common metal-heme centres found in metalloenzymes (Table 6.2) [39–73].

Contrary to type I is metal compound nanoparticles, which mimic the metal-heme redox centre of metalloenzymes, and type II nanozymes are metal nanoparticles that catalyse the same processes as natural enzymes [77–79]. The metals used to make these metal nanoparticles naturally stimulate a variety of heterogeneous reactions. These metals include, but are not limited to, gold, silver, platinum, palladium, and iridium [83–85]. According to Rossi and colleagues, under specific circumstances unprotected "naked" nanoparticles with a diameter of 3.6 nm in the gold nanoparticles may behave as a glucose oxidase mimic by initially catalysing glucose oxidase-like processes in the presence of high glucose [86]. Additionally, peroxidase activity was imitated by gold nanoparticles [87, 88].

Gao and colleagues [89] found that silver, gold, palladium, and platinum nanoparticles have peroxidase-like activities at acidic pH and catalase-like activities at basic pH. Nie and colleagues [90] subsequently investigated the pH-switchable phenomenon using 1–2 nm platinum nanoparticles and found that under basic conditions, catalase-like activity was evident whereas under acidic conditions, peroxidase-like activity predominated [81, 82].

**Table 6.2:** Examples of metallozymes.

| Centre metal | | Enzyme |
|---|---|---|
| Zinc | Zn | Organophosphate hydrolase, alcohol dehydrogenase, carbonic anhydrase |
| Iron | Fe | Cytochrome oxidase, peroxidase, catalase |
| Manganese | Mn | Hexokinase, enolase |
| Copper | Cu | Laccase, lysyl oxidase, tyrosinase |
| Cobalt | Co | Dipeptidase |

The capacity of metal-based nanozymes to create alloys with different elemental compositions is another noteworthy quality [91, 92]. As a result, alloy compositions, which are categorized as type III nanozymes here, can be altered to tune the enzyme-mimicking activities [93, 96–99]. They suggested that the electronic structure of alloying was responsible for the composition-dependent activity. To further improve multi-enzymatic activities, Yin and colleagues [94] showed an additional rise in multi-enzymatic activities using Au–Pt bimetallic nanoparticles to exhibit peroxidase, oxidase, and catalase-like activities by altering the Pt and Au molar ratio. Ir–Pd nanocubes with increased peroxidase activity were created by covering Pd nanocubes with an atomic layer of Ir [95]. Since the Ir–Pd(100) surface had larger adsorption energy than the Pd(100), it was predicted that the hydrogen peroxide's breakdown into hydroxyl radicals would be more energy-efficient. Kim et al. [100] co-doped graphene oxide with nitrogen and boron to further reduce the bandgap and improve the peroxidase-like action, resulting in considerably better catalytic performance than undoped graphene oxide. Ren et al.'s [101] use of graphene oxide quantum dots allowed them to show both peroxidase- and catalase-like activity [102–105].

These restrictions restrict the broad range of uses for these typical nanozymes. New methods for overcoming these limitations have therefore been developed, such as the spatial or three-dimensional structural mimicry of actual enzyme active regions [116, 117]. The geometry of pre-existing metal-binding centres, peripheral-binding sites, or the constrained and empty space at the heart of natural enzymes is mimicked to construct these structural imitators (type IV nanozymes) [107–111]. In this design, the transition metal nodes holding the MOFs themselves can serve as biomimetic catalysts while the high porosity structure created by the metal organic linkers can serve as substrate-binding sites. MOFs exhibit great catalytic effectiveness because of their highly specialized surface areas, exposed active sites, and tuneable pore sizes [106] [114, 115].

There were both peroxidase and catalase in the Co/2Fe-MOF, a bimetallic-MOF. Furthermore, Min and colleagues [112] demonstrated that $CeO_2$-MOF functions as a hydrolase mimic by releasing a chemical link with water. Similar to MOFs that have been successfully used as reactive oxygen species (ROS) scavengers, Prussian Blue nanoparticles can behave as multi-enzyme catalase-like activities, peroxidase, and superoxide dismutase, among others, are mimics [113].

## 6.3 Use cases for CD-based nanozymes

Carbon-based materials are essential for the advancement of material science. Modern industrial carbon (like carbon fibres and graphite), conventional industrial carbon (like activated carbon and carbon black), and innovative carbon nanomaterials (like graphene and carbon nanotubes) are all forms of carbon nanotubes. Critical claims in several industries, such as biomedicine, catalysis, optoelectronics, and anti-counterfeiting, are made possible by these features [118–125].

### 6.3.1 Biomedical application of CD-based nanozymes

It should come as no surprise that biomedicine is one of the most exciting and well-reported uses of CDs. Studies on the in vitro cytotoxicity of a few different cell lines show that CDs have little to no toxicity and good biocompatibility even at large dosages [170, 195]. CDs are therefore appropriate for biomedical applications. Because they are inexpensive, compact size, customizable surface functionalities, high photostability, and potential as a replacement for conventional fluorescent materials in illness detection, therapy, and healthcare supplements, CDs are also a prospective alternative to these materials [128, 130–133].

The CD biomedical applications covered in this section include nano-medicine, phototherapy, drug/gene delivery, and bio-imaging. In biological applications, CD toxicity and biocompatibility are crucial elements. Most CDs could either break down or be eliminated directly from the organism and were generally non-toxic at low quantities (like 10 g $mL^{-1}$) (181). CD toxicity, however, varies for their quantity, molecular makeup, and size. Increased amounts may interfere with organ development and cause cell damage (188). Furthermore, because ROS are produced, CDs may not be poisonous in the dark but become hazardous when exposed to light (186) [127].

### 6.3.2 Biomedicine application of CD-based nanozymes

There is a significant market for affordable, durable, and efficient antimicrobials. The above-mentioned materials can be substituted with CD-based nanozymes due to their distinct electrical, optical, thermal, and mechanical capabilities. Some CDs with nitrogen doping were made by Zhang et al. [32] to imitate oxidase action. Such CDs can quickly replicate the oxidation reaction and successfully stop *Salmonella* and *Escherichia coli* (*E. coli*) from growing (215). However, it did have anti-microbial action [136, 138–140].

### 6.3.3 Bio-imaging and bio-detection of CD-based nanozymes

Biological events are immediately and painlessly observed using probes and detectors during the bio-imaging process. Images of plant tissue, microorganisms, and cells have been captured on numerous CDs [162].

### 6.3.4 Detection of $O_2$

Most $O_2$ sensors have, up until recently, picked up superoxide anion generated by cells.

### 6.3.5 Detection of $H_2O_2$

$H_2O_2$ is a significant messenger molecule as a result of different biological processes. The high peroxidase-like activity of GQDs is the cause of this. (Table 6.3) When employed as designed, the $H_2O_2$ sensor demonstrated efficient electrochemical catalytic abilities for $H_2O_2$ breakdown. A hybrid FePt and graphene oxide nanozyme with a 2.2 M [132] detection limit was also developed by Chen et al. [129] for the sensitive detection of $H_2O_2$ [137]. Using Pt/Pd nanodendrites based on graphene oxide nanozymes, Chen et al. [129] reported colorimetric detection of $H_2O_2$ like this.

**Table 6.3:** Summary of the application of CD-based nanozymes in detection [126].

| Detection method | Sample | Detection limit | Linear range | References [130] | Years |
|---|---|---|---|---|---|
| Colorimetric detection | $H_2O_2$ | 0.16 µM | 5–500 µM, 500 µM–4 mM | [128] | 2016 |
| – | Glutathione | – | 0.058 µM | [129] | 2019 |
| – | – | 0.2 µM | 0.5–25 µM | [132] | 2020 |
| – | Ascorbic acid | 9 nM | 50–2500 nM | [133] | 2019 |
| – | – | 0.14 µM | 1.0–105 µM | [134] | 2020 |
| – | Uric acid | 0.64 µM | 2–150 µM | [137] | 2020 |
| – | Pyrophosphate | 4.29 nM | – | [139] | 2020b |

**Table 6.3** (continued)

| Detection method | Sample | Detection limit | Linear range | References [130] | Years |
|---|---|---|---|---|---|
| | Ion | | | | |
| Collaborative detection by colorimetric and fluorescence | $H_2O_2$ | 0.11 µM (colorimetric method) | – | Su et al. | 2020 |
| Methods | | 0.15 µM (fluorescence method) | | | |
| – | – | 0.35 µM | 1.67 µM–2.01 mM | Yang et al. | 2021b |
| – | Glucose | 0.15 µM (colorimetric method fluorescence method) | – | Su et al. | 2020 |
| – | Xanthine | 0.11 µM (colorimetric method) | – | Su et al. | 2020 |
| | | 0.12 µM (fluorescence method) | | | |
| – | Hydroquinone ($H_2Q$) | 1 µM | 0.05–20 mM | Ren et al. | 2015 |
| | | | 1–30 mM | | |
| – | Cholesterol | 0.49 µM | 1.66 µM | Yang et al. | 2021b |
| | | | 1.65 mM | | |
| Collaborative detection by colorimetric and SERS | Uric acid | 1–500 µM (colorimetric method) | – | Wang et al. | 2019 |
| methods | | 0.01–500 µM | | | |
| | | SERS method | | | |
| – | Glucose | –7 5 × 10 M | – | Gan et al. | 2021 |
| Double emission carbon spot detection | O-Phenylenediamine | 0.58 µM | – | Mathivanan et al. | 2020 |
| – | $H_2O_2$ | 0.27 µM | – | Mathivanan et al. | 2020 |
| – | Hydroquinone ($H_2Q$) | 0.04 µM | 1.0–75 µM | Wang et al. | 2019 |

### 6.3.6 Detection of glucose and glutathione

As was previously noted, $H_2O_2$ is a crucial signalling molecule in pathological and metabolic processes, particularly in glucose metabolism. In the absence of oxygen, glucose can be catalysed to create gluconolactone and $H_2O_2$. As a result, glucose and glutathione (GSH) can be measured indirectly by checking the amount of $H_2O_2$ that has been released by enzymes (184, 189, 190, 203).

### 6.3.7 Detection of proteins

The tailored immune sorbent assay to find protein molecules has also used carbon nanozymes. The first method for the extremely sensitive detection of carcino embryonic antigen was created by [141].

### 6.3.8 Detection of nucleic acids

Both applications in nucleic acid detection and uses for sensitive nanozyme assays have been discovered by researchers.

### 6.3.9 Detection of cancer cells

Recently, carbon nanozymes have become an intriguing cancer cell monitoring and detection assay [141].

### 6.3.10 Bio-sensing of CD-based nanozymes

Nanozymes have gained popularity as an ideal and crucial component for biosensors since they are less expensive, more stable, and easier to synthesize than protein enzymes. Potential biosensors have been investigated using $MoS_2$ nanoribbons, ferromagnetic nanoparticles, AuNP@$MoS_2$QD gold nanoparticles, and other inorganic nanomaterials with a variety of enzymatic activity [134].

### 6.3.11 The biosensor colorimetric

The cholesterol oxidase and cascade colorimetric biosensor demonstrated excellent selectivity and high sensitivity to the target. The detection cut-off was reduced to 7 mM (219).

### 6.3.12 Sensors electrochemical of CD-based nanozymes

Electrochemical sensors have several advantages, including linear output, good resolution, low power consumption, repeatability, and precision. A contemporary topic is the application of nanozymes based on CDs in electrochemical sensors [142–146].

### 6.3.13 The high catalytic activity of CD-based nanozymes

The copper based oxidase, superoxide oxidase, peroxidase, and catalase reactions are basic nanozyme catalytic reactions. Despite the variety and vast range of catalytic abilities displayed by natural enzymes, creating nanozymes for novel enzyme reactions is incredibly challenging [135]. However, a 200 °C high-temperature alkaline solution with a pH of 8.5 is the ideal reaction environment for this nanozyme. Its employment in biological systems is greatly constrained by these extreme reaction conditions [147–151].

### 6.3.14 Environmental application of CD-based nanozymes

Similar to real enzymes, environmental conditions such as pH levels, ionic strengths, temperatures, and light have a significant impact on the catalytic activity of carbon nanozymes [137]. Additionally, beyond 60 °C, HRP's catalytic activity was almost completely suppressed while N-GQDs continued to exhibit strong catalytic capabilities. Carbon nanoparticles' distinctive optical features also make carbon nanozymes the most light-sensitive compound. Carbon nanoparticles' enzyme-like activity can thus be controlled by light [160].

### 6.3.15 Oxidase/laccase-like activity

Nanozymes have a number of benefits over natural oxidases in terms of use, production, and catalytic activity. Due to their capacity to catalyse oxidation events without the presence of $H_2O_2$, several oxidase-like nanozymes have been identified. High catalytic activity has been observed for metal nanoparticles, nanocomposite carbon materials, and MOx [56].

Shamsipur et al. [161] studied carbon quantum dots (CD-1) with peroxidase activity to track GSH levels in human blood. $Na_2EDTA.2H_2O$ was used as the starting material, and it was heated for two hours at 40 °C under argon (Figure 6.4). When TMB is added to TMBox, CD-1, a peroxidase, can be used to catalyse the conversion of TMB to $H_2O_2$, which changes the colour of the substance from colourless to blue. $H_2O_2$ and TMB I and II had largely neutral reactions lacking CD-1. At wavelengths 370, 450, and 653, three different emission peaks were seen when CD-1 was introduced [152–156].

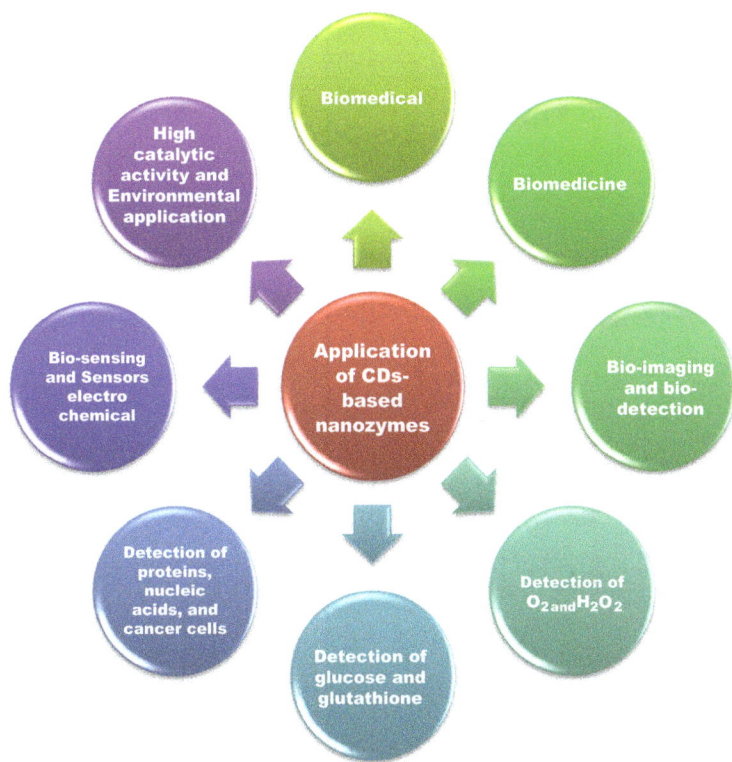

**Figure 6.4:** Applications of CDs.

To detect $H_2O_2$ at the nanomolar level, Yousefinejad et al. [162] created new Fe-CD-2 as a peroxidase mimic. To create Fe-CD-2, CD-2, and $Fe_3O_4$ NPs were combined in an acidic environment at a 4:1 ratio, and CD-2 was created utilizing carbon soot and nitric acid as precursors (Figure 6.5). Fe-CD-2 was identified by transmission electron microscopy pictures as a hemisphere with a size of roughly 10 nm. With $H_2O_2$, Fe-CD-2 was able to catalyse the oxidation of TMB to TMBox, which resulted in a colour shift from colourless to blue. Under the same circumstances, TMBox displayed more absorbance with Fe-CD-2 than with CD-2 and $Fe_3O_4$ NPs alone. The absorbance displayed a remarkable linear relationship with the $H_2O_2$ concentration starting at 10 nmol $L^{-1}$ in the presence of TMB and Fe-CD-2 to 1 mmol $L^{-1}$, with a remarkably low detection threshold of just 1 nmol $L^{-1}$.

To detect chromium ions [Cr(III)], Li et al. [163] reported copper-doped CDs (CD-7) as a peroxidase mimic. $Na_2[Cu(EDTA)]$ was cooked at 27 °C for 2 h in a tube furnace to create CD-7 (Figure 6.7). OPD was oxidized in DAP in the presence of CD-7 by $H_2O_2$, which released yellow fluorescence at 564 nm under excitation of 421 nm. A new absorption peak at 400 nm formed once Cr(III) was introduced, and the produced DAP's fluorescence intensity rapidly reduced (Table 6.4).

**Table 6.4:** Kinetic parameters of recent studies on CDs, doped-CDs, and CD-based hybrid nanozymes with oxidase/laccase-like activity and their application in detection systems.

| CDs/CD-hybrid | Enzyme-like | Substrates | Michaelis–Menten parameters | | Application | | | References |
|---|---|---|---|---|---|---|---|---|
| | | | $K_m$ (mM) | $V_{max}$ ($10^{-7}$ M s$^{-1}$) | Detection | Linear range (µM) | LOD (µM) | |
| Ce$^{3+}$-CDs | Oxidase | TMB | 0.28 (pH 4), 0.23 (pH 7) | 0.28 (pH 4) and 0.239 (pH 7) | NA | NA | NA | Li et al. [58] |
| Cu-CDs | Laccase | TMB | NA | NA | Hydroquinone | 1,000–20,000 | 980 | Ren et al. [67] |
| Ce$^{4+}$-GQDs | Oxidase | OPD | NA | NA | Aledronate Sodium | 0.1–1.0 1.0–100 | 0.033 | Xia et al. [65] |
| MnO$_2$-CDs | Oxidase | TMB | NA | NA | Glutathione | 1.0–100 | 0.27 | Cao et al. [68] |
| CD-Cu$_4$O$_3$ | Oxidase | TMB | 10.69 | 209.3 | | | | Li et al. [58] |
| Ag/His-GQD | Oxidase | TMB | 0.6 | 2.25 | D-Penicillamine | 0.1–1.25 | 0.034 | Dan et al. [56] |
| MnO$_2$-CQDs | Oxidase | OPD | NA | NA | 6-Mercaptopurine | 0.46–100 | 0.14 | An et al. [66] |
| Mn-CDs | Oxidase | TMB | NA | NA | Ascorbic acid | 0.05–2.5 | 0.009 | Zhuo et al. [60] |

Note: NR represents not reported.

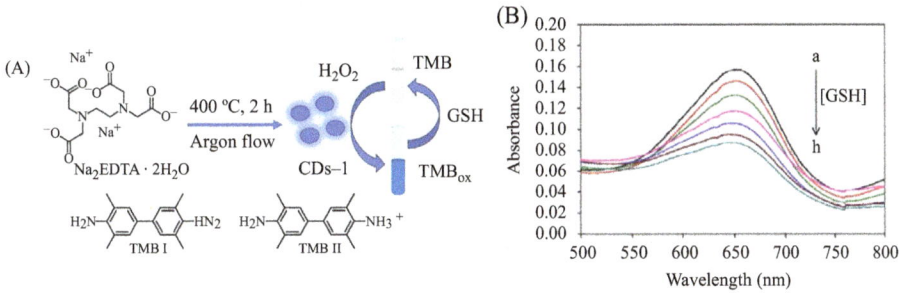

**Figure 6.5:** (A) The method of making CD-1 and the theory behind how it reacts with TMB I and II in the presence of $H_2O_2$ before reacting with GSH. (B) TMBox absorption as GSH increases (copying is permitted [161]. Copyright 2014, Elsevier).

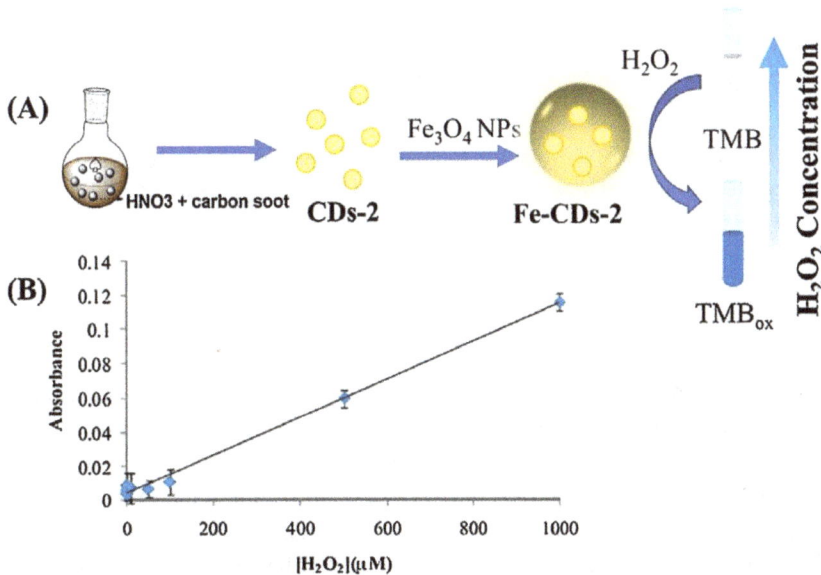

**Figure 6.6:** (A) The Fe-CD-2 synthesis pathway and Fe-CD-2 catalysed TMB oxidation by $H_2O_2$. (B) TMBox absorbance at 653 nm as a function of $H_2O_2$ concentration (mM: mmol $L^{-1}$) (copying is permitted [162]. Copyright 2017, Elsevier).

A nanozyme (CD-15) that may release DNAzyme to boost catalytic efficiency in the presence of $K^+$ was described by Li et al. [164] in detail. Corn starch and ethylenediamine were microwaved at 14 °C for 10 min to create CD-15 (Figure 6.8). Apt-CDs15 decreased CD-15's fluorescence when it was coupled with an aptamer (Apt), which was explained by the inhibition of catalytic activity. Apt-CD-15 developed hemin-containing DNAzyme (HM) to increase CD-15's catalytic activity when $K^+$ is present. In the presence of $H_2O_2$, CD-15 and DNAzyme accelerated the production of TMBox, which displayed a

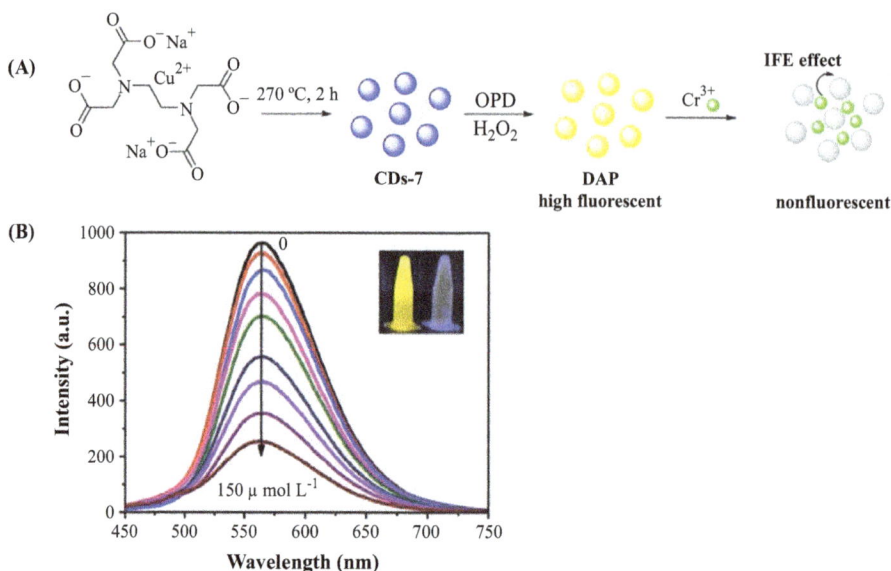

**Figure 6.7:** (A) Mechanism of CD-7's synthesis and detection with peroxidase activity for Cr (III). (B) Fluorescence intensity of DAP as Cr(III) concentration is increased (copying is permitted [163]. Copyright 2020, Elsevier).

fluorescence signal at 403 nm and an absorption band at 563 nm. As the concentration of Apt-CD-15 with $K^+$ was raised, the fluorescence intensity of TMBox gradually rose [157–159].

Novel copper-doped CDs (CD-18) were described by Ren et al. [165] as an oxidase mimic for tracking hydroquinone ($H_2Q$) levels. Cu ($NO_3$)$_2$ and sodium salt of poly (methacrylic acid) were used as raw materials in the hydrothermal process to create CD-18 (Figure 6.9). In a wide pH range, CD-18 showed outstanding photostability and stability and released blue fluorescence at 460 nm (3.0–13.5). The colourless compound *p*-phenylenediamine (PPD) turned brown after 30 min after being added to CD-18, which promoted the oxidation of PPD and produced a distinct absorption peak at 495 nm. The ideal pH was 8.5, and CD-18 demonstrated greater temperature stability than laccase. PPD and the oxidation product of PPD on the surface of CD-18 could be removed by flocculation between polyacrylamide (PAM+) and CD-18 when PAM+ was applied. Additionally, CD-18 responded to $H_2Q$ by changing its hue from colourless to yellow in less than 10 min. In carbonate buffer (pH 9.2), the relative fluorescence quenching ($I/I_0$) demonstrated a remarkable linear relationship with the $H_2Q$ concentration in the range of 0.05–2 mmol $L^{-1}$. The fluorescence intensity at 460 nm decreased with increasing $H_2Q$ concentration. The detection limit was 1 mmol $L^{-1}$, which is substantially less than the Environmental Protection Agency's 3 mmol $L^{-1}$ permissible thresholds. Additionally, CD-18 responded to $H_2Q$ by changing its hue from colourless to yellow in less than 10 min. In carbonate buffer (pH 9.2), the relative fluorescence quenching ($I/I_0$) demonstrated a remarkable linear

**(A)**

ethylenediamine

Au(III)

CDs-15

**(B)**

**(C)**

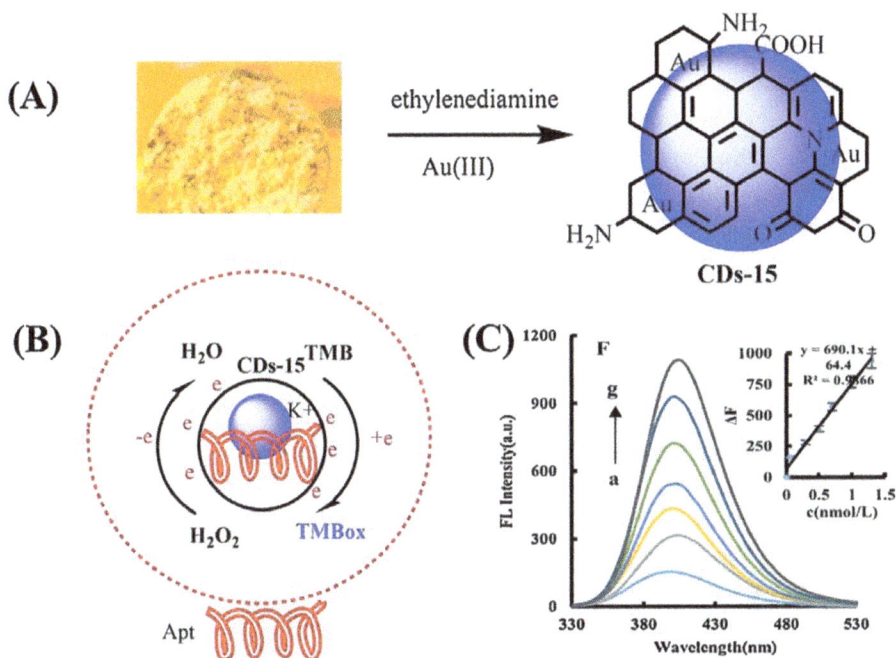

**Figure 6.8:** (A) The CD-15 synthesis method. (B) The process by which TMB is oxidized when Apt-CD-15 and K⁺ are present. (C) Fluorescence intensity of the TMBox as K⁺ concentration increases (reproduced with permission [164]. Copyright 2020, Elsevier).

relationship with the $H_2Q$ concentration in the range of 0.05–2 mmol $L^{-1}$. The fluorescence intensity at 460 nm decreased with increasing $H_2Q$ concentration. The detection limit was 1 mmol $L^{-1}$, which is substantially less than the Environmental Protection Agency's 3 mmol $L^{-1}$ permissible thresholds.

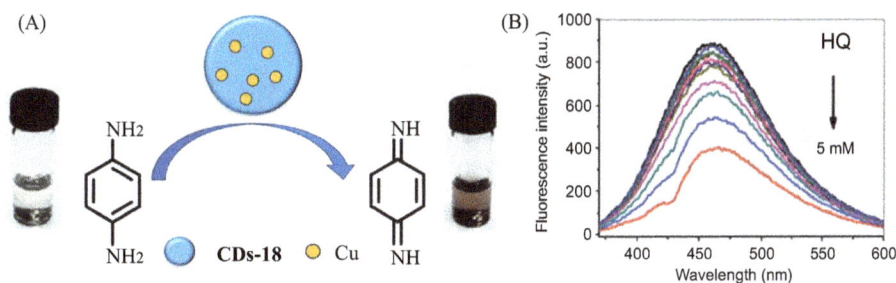

**(A)**

**(B)**

**Figure 6.9:** (A) Schematic representation of CD-18 reacting to PPD. (B) Variation in CD-18's fluorescence intensity as $H_2Q$ concentration (mM: mmol $L^{-1}$) was increased in the carbonate buffe (copied with permission [165]. Copyright 2018, Royal Society of Chemistry).

Li et al. [166] reported CD-21 with the $Ce^{3+}$-driven pH-dependent photo-oxidation feature. Under visible light, in the presence of oxygen, CD-21 converted colourless TMB to blue TMBox at the natural pH (Figure 6.10). With rising pH, the $K_m$ value of CD-21-$Ce^{3+}$ + EDTA•2Na decreased more than that of CDs-21, indicating that CD-21-$Ce^{3+}$ + EDTA•2Na has a higher affinity for TMB. TMB was converted to TMBox under acidic conditions by the $1O^2$ produced from CD-21 in the presence of light. In contrast, because the $Ce^{3+}/Ce^{4+}$ potential diminishes with rising pH, under the neutral pH condition, $1O_2$, which was photo-generated by CD-21-$Ce^{3+}$ + EDTA 2Na under illumination, oxidized $Ce^{3+}$ to $Ce^{4+}$ and TMB was then further oxidized by $Ce^{4+}$. Additionally, the main factor in the oxidation of TMB in a pH-neutral environment was the oxidation product $Ce^{4+}$ which was stabilized by EDTA.2Na to stop the production of cerium hydroxide.

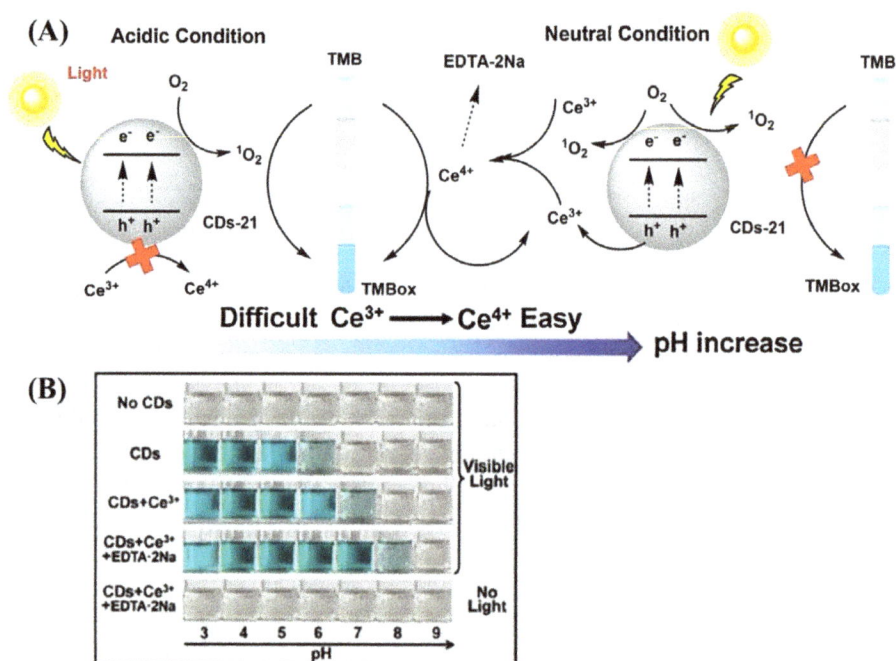

Figure 6.10: (A) The process by which CD-21-$Ce^{3+}$ + EDTA 2Na oxidizes TMB in ambient light. (B) Variations in TMB's colour under various circumstances (copied with permission [166]. Copyright 2020, Elsevier).

To scavenge for free radicals, Dehvari et al. [167] created peroxidase-active CDs (CD-25). By employing citric acid trisodium salt, N-acetyl-L-cysteine, manganese acetate, and $Na_2S$ as starting ingredients, CD-25 was created using the microwave hydrothermal technique (Figure 6.11). Because the oxidation of TMB was catalysed by the production of OH, the solution changed from colourless to blue when TMB was treated with $H_2O_2$ and $Fe^{2+}$. The distinctive absorption peak of ox-TMB also developed at 652 nm. The elimination of OH by CD-25 was the reason for the system's considerable decrease in absorbance at 652 nm when CD-25 was added. Furthermore, CD-25 had a significant ability for scavenging reactive nitrogen species; at 100 mg mL$^{-1}$, CD-25 demonstrated a 98% scavenging effect for ONOO. As soon as the CD-25 surface was in contact with hyaluronic acid (HA), the increase in surface negative functional groups significantly increased the water solubility of HA-CD-25.

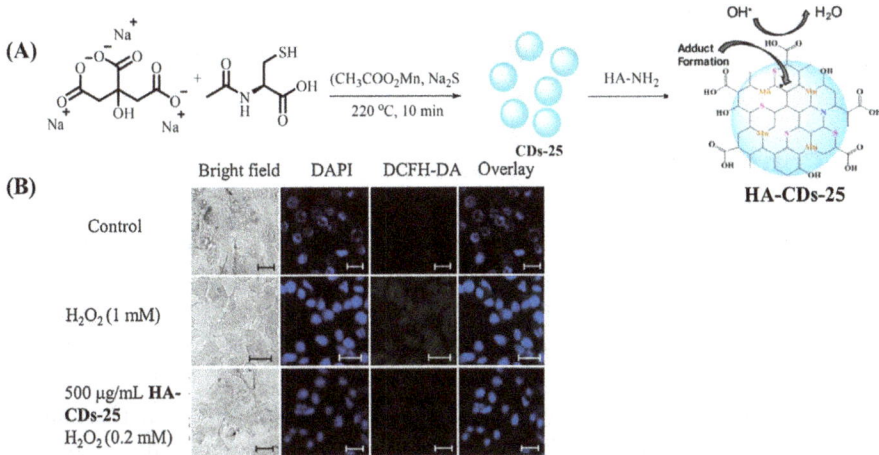

**Figure 6.11:** (A) The method by which HA-CD-25 is synthesized and the associated mechanism for lowering ROS. (B) Photographs of $B_{16}F_1$ cells after they had been exposed to DCFH-DA, $H_2O_2$ (mM: mmol L$^{-1}$), and HA-CD-25 (reproduced with permission [167]. Copyright 2020, Elsevier).

## 6.4 Current issues and proposed fixes

However, there are still a number of obstacles to be overcome:
1. The link between structure and activity has to be better understood. As was previously mentioned, a number of carbon nanostructures have enzyme catalytic activity. However, the structure–activity relationship does not appear to follow any overall trend. Each system displays unique characteristics when the catalytic ability is evaluated, and these characteristics are explained by particular mechanisms. The hypothesized processes are typically diverse, which make it difficult to develop CDs nanozymes in an efficient manner.
2. A number of variables, including the raw materials used to make CDs, influence the activity of nanozymes.
3. Although CDs are biocompatible, a thorough assessment of their biosafety and biocompatibility should be conducted. Even when doped with additional elements like nitrogen, CDs often exhibit no cytotoxicity or very little cytotoxicity [80].

4. The selection of nanozymes is constrained. As the bulk of nanozymes are exclusively concerned with replicating the catalytic functions of natural oxidoreductases, the present research advocates for a rise in varied nanozymes. The abilities of semi-artificial enzymes will be mimicked by the next generation of nanozymes.
5. Nanozymes with multiple functions may be complicated. Natural enzymes typically follow the tenets of coordination chemistry to catalyse extremely precise reactions requiring a metal coordination site. Therefore, applying coordination chemistry principles could facilitate the rational design of nanozymes [76].

## 6.5 Conclusions

Significant improvements have been made in synthetic methods, structures, characteristics, mechanistic studies, and application development in CD research. Additionally, a size or polarity-based purification technique should be established. Particularly, the growth of characterizing the formation of CDs is essential for enabling the controlled synthesis of CDs with unique nanostructures. Additionally, this outlook discusses CDs, a novel class of carbon-based non-materials, and how they exhibit exceptional biocompatibility, distinctive optical properties, low cost, easy modification, and fictionalization as well as having significant and extraordinary potential for a wide range of applications. But more research needs to be done on the actual application of CD-based materials.

Due to their versatility, adaptability, and design ability, CDs can be used in a variety of vibrant applications. Despite numerous researches, there are still substantial CD problems that need to be resolved. Making CDs with clearly defined structures and morphologies and understanding the specific reaction mechanism of preparations remain challenging tasks. The CDs have not yet been fully described, particularly about to photogene rated charge separations, toxicity, and biocompatibility. One might get excited about the enhanced performance and novel uses of CDs by harnessing the tremendous design potential of carbon and organics.

## References

[1]   Breaker RR. DNA enzymes. Nat Biotechnol. 1997, 15, 427–431.
[2]   Moore JC, Robins K. Engineering the third wave of biocatalysis. Nature. 2012, 485, 185–194.
[3]   Genet JP. Asymmetric catalytic hydrogenation. Design of new Ru catalysts and chiral ligand: From laboratory to industrial applications. Acc Chem Res. 2003, 36, 908–918.
[4]   Behrens M, Studt F, Kasatkin I, Kühl S, Hävecker M, Abild-Pedersen F, Zander S, Girgsdies F, Kurr P, Kniep BL. The active site of methanol synthesis over Cu/ZnO/Al$_2$O$_3$ industrial catalysts. Science. 2012, 336, 893–897.
[5]   Meunier B, de Visser SP, Shaik S. Mechanism of oxidation reactions catalyzed by cytochrome P450 enzymes. Chem Rev. 2004, 104, 3947–3980.

[6]   Kirby AJ. Efficiency of proton transfer catalysis in models and enzymes. Acc Chem Res. 1997, 30, 290–296.

[7]   Haseloff J, Gerlach WL. Simple RNA enzymes with new and highly specific endoribonuclease activities. Nature. 1988, 334, 585–591.

[8]   Hubatsch I, Ridderström M, Mannervik B. Human glutathione transferase A4-4: An alpha class enzyme with high catalytic efficiency in the conjugation of 4-hydroxynonenal and other genotoxic products of lipid peroxidation. Biochem J. 1998, 330, 175–179.

[9]   Posorske LH. Industrial-scale application of enzymes to the fats and oil industry. J Am Oil Chem Soc. 1984, 61, 1758–1760.

[10]  Choct M. Enzymes for the feed industry: Past, present and future. World's Poult Sci J. 2006, 62, 5–16.

[11]  Abuchowski A, Kazo GM, Verhoest CRJ, Van ET, Kafkewitz D, Nucci ML, Viau AT, Davis FF. Cancer therapy with chemically modified enzymes. I. Antitumor properties of polyethylene glycol-asparaginase conjugates. Cancer Biochem Biophys. 1984, 7, 175–186.

[12]  Gurung N, Ray S, Bose S, Rai V. A broader view: Microbial enzymes and their relevance in industries, medicine, and beyond. Biomed Res Int. 2013, 2013, 1–18.

[13]  Wang XY, Guo WJ, Hu YH, Wu JJ, Wei H. Nanozymes: Next Wave of Artificial Enzymes, Springer, 2016.

[14]  Yan XY. Nanozyme: A new type of artificial enzyme. Prog Biochem Biophys. 2018, 45, 101–104.

[15]  Breslow R. Biomimetic chemistry and artificial enzymes: Catalysis by design. Acc Chem Res. 1995, 28, 146–153.

[16]  Motherwell WB, Bingham MJ, Six Y. Recent progress in the design and synthesis of artificial enzymes. Tetrahedron. 2001, 57, 4663–4686.

[17]  Kirby AJ. Enzyme mimics. Angew Chem Int Ed Engl. 1994, 33, 551–553.

[18]  Ali SS, Hardt JI, Quick KL, Sook Kim-Han J, Erlanger BF, Huang TT, Epstein CJ, Dugan LL. A biologically effective fullerene (C60) derivative with superoxide dismutase mimetic properties. Free Radic Biol Med. 2004, 37, 1191–1202.

[19]  Kataky R, Morgan E. Potential of enzyme mimics in biomimetic sensors: A modified-cyclodextrin as a dehydrogenase enzyme mimic. Biosens Bioelectron. 2003, 18, 1407–1417.

[20]  Kirkorian K, Ellis A, Twyman LJ. Catalytic hyperbranched polymers as enzyme mimics; exploiting the principles of encapsulation and supramolecular chemistry. Chem Soc Rev. 2012, 41, 6138–6159.

[21]  Liu L, Breslow R. Dendrimeric pyridoxamine enzyme mimics. J Am Chem Soc. 2003, 125, 12110–12111.

[22]  Anderson HL, Sanders JKM. Enzyme mimics based on cyclic porphyrin oligomers: Strategy, design and exploratory synthesis. J Chem Soc Perkin Trans 1. 1995, 0, 2223–2229.

[23]  Romanovsky BV. Transition metal complexes in inorganic polymers as enzyme mimics. Macromol Symp. 1994, 80, 185–192.

[24]  Gong L, Zhao ZL, Lv YF, Huan SY, Fu T, Zhang XB, Shen GL, Yu RQ. DNAzyme-based biosensors and nanodevices. Chem Commun. 2015, 51, 979–995.

[25]  Gao LZ, Zhuang J, Nie L, Zhang JB, Zhang Y, Gu N, Wang TH, Feng J, Yang DL, Perrett S, et al. Intrinsic peroxidase-like activity of ferromagnetic nanoparticles. Nat Nanotechnol. 2007, 2, 577–583.

[26]  Wei H, Wang EK. Nanomaterials with enzyme-like characteristics (nanozymes): Next-generation artificial enzymes. Chem Soc Rev. 2013, 42, 6060–6093.

[27]  Lin YH, Ren JS, Qu XG. Catalytically active nanomaterials: A promising candidate for artificial enzymes. Acc Chem Res. 2014, 47, 1097–1105.

[28]  Wu JJ, Wang XY, Wang Q, Lou ZP, Li SR, Zhu YY, Qin L, Wei H. Nanomaterials with enzyme-like characteristics (nanozymes): Next-generation artificial enzymes (II). Chem Soc Rev. 2019. doi: 10.1039/C8CS00457A.

[29]  Ragg R, Tahir MN, Tremel W. Solids go bio: Inorganic nanoparticles as enzyme mimics. Eur J Inorg Chem. 2016, 2016, 1906–1915.

[30]  Wang H, Wan KW, Shi XH. Recent advances in nanozyme research. Adv Mater. 2018, 1805368.

[31]    Zheng XX, Liu Q, Jing C, Li Y, Li D, Luo WJ, Wen YQ, He Y, Huang Q, Long YT, et al. Catalytic gold nanoparticles for nanoplasmonic detection of DNA hybridization. Angew Chem Int Ed. 2011, 50, 11994–11998.

[32]    Asati A, Santra S, Kaittanis C, Nath S, Perez JM. Oxidase-like activity of polymer-coated cerium oxide nanoparticles. Angew Chem Int Ed. 2009, 48, 2308–2312.

[33]    Vernekar AA, Sinha D, Srivastava S, Paramasivam PU, D'Silva P, Mugesh G. An antioxidant nanozyme that uncovers the cytoprotective potential of vanadia nanowires. Nat Commun. 2014, 5, 5301.

[34]    Song Y, Zhu S, Yang B. Bioimaging based on fluorescent carbon dots. Rsc Advances. 2014; 4(52):27184–200.

[35]    Natalio F, André R, Hartog AF, Stoll B, Jochum KP, Wever R, Tremel W. Vanadium pentoxide nanoparticles mimic vanadium haloperoxidases and thwart biofilm formation. Nat Nanotechnol. 2012, 7, 530–535.

[36]    Sun HJ, Gao N, Dong K, Ren JS, Qu XG. Graphene quantum dots-band-aids used for wound disinfection. ACS Nano. 2014, 8, 6202–6210.

[37]    Pogacean F, Socaci C, Pruneanu S, Biris AR, Coros M, Magerusan L, Katona G, Turcu R, Borodi G. Graphene based nanomaterials as chemical sensors for hydrogen peroxide – a comparison study of their intrinsic peroxidase catalytic behavior. Sens Actuators B. 2015, 213, 474–483.

[38]    Ragg R, Natalio F, Tahir MN, Janssen H, Kashyap A, Strand D, Strand S, Tremel W. Molybdenum trioxide nanoparticles with intrinsic sulfite oxidase activity. ACS Nano. 2014, 8, 5182–5189.

[39]    Liu BW, Liu JW. Surface modification of nanozymes. Nano Res. 2017, 10, 1125–1148.

[40]    Zhang W, Hu SL, Yin JJ, He WW, Lu W, Ma M, Gu N, Zhang Y. Prussian blue nanoparticles as multienzyme mimetics and reactive oxygen species scavengers. J Am Chem Soc. 2016, 138, 5860–5865.

[41]    Bleeker EAJ, de Jong WH, Geertsma RE, Groenewold M, Heugens EHW, Koers-Jacquemijns M, van de Meent D, Popma JR, Rietveld AG, Wijnhoven SWP, et al. Considerations on the EU definition of a nanomaterial: Science to support policy making. Regul Toxicol Pharmacol. 2013, 65, 119–125.

[42]    Maynard AD. Don't define nanomaterials. Nature. 2011, 475, 31.

[43]    Zhou YB, Liu BW, Yang RH, Liu JW. Filling in the gaps between nanozymes and enzymes: Challenges and opportunities. Bioconjugate Chem. 2017, 28, 2903–2909.

[44]    Tonga GY, Jeong Y, Duncan B, Mizuhara T, Mout R, Das R, Kim ST, Yeh YC, Yan B, Hou S, et al. Supramolecular regulation of bioorthogonal catalysis in cells using nanoparticle embedded transition metal catalysts. Nat Chem. 2015, 7, 597–603.

[45]    Wang FM, Zhang Y, Du Z, Ren JS, Qu XG. Designed heterogeneous palladium catalysts for reversible light-controlled bioorthogonal catalysis in living cells. Nat Commun. 2018, 9, 1209.

[46]    Gupta A, Das R, Tonga GY, Mizuhara T, Rotello VM. Charge-switchable nanozymes for bioorthogonal imaging of biofilm associated infections. ACS Nano. 2018, 12, 89–94.

[47]    Qiu H, Pu F, Ran X, Liu CQ, Ren JS, Qu XG. Nanozyme as artificial receptor with multiple readouts for pattern recognition. Anal Chem. 2018, 90, 11775–11779.

[48]    Sharma TK, Ramanathan R, Weerathunge P, Mohammad Taheri M, Daima HK, Shukla R, Bansal V. Aptamer-mediated 'turn-off/turn-on' nanozyme activity of gold nanoparticles for kanamycin detection. Chem Commun. 2014, 50, 15856–15859.

[49]    Tian L, Qi JX, Oderinde O, Yao C, Song W, Wang YH. Planar intercalated copper (II) complex molecule as small molecule enzyme mimic combined with $Fe_3O_4$ nanozyme for bienzyme synergistic catalysis applied to the microRNA biosensor. Biosens Bioelectron. 2018, 110, 110–117.

[50]    Huang YY, Ran X, Lin YH, Ren JS, Qu XG. Self assembly of an organic-inorganic hybrid nanoflower as an efficient biomimetic catalyst for self-activated tandem reactions. Chem Commun. 2015, 51, 4386–4389.

[51]    Gao LZ, Yan XY. Nanozymes: An emerging field bridging nanotechnology and biology. Sci China: Life Sci. 2016, 59, 400–402.

[52]   Duan DM, Fan KL, Zhang DX, Tan SG, Liang MF, Liu Y, Zhang JL, Zhang PH, Qiu XG, Kobinger GP, et al. Nanozyme-strip for rapid local diagnosis of Ebola. Biosens Bioelectron. 2015, 74, 134–141.

[53]   Huang YY, Lin YH, Pu F, Ren JS, Qu XG. The current progress of nanozymes in disease treatments. Prog Biochem Biophys. 2018, 45, 256–267.

[54]   Yang BW, Chen Y, Shi JL. Nanozymes in catalytic cancer theranostics. Prog Biochem Biophys. 2018, 45, 237–255.

[55]   Tang Y, Qiu ZY, Xu ZB, Gao LZ. Antibacterial mechanism and applications of nanozymes. Prog Biochem Biophys. 2018, 45, 118–128.

[56]   Chen ZW, Wang ZZ, Ren JS, Qu XG. Enzyme mimicry for combating bacteria and biofilms. Acc Chem Res. 2018, 51, 789–799.

[57]   Niu JS, Sun YH, Wang FM, Zhao CQ, Ren JS, Qu XG. Photomodulated nanozyme used for a Gram-selective antimicrobial. Chem Mater. 2018, 30, 7027–7033.

[58]   Wu JJ, Li SR, Wei H. Integrated nanozymes: Facile preparation and biomedical applications. Chem Commun. 2018, 54, 6520–6530.

[59]   Popov AL, Popova NR, Tarakina NV, Ivanova OS, Ermakov AM, Ivanov VK, Sukhorukov GB. Intracellular delivery of antioxidant $CeO_2$ nanoparticles via polyelectrolyte microcapsules. ACS Biomater Sci Eng. 2018, 4, 2453–2462.

[60]   Batrakova EV, Li S, Reynolds AD, Mosley RL, Bronich TK, Kabanov AV, Gendelman HE. A macrophage-nanozyme delivery system for Parkinson's disease. Bioconjugate Chem. 2007, 18, 1498–1506.

[61]   Li W, Liu CQ, Guan YJ, Ren JS, Qu XG, Liu C. Manganese dioxide nanozymes as responsive cytoprotective shells for individual living cell encapsulation. Angew Chem Int Ed. 2017, 56, 13661–13665.

[62]   Liang MM, Fan KL, Pan Y, Jiang H, Wang F, Yang DL, Lu D, Feng J, Zhao JJ, Yang L, et al. $Fe_3O_4$ magnetic nanoparticle peroxidase mimetic-based colorimetric assay for the rapid detection of organo phosphorus pesticide and nerve agent. Anal Chem. 2013, 85, 308–312.

[63]   Hu YH, Cheng HJ, Zhao XZ, Wu JJ, Muhammad F, Lin SC, He J, Zhou LQ, Zhang CP, Deng Y, et al. Surface enhanced Raman scattering active gold nanoparticles with enzyme mimicking activities for measuring glucose and lactate in living tissues. ACS Nano. 2017, 11, 5558–5566.

[64]   Luo WJ, Zhu CF, Su S, Li D, He Y, Huang Q, Fan CH. Self-catalyzed, self-limiting growth of glucose oxidase-mimicking gold nanoparticles. ACS Nano. 2010, 4, 7451–7458.

[65]   Tao Y, Ju EG, Ren JS, Qu XG. Polypyrrole nanoparticles as promising enzyme mimics for sensitive hydrogen peroxide detection. Chem Commun. 2014, 50, 3030–3032.

[66]   Zhang Y, Wang FM, Liu CQ, Wang ZZ, Kang LH, Huang YY, Dong K, Ren JS, Qu XG. Nanozyme decorated metal-organic frameworks for enhanced photodynamic therapy. ACS Nano. 2018, 12, 651–661.

[67]   Fan J, Yin JJ, Ning B, Wu XC, Hu Y, Ferrari M, Anderson GJ, Wei JY, Zhao YL, Nie GJ. Direct evidence for catalase and peroxidase activities of ferritin-platinum nanoparticles. Biomaterials. 2011, 32, 1611–1618.

[68]   Ge CC, Fang G, Shen XM, Chong Y, Wamer WG, Gao XF, Chai ZF, Chen CY, Yin JJ. Facet energy versus enzyme-like activities: The unexpected protection of palladium nanocrystals against oxidative damage. ACS Nano. 2016, 10, 10436–10445.

[69]   Karakoti AS, Singh S, Kumar A, Malinska M, Kuchibhatla SVNT, Wozniak K, Self WT, Seal S. PEGylated nanoceria as radical scavenger with tunable redox chemistry. J Am Chem Soc. 2009, 131, 14144–14145.

[70]   Li YY, He X, Yin JJ, Ma YH, Zhang P, Li JY, Ding YY, Zhang J, Zhao YL, Chai ZF, et al. Acquired superoxide scavenging ability of ceria nanoparticles. Angew Chem Int Ed. 2015, 54, 1832–1835.

[71]   Kim CK, Kim T, Choi IY, Soh M, Kim D, Kim YJ, Jang H, Yang HS, Kim JY, Park HK, et al. Ceria nanoparticles that can protect against ischemic stroke. Angew Chem Int Ed. 2012, 51, 11039–11043.

[72] Huang YY, Liu Z, Liu CQ, Ju EG, Zhang Y, Ren JS, Qu XG. Self-assembly of multi-nanozymes to mimic an intracellular antioxidant defense system. Angew Chem Int Ed. 2016, 55, 6646–6650.

[73] Lin YH, Li ZH, Chen ZW, Ren JS, Qu XG. Mesoporous silica-encapsulated gold nanoparticles as artificial enzymes for self-activated cascade catalysis. Biomaterials. 2013, 34, 2600–2610.

[74] Holm RH, Kennepohl P, Solomon EI. Structural and functional aspects of metal sites in biology. Chem Rev. 1996, 96, 2239–2314. doi: 10.1021/cr9500390.

[75] Palizban AA, Sadeghi-Aliabadi H, Abdollahpour F. Effect of cerium lanthanide on Hela and MCF-7 cancer cell growth in the presence of transferrin. Res Pharm Sci. 2010, 5, 119–125. doi: doi.

[76] Celardo I, Pedersen JZ, Traversa E, Ghibelli L. Pharmacological potential of cerium oxide nanoparticles. Nanoscale. 2011, 3, 1411–1420. doi: 10.1039/c0nr00875c.

[77] Baldim V, Bedioui F, Mignet N, Margaill I, Berret JF. The enzyme-like catalytic activity of cerium oxide nanoparticles and its dependency on Ce3+ surface area concentration. Nanoscale. 2018, 10, 6971–6980. doi: 10.1039/c8nr00325d.

[78] Mu JS, Wang Y, Zhao M, Zhang L. Intrinsic peroxidase-like activity and catalase-like activity of $Co_3O_4$ nanoparticles. Chem Commun. 2012, 48, 2540–2542. doi: 10.1039/c2cc17013b.

[79] Lin SB, Wang YY, Chen ZZ, Li LB, Zeng JF, Dong QR, Wang Y, Chai ZF. Biomineralized enzyme-like cobalt sulfide nanodots for synergetic phototherapy with tumor multimodal imaging navigation. ACS Sustain Chem Eng. 2018, 6, 12061–12069. doi: 10.1021/acssuschemeng.8b02386.

[80] Chen W, Chen J, Liu AL, Wang LM, Li GW, Lin XH. Peroxidase-like activity of cupric oxide nanoparticle. Chem-CatChem. 2011, 3, 1151–1154. doi: 10.1002/cctc.201100064.

[81] Wan Y, Qi P, Zhang D, Wu JJ, Wang Y. Manganese oxide nanowire-mediated enzyme-linked immunosorbent assay. Biosens Bioelectron. 2012, 33, 69–74. doi: 10.1016/j.bios.2011.12.033.

[82] Pijpers IAB, Cao S, Llopis-Lorente A, Zhu J, Song S, Joosten RRM, Meng F, Friedrich H, Williams DS, Sánchez S, et al. Hybrid biodegradable nanomotors through compartmentalized synthesis. Nano Lett. 2020, 20, 4472–4480. doi: 10.1021/acs.nanolett.0c01268.

[83] He WW, Zhou YT, Warner WG, Hu XN, Wu XC, Zheng Z, Boudreau MD, Yin JJ. Intrinsic catalytic activity of Au nanoparticles with respect to hydrogen peroxide decomposition and superoxide scavenging. Biomaterials. 2013, 34, 765–773. doi: 10.1016/j.biomaterials.2012.10.010.

[84] Liu CP, Wu TH, Lin YL, Liu CY, Wang S, Lin SY. Tailoring enzyme-like activities of gold nanoclusters by polymeric tertiary amines for protecting neurons against oxidative stress. Small. 2016, 12, 4127–4135. doi: 10.1002/smll.201503919.

[85] Long R, Huang H, Li YP, Song L, Xiong YJ. Palladium-based nanomaterials: A platform to produce reactive oxygen species for catalyzing oxidation reactions. Adv Mater. 2015, 27, 7025–7042. doi: 10.1002/adma.201502068.

[86] Comotti M, Della Pina C, Matarrese R, Rossi M. The catalytic activity of "naked" gold particles. Angew Chem Int Ed. 2004, 43, 5812–5815. doi: 10.1002/anie.200460446.

[87] Wang S, Chen W, Liu AL, Hong L, Deng HH, Lin XH. Comparison of the peroxidase-like activity of unmodified, amino-modified, and citrate-capped gold nanoparticles. ChemPhysChem. 2012, 13, 1199–1204. doi: 10.1002/cphc.201100906.

[88] Jv Y, Li BX, Cao R. Positively-charged gold nanoparticles as peroxidiase mimic and their application in hydrogen peroxide and glucose detection. Chem Commun. 2010, 46, 8017–8019. doi: 10.1039/c0cc02698k.

[89] Li JN, Liu WQ, Wu XC, Gao XF. Mechanism of pH-switchable peroxidase and catalase-like activities of gold, silver, platinum and palladium. Biomaterials. 2015, 48, 37–44. doi: 10.1016/j.biomaterials.2015.01.012.

[90] Fan J, Yin JJ, Ning B, Wu XC, Hu Y, Ferrari M, Anderson GJ, Wei JY, Zhao YL, Nie GJ. Direct evidence for catalase and peroxidase activities of ferritin-platinum nanoparticles. Biomaterials. 2011, 32, 1611–1618. doi: 10.1016/j.biomaterials.2010.11.004.

[91]   Shen XM, Liu WQ, Gao XJ, Lu ZH, Wu XC, Gao XF. Mechanisms of oxidase and superoxide
       dismutation-like activities of gold, silver, platinum, and palladium, and their alloys: A general way to
       the activation of molecular oxygen. J Am Chem Soc. 2015, 137, 15882–15891. doi: 10.1021/
       jacs.5b10346.
[92]   Xu Y, Chen L, Wang XC, Yao WT, Zhang Q. Recent advances in noble metal based composite
       nanocatalysts: Colloidal synthesis, properties, and catalytic applications. Nanoscale. 2015, 7,
       10559–10583. doi: 10.1039/c5nr02216a.
[93]   He WW, Wu XC, Liu JB, Hu XN, Zhang K, Hou SA, Zhou WY, Xie SS. Design of AgM bimetallic alloy
       nanostructures (M = Au, Pd, Pt) with tunable morphology and peroxidase-like activity. Chem Mater.
       2010, 22, 2988–2994. doi: 10.1021/cm100393v.
[94]   He WW, Liu Y, Yuan JS, Yin JJ, Wu XC, Hu XN, Zhang K, Liu JB, Chen CY, Ji YL, et al. Au@Pt nanostruc-
       tures as oxidase and peroxidase mimetics for use in immunoassays. Biomaterials. 2011, 32,
       1139–1147. doi: 10.1016/j.biomaterials.2010.09.040.
[95]   Xia XH, Zhang JT, Lu N, Kim MJ, Ghale K, Xu Y, McKenzie E, Liu JB, Yet HH. Pd-Ir core-shell
       nanocubes: A type of highly efficient and versatile peroxidase mimic. ACS Nano. 2015, 9,
       9994–10004. doi: 10.1021/acsnano.5b03525.
[96]   Garg B, Bisht T. Carbon nanodots as peroxidase nanozymes for biosensing. Molecules. 2016, 21,
       1653. doi: 10.3390/molecules21121653.
[97]   Sun H, Zhou Y, Ren J, Qu X. Carbon nanozymes: Enzymatic properties, catalytic mechanism, and
       applications. Angew Chem Int Ed. 2018, 57, 9224–9237. doi: 10.1002/anie.201712469.
[98]   Shi W, Wang Q, Long Y, Cheng Z, Chen S, Zheng H, Huang Y. Carbon nanodots as peroxidase
       mimetics and their applications to glucose detection. Chem Commun. 2011, 47, 6695–6697. doi:
       10.1039/C1CC11943E.
[99]   Song YJ, Qu KG, Zhao C, Ren JS, Qu XG. Graphene oxide: Intrinsic peroxidase catalytic activity and its
       application to glucose detection. Adv Mater. 2010, 22, 2206–2210. doi: 10.1002/adma.200903783.
[100]  Kim MS, Cho S, Joo SH, Lee J, Kwak SK, Kim MIN. B-codoped graphene: A strong candidate to
       replace natural peroxidase in sensitive and selective bioassays. ACS Nano. 2019, 13, 4312–4321. doi:
       10.1021/acsnano.8b09519.
[101]  Ren CX, Hu XG, Zhou QX. Graphene oxide quantum dots reduce oxidative stress and inhibit
       neurotoxicity in vitro and in vivo through catalase-like activity and metabolic regulation. Adv Sci.
       2018, 5, 1700595. doi: 10.1002/advs.201700595.
[102]  Wu XC, Zhang Y, Han T, Wu HX, Guo SW, Zhang JY. Composite of graphene quantum dots and Fe₃O₄
       nanoparticles: Peroxidase activity and application in phenolic compound removal. RSC Adv. 2014, 4,
       3299–3305. doi: 10.1039/c3ra44709j.
[103]  Dong YM, Zhang JJ, Jiang PP, Wang GL, Wu XM, Zhao H, Zhang C. Superior peroxidase mimetic
       activity of carbon dots-Pt nanocomposites relies on synergistic effects. New J Chem. 2015, 39,
       4141–4146. doi: 10.1039/c5nj00012b.
[104]  Zheng C, Ke WJ, Yin TX, An XQ. Intrinsic peroxidase-like activity and the catalytic mechanism of
       gold@carbon dots nanocomposites. RSC Adv. 2016, 6, 35280–35286. doi: 10.1039/c6ra01917j.
[105]  Chen QM, Liang CH, Zhang XD, Huang YM. High oxidase-mimic activity of Fe nanoparticles
       embedded in an N-rich porous carbon and their application for sensing of dopamine. Talanta. 2018,
       182, 476–483. doi: 10.1016/j.talanta.2018.02.032.
[106]  Farha OK, Shultz AM, Sarjeant AA, Nguyen ST, Hupp JT. Active-site-accessible, porphyrinic
       metal-organic frame-work materials. J Am Chem Soc. 2011, 133, 5652–5655. doi: 10.1021/ja111042f.
[107]  Liu YL, Zhao XJ, Yang XX, Li YF. A nanosized metal-organic framework of Fe-MIL-88NH2 as a novel
       peroxidase mimic used for colorimetric detection of glucose. Analyst. 2013, 138, 4526–4531. doi:
       10.1039/c3an00560g.

[108] Wang CH, Gao J, Cao YL, Tan HL. Colorimetric logic gate for alkaline phosphatase based on copper (II)-based met-al-organic frameworks with peroxidase-like activity. Anal Chim Acta. 2018, 1004, 74–81. doi: 10.1016/j.aca.2017.11.078.

[109] Chen JY, Shu Y, Li HL, Xu Q, Hu XY. Nickel metal-organic framework 2D nanosheets with enhanced peroxidase nanozyme activity for colorimetric detection of $H_2O_2$. Talanta. 2018, 189, 254–261. doi: 10.1016/j.talanta.2018.06.075.

[110] Li HP, Liu HF, Zhang JD, Cheng YX, Zhang CL, Fei XY, Xian YZ. Platinum nanoparticle encapsulated met-al-organic frameworks for colorimetric measurement and facile removal of mercury(II). ACS Appl Mater Inter. 2017, 9, 40716–40725. doi: 10.1021/acsami.7b13695.

[111] Yang HG, Yang RT, Zhang P, Qin YM, Chen T, Ye FG. A bimetallic (Co/2Fe) metal-organic framework with oxidase and peroxidase mimicking activity for colorimetric detection of hydrogen peroxide. Microchim Acta. 2017, 184, 4629–4635. doi: 10.1007/s00604-017-2509-4.

[112] Xu HM, Liu M, Huang XD, Min QH, Zhu JJ. Multiplexed quantitative MALDI MS approach for assessing activity and inhibition of protein kinases based on postenrichment dephosphorylation of phosphopeptides by metal-organic frame-work-templated porous $CeO_2$. Anal Chem. 2018, 90, 9859–9867. doi: 10.1021/acs.analchem.8b01938.

[113] Zhang W, Hu SL, Yin JJ, He WW, Lu W, Ma M, Gu N, Zhang Y. Prussian blue nanoparticles as multienzyme mimetics and reactive oxygen species scavengers. J Am Chem Soc. 2016, 138, 5860–5865. doi: 10.1021/jacs.5b12070.

[114] Lin YH, Ren JS, Qu XG. Catalytically active nanomaterials: A promising candidate for artificial enzymes. Acc Chem Res. 2014, 47, 1097–1105. doi: 10.1021/ar400250z.

[115] Ghosh S, Roy P, Karmodak N, Jemmis ED, Mugesh G. Nanoisozymes: Crystal-facet-dependent enzyme-mimetic activity of $V_2O_5$ nanomaterials. Angew Chem Int Edit. 2018, 57, 4510–4515. doi: 10.1002/anie.201800681.

[116] Huang L, Chen JX, Gan LF, Wang J, Dong SJ. Single-atom nanozymes. Sci Adv. 2019, 5, eaav5490. doi: 10.1126/sciadv.aav5490.

[117] Benedetti TM, Andronescu C, Cheong S, Wilde P, Wordsworth J, Kientz M, Tilley RD, Schuhmann W, Gooding JJ. Electrocatalytic nanoparticles that mimic the three-dimensional geometric architecture of enzymes: Nanozymes. J Am Chem Soc. 2018, 140, 13449–13455. doi: 10.1021/jacs.8b08664.

[118] Chen YJ, Ji SF, Wang YG, Dong JC, Chen WX, Li Z, Shen RA, Zheng LR, Zhuang ZB, Wang DS, et al. Isolated single iron atoms anchored on N-doped porous carbon as an efficient electrocatalyst for the oxygen reduction reaction. Angew Chem Int Ed. 2017, 56, 6937–6941. doi: 10.1002/anie.201702473.

[119] Huang XY, Groves JT. Oxygen activation and radical transformations in heme proteins and metalloporphyrins. Chem Rev. 2018, 118, 2491–2553. doi: 10.1021/acs.chemrev.7b00373.

[120] Jiao L, Wu JB, Zhong H, Zhang Y, Xu WQ, Wu Y, Chen YF, Yan HY, Zhang QH, Gu WL, et al. Densely isolated $FeN_4$ sites for peroxidase mimicking. ACS Catal. 2020, 10, 6422–6429. doi: 10.1021/acscatal.0c01647.

[121] Asati A, Santra S, Kaittanis C, Nath S, Perez JM. Oxidase-like activity of polymer-coated cerium oxide nanoparticles. Angew Chem. 2009, 121(13), 2344–2348.

[122] Bao YW, Hua XW, Ran HH, Zeng J, Wu FG. Metal-doped carbon nanoparticles with intrinsic peroxidase-like activity for colorimetric detection of $H_2O_2$ and glucose. J Mat Chem B. 2019, 7(2), 296–304.

[123] Bing W, Sun H, Yan Z, Ren J, Qu X. Programmed bacteria death induced by carbon dots with different surface charge. Small. 2016, 12(34), 4713–4718.

[124] Bouzas-Ramos D, Cigales Canga J, Mayo JC, Sainz RM, Ruiz Encinar J, Costa-Fernandez JM. Carbon quantum dots codoped with nitrogen and lanthanides for multimodal imaging. Adv Funct Mater. 2019, 29(38), 1903884.

[125] Cabaleiro-Lago C, Quinlan-Pluck F, Lynch I, Lindman S, Minogue AM, Thulin E, Linse S. Inhibition of amyloid β protein fibrillation by polymeric nanoparticles. J Am Chem Soc. 2008, 130(46), 15437–15443.

[126] Jin J, Linlin L, Zhang L, Luan Z, Xin* S, Song K. Progress in the Application of Carbon Dots-Based Nanozymes. Front Chem. 2021, 9, 1–8.

[127] Cai X, Wang Z, Zhang H, Li Y, Chen K, Zhao H, Lan M. Carbon-mediated synthesis of shape-controllable manganese phosphate as nanozymes for modulation of superoxide anions in HeLa cells. J Mat Chem B. 2019, 7(3), 401–407.

[128] Chandra A, Deshpande S, Shinde DB, Pillai VK, Singh N. Mitigating the cytotoxicity of graphene quantum dots and enhancing their applications in bioimaging and drug delivery. ACS Macro Lett. 2014, 3(10), 1064–1068.

[129] Chen K, Chou W, Liu L, Cui Y, Xue P, Jia M. Electrochemical sensors fabricated by electrospinning technology: An overview. Sensors. 2019, 19(17), 3676.

[130] Chen M, Yang B, Zhu J, Liu H, Zhang X, Zheng X, Liu Q. FePt nanoparticles-decorated graphene oxide nanosheets as enhanced peroxidase mimics for sensitive response to H2O2. Mater Sci Eng C. 2018, 90, 610–620.

[131] Chen Q, Li S, Liu Y, Zhang X, Tang Y, Chai H, Huang Y. Size-controllable Fe-N/C single-atom nanozyme with exceptional oxidase-like activity for sensitive detection of alkaline phosphatase. Sensors and Actuators B. Chemical. 2020, 305, 127511.

[132] Chen X, Su B, Cai Z, Chen X, Oyama M. PtPd nanodendrites supported on graphene nanosheets: A peroxidase-like catalyst for colorimetric detection of $H_2O_2$. Sensors and Actuators B. Chemical. 2014, 201, 286–292.

[133] Chiti F, Dobson CM. Protein misfolding, functional amyloid, and human disease. Annu Rev Biochem. 2006, 75(1), 333–366.

[134] Ding H, Cai Y, Gao L, Liang M, Miao B, Wu H, Nie G. Exosome-like nanozyme vesicles for H2O2-responsive catalytic photoacoustic imaging of xenograft nasopharyngeal carcinoma. Nano Letters. 2018, 19(1), 203–209.

[135] Ding Y, Liu H, Gao LN, Fu M, Luo X, Zhang X, Zeng RC. Fe-doped Ag2S with excellent peroxidase-like activity for colorimetric determination of $H_2O_2$. J Alloys Compd. 2019, 785, 1189–1197.

[136] Du J, Qi S, Chen J, Yang Y, Fan T, Zhang P, Zhu C. Fabrication of highly active phosphatase-like fluorescent cerium-doped carbon dots for in situ monitoring the hydrolysis of phosphate diesters. Rsc Adv. 2020, 10(68), 41551–41559.

[137] Fan K, Xi J, Fan L, Wang P, Zhu C, Tang Y, Gao L. In vivo guiding nitrogen-doped carbon nanozyme for tumor catalytic therapy. Nat Commun. 2018, 9(1), 1440.

[138] Fan Z, Zhou S, Garcia C, Fan L, Zhou J. pH-Responsive fluorescent graphene quantum dots for fluorescence-guided cancer surgery and diagnosis. Nanoscale. 2017, 9(15), 4928–4933.

[139] Fan Z, Zhou S, Garcia C, Fan L, Zhou J. pH-Responsive fluorescent graphene quantum dots for fluorescence-guided cancer surgery and diagnosis. Nanoscale. 2017, 9(15), 4928–4933.

[140] Fasciani C, Silvero MJ, Anghel MA, Arguello GA, Becerra MC, Scaiano JC. Aspartame-stabilized gold–silver bimetallic biocompatible nanostructures with plasmonic photothermal properties, antibacterial activity, and long-term stability. J Am Chem Soc. 2014, 136(50), 17394–17397.

[141] Fischbach MA, Walsh CT. Antibiotics for emerging pathogens. Science. 2009, 325(5944), 1089–1093.

[142] Gao L, Zhuang J, Nie L, Zhang J, Zhang Y, Gu N, Yan X. Intrinsic peroxidase-like activity of ferromagnetic nanoparticles. Nature Nanotechnol. 2007, 2(9), 577–583.

[143] Geng X, Sun Y, Li Z, Yang R, Zhao Y, Guo Y, Qu L. Retrosynthesis of tunable fluorescent carbon dots for precise long-term mitochondrial tracking. Small. 2019, 15(48), 1901517.

[144] Glabe CG. Common mechanisms of amyloid oligomer pathogenesis in degenerative disease. Neurobiol Aging. 2006, 27(4), 570–575.

[145] Guo J, Wang Y, Zhao M. Target-directed functionalized ferrous phosphate-carbon dots fluorescent nanostructures as peroxidase mimetics for cancer cell detection and ROS-mediated therapy. Sens Actuators B Chem. 2019, 297, 126739.

[146] Han X, Park J, Wu W, Malagon A, Wang L, Vargas E, Leblanc RM. A resorcinarene for inhibition of Aβ fibrillation. Chemical Science. 2017, 8(3), 2003–2009.

[147] Honarasa F, Kamshoori FH, Fathi S, Motamedifar Z. Carbon dots on $V_2O_5$ nanowires are a viable peroxidase mimic for colorimetric determination of hydrogen peroxide and glucose. Mikrochim Acta. 2019, 186(4), 1–7.

[148] Honarasa F, Keshtkar S, Eskandari N, Eghbal M. Catalytic and electrocatalytic activities of $Fe_3O_4$/$CeO_2$/C-dot nanocomposite. Chem Papers. 2021, 75(6), 2371–2378.

[149] Hu S, Zhang W, Li N, Chang Q, Yang J. Integrating biphase γ-and α-$Fe_2O_3$ with carbon dots as a synergistic nanozyme with easy recycle and high catalytic activity. Appl Surf Sci. 2021, 545, 148987.

[150] Hua XW, Bao YW, Wu FG. Fluorescent carbon quantum dots with intrinsic nucleolus-targeting capability for nucleolus imaging and enhanced cytosolic and nuclear drug delivery. ACS Appl Mater Interfaces. 2018, 10(13), 10664–10677.

[151] Ji Z, Arvapalli DM, Zhang W, Yin Z, Wei J. Nitrogen and sulfur co-doped carbon nanodots in living EA. hy926 and A549 cells: Oxidative stress effect and mitochondria targeting. J Mater Sci. 2020, 55(14), 6093–6104.

[152] Kohanski MA, Dwyer DJ, Collins JJ. How antibiotics kill bacteria: From targets to networks. Nat Rev Microb. 2010, 8(6), 423–435.

[153] Kong W, Liu J, Liu R, Li H, Liu Y, Huang H, Kang Z. Quantitative and real-time effects of carbon quantum dots on single living HeLa cell membrane permeability. Nanoscale. 2014, 6(10), 5116–5120.

[154] Lan M, Zhao S, Zhang Z, Yan L, Guo L, Niu G, Zhang W. Two-photon-excited near-infrared emissive carbon dots as multifunctional agents for fluorescence imaging and photothermal therapy. Nano Res. 2017, 10(9), 3113–3123.

[155] Leidinger P, Treptow J, Hagens K, Eich J, Zehethofer N, Schwudke D, Feldmann C. Isoniazid@ $Fe_2O_3$ nanocontainers and their antibacterial effect on tuberculosis mycobacteria. Angew Chem Int Ed. 2015, 54(43), 12597–12601.

[156] Leidinger P, Treptow J, Hagens K, Eich J, Zehethofer N, Schwudke D, Feldmann C. Isoniazid@ $Fe_2O_3$ nanocontainers and their antibacterial effect on tuberculosis mycobacteria. Angew Chem Int Ed. 2015, 54(43), 12597–12601.

[157] Li B, Chen D, Nie M, Wang J, Li Y, Yang Y. Carbon Dots/$Cu_2O$ Composite with Intrinsic High Protease-Like Activity for Hydrolysis of Proteins under Physiological Conditions. Part Part Syst Charact. 2018, 35(11), 1800277.

[158] Li C, Mezzenga R. The interplay between carbon nanomaterials and amyloid fibrils in bio-nanotechnology. Nanoscale. 2013, 5(14), 6207–6218.

[159] Li F, Chang Q, Li N, Xue C, Liu H, Yang J, Wang H. Carbon dots-stabilized $Cu_4O_3$ for a multi-responsive nanozyme with exceptionally high activity. Chem Eng J. 2020, 394, 125045.

[160] Li H, Huang J, Song Y, Zhang M, Wang H, Lu F, Kang Z. Degradable carbon dots with broad-spectrum antibacterial activity. ACS Appl Mater Interfaces. 2018, 10(32), 26936–26946.

[161] Shamsipur M, Safavi A, Mohammadpour Z. Indirect colorimetric detection of glutathione based on its radical restoration ability using carbon nanodots as nanozymes, Sens. Actuators B Chem. 2014, 199, 463–469.

[162] Yousefinejad S, Rasti H, Hajebi M, Kowsarib M, Sadravib S, Honarasab F. Design of C-dots/$Fe_3O_4$ magnetic nanocomposite as an efficient new nanozyme and its application for determination of $H_2O_2$ in nanomolar level, Sens. Actuators B Chem. 2017, 247, 691–696.

[163] Li Q, Yang D, Yang Y. Spectrofluorimetric determination of Cr(VI) and Cr(III) by quenching effect of Cr(III) based on the Cu-CDs with peroxidase-mimicking activity, Spectrochim. Acta A Mol Biomol Spectrosc. 2021, 244, 118882.

[164]  Li C, Liu Q, Wang X, Luo Y, Jiang Z. An ultrasensitive K+ fluorescence/absorption di-mode assay
based on highly co-catalysis carbon dot nanozyme and DNAzyme. Microchem J. 2020, 159, 105508.

[165]  Ren X, Liu J, Ren J, Tang F, Meng X. One-pot synthesis of active copper-containing carbon dots with
laccase-like activities. Nanoscale. 2015, 7, 19641–19646.

[166]  Li S, Pang E, Gao C, Chang Q, Hu S, Li. N. Cerium-mediated photooxidation for tuning pH-
dependent oxidase-like activity. Chem Eng J. 2020, 397, 125471.

[167]  Dehvaria K, Hui S, Jin C, Lina S, Mekonnen W, Yong-Chien G, Jia-Yaw L. Heteroatom doped carbon
dots with nanoenzyme like properties as theranostic platforms for free radical scavenging, imaging,
and chemotherapy. Changac Acta Biomaterialia. 2020, 114, 343–357.

Muhammad Alamgeer

# Chapter 7
# Carbon dots in food safety detection

**Abstract:** Carbon dots have received a portion of courtesy over the past as a promising novel nanomaterial with unique visual and functional physicochemical features, high compatibility, low cost, and high sensitivity. This study aims at discussing current discovery principles, such as fluorescence extraction and recovery procedures using carbon dots and other nanobiological technologies. As synthetic and corrective methods for creating carbon dots, Surface and surface methods, as well as techniques for utilizing surface passivation and heteroatom doping, are presented as synthetic and corrective ways of producing carbon dots. This chapter discusses about their uses in the food industry to identify nutrients, prohibited substances, pathogenic bacteria, and toxins in food materials. Finally, concerns or obstacles need to be addressed, and the problems that need to be overcome or resolved are highlighted, as well as other innovative ways of combining carbon dots and other materials to produce stable and special nanosensors in many fields. Although carbon-based sensors have shown strength in the food sensor industry because food samples contain complex compounds that can cause disruption, additional innovations to mix carbon dots and other materials are needed to create sensitive and selective sensing probes.

**Keywords:** Carbon dots, food security, principles of sustainable development, governance, human development, planetary health

## 7.1 Introduction

Carbon dots (CD) received a lot of attention as their structures are different from the one published in 2004. The advantages of CD include high water solubility, easy source modification, chemical stability, high resistance to photobleaching, low toxicity, and high biocompatibility [1–5]. CD nanostructures are considered promising candidates for fluorescent sensors, bioimaging, and energy modification. Tens of thousands of precursors were used to create CDs using hundreds of methods. The raw materials used range from commercial chemicals to natural products. Blending methods include hydrothermal treatment, microwave heating, ultrasonic processing, UV irradiation, and electrochemical oxidation [6–7]. The fluorescence release of CDs is mainly due to the main structures and additional conditions. From the machine, CD fluorescence refers

**Muhammad Alamgeer,** Khwaja Fareed University of Engineering and Information Technology, Rahim Yar Khan, Pakistan

https://doi.org/10.1515/9783110799958-007

to electrons and holes due to the effect of quantum confinement on the surface of CDs. Therefore, we get a strong release of prepared CDs in small sizes to achieve a large amount of specific surface area, that is, the ratio of carbon particle surface to the particle volume [8–10]. Food hygiene is the condition and measure needed to confirm diet care from manufacture to ingesting. It is a vital obligation of any diet course that the nourishment bent must be harmless to use. Food security is a fundamental need, but some risks can be overlooked in developing effective and efficient procedures. Food security remains a major concern due to the outbreak of food-borne diseases that cost people, the nourishment commerce, and the budget enormously. In England and Wales, the quantity of nutriment exterminating reports gradually increased from around 15,000 in the premature 1980s to a crowning of over 60,000 in 1996 [11–15].

## 7.1.1 Time

Past was foremost written, and many of the nutrition protection issues we face currently are not newfangled. While governments around the world are committed to improving nutritional safety, the emergence of food-borne diseases continues to remain a chief municipal health problem in both advanced and emergent countries [16–20].

Nutrition contamination and recontamination are most likely at lower socioeconomic levels due to bad environmental conditions, personal hygiene, low or insufficient water supplies, and poor sanitation and food supply maintenance. Nutriment safety risks are impurities that can make diet manufacturing hazardous in the production process. A lack of proper food hygiene can result in foodborne diseases and even death. In today's world, contaminated food is the most prevalent cause and a major contributor to intestinal sickness, malnutrition, disease resistance, and productivity loss. Intestinal illnesses produced by nourishment pollution may be evaded to a substantial degree if nutrition cleanliness procedures are trailed at all points of diet gaining, stowage, training, and ingesting. More study has been conducted in the current centuries on the synthesis, alteration, and request of Carbon dots based nanomaterials. Carbon-built nanomaterials of diverse morphologies have been effectively synthesized and exploited in a variety of study fields. In general, the local structure of carbon dots based nanomaterials is defined by the heterocyclic structure, nature of C=C bonds, leading to extraordinary chemical and electrical properties. These advantages promote the broad application of this type of nanomaterial in a variety of fields, including environmental monitoring, energy conservation, and life science. Carbon-founded nanomaterials have been utilized to create extremely efficient sensors for testing diet safety, as well as for the generation, detection, and development of sensory signals. The complex research of novel carbon sources, such as graphene and carbon marks, has increased carbon-based sensitivity and their potential for advancement in the creation of diet security strategies with tall accuracy, low distortion, and ease of use. This research examines several features of carbon-based nanomaterials and the methodologies used to determine food safety. A recent study in the design and building of higher performance food safety encampments is discussed in great detail. This chapter outlines the research and development direction of carbon nanomaterial chemistry and biosensors for detecting and analysing pesticide, veterinary, and pharmaceutical residues [26–31].

## 7.2 Food contamination during processing

The presence of unsolicited substances such as dust and particles during manufacture and conveyance is called adulteration. The term "contaminated" refers to unsolicited substances found in a diet. This contamination affects the quality of the product or process. The food poisons are basic reason for the pathogenic diseases. Small capacity taster handling techniques have been developed and offered as effective approaches for food inspection, minimizing matrix conflicts and enabling the analysis to be focused on the sample [32–35].

However, identifying contaminants that may arise through food production, processing, or packaging remains stimulating. Information about possible contamination at each stage of food processing is important. The following sections recognize the foremost contaminants at each stage, control means, and customs to prevent or reduce food intake. This information is important in determining the source of contaminants in the final diet. Malnutrition can be associated with 97% of all food-borne diseases. The likelihood of food contamination and recontamination is particularly high at lower socioeconomic levels due to poor environmental conditions, personal hygiene, low standards, scarce water supply, poor sanitation, and food supply. Food safety hazards are contaminants that can make food production unsafe. The absence of appropriate nourishment cleanliness can be the main source for food poisoning and purchaser's death. In the modern world, an unclean diet is the main contributor to digestive problems, undernutrition, disease resistance, and decreased productivity. To a large extent, intestinal infections caused by food contamination can be prevented if safe food hygiene procedures are followed in various stages of food purchase, storage, preparation, and consumption. The World Health Organization (WHO) has long recognized the need to educate food handlers about their responsibility for food safety. In the 1990s, WHO developed the ten first-rate rule for harmless diet training and introduced the five keys to safe food in 2001. Knowing the rank of safe food for hominid fitness, the WHO has chosen the theme of World Wellbeing Day Nutrition Security 2015 to ensure food security from one farm to another. Food contamination processing is evident in the Pace scopes of diet dispensation in (Figure 7.1). The presence of uninvited substances such as dust and particles during production and transportation is called contamination. The term "contaminated" refers to unwanted substances found in food [36–40]. This contamination affects the quality of the product or process. Nutrition contamination of infectious or chemical origin is a thoughtful problem for consumers. Taster handling diplomacies, such as minor-bulk extraction procedures, have been developed to provide powerful sustenance scrutiny tools that abolish matrix block and tolerate sample-engrossed analysis. However, identifying contaminants that may transpire during the production, prosing, or packaging of nutriment remains challenging. Information about possible contamination at each stage of nutrition processing is important. The following sections describe how to detect, control, prevent, or reduce food intake at each stage. This information is important in determining the source of pollutants in the last diet. Outside of the evolution of the food contamination industry, increased pesticide use or accumulated activities can contribute to food contamination. The major route of nourishment contamination is through the use of manures and insecticides, as they can be the reason for fitness glitches if eaten by humans [41–45].

## 7.2.1 External pollution

Industrialization, advances in pesticide use, or urban activities can all worsen food corruption. Manures and insecticides are the most prevalent sources of food contamination, and their use by humans can have negative health effects. Since heavy metals are present in the atmosphere, soil, and water, they can be extremely dangerous. Now let's talk about the meal. Research on heavy metals was conducted on a range of foods, including tea, honey, spinach, potatoes, and salmon.The primary methods used in the analysis of heavy metals include flame atomic absorption spectrometry, graphite furnace atomic absorption spectrometry, cold vapor atomic absorption spectrometry, inductively coupled plasma atomic emission a, and inductive spectrometry plasma mass spectrometry. To evaluate antimicrobial residues in foods like meat, eggs, or milk contamination of food during transportation, liquid chromatography techniques like the microbial inhibition plate test have been developed. Additionally, food contamination can happen while in transportation. It can be produced by the waterfalls from gasoline and diesel vehicles or by the pollutants from a vehicle used to transport food. Food safety may be jeopardized by any of these several contaminants. In 1999, severe illness was brought on by moldy pallets used for storing and delivering food packaging in the European Economic Community. The opposing pollution from cleaning chemicals or other sources has also repeatedly hurt a long-distance cruise ship. Research is a great illustration of how high theoretically barrier materials can be used to prevent food contamination with naphthalene, methyl bromide, toluene, ethylbenzene, and ortho par xylenes. Contamination from classes on household tasks: Cleaning and disinfection are essential to reducing diet adulteration because they eliminate the presence of potentially dangerous germs throughout the distribution of nutrition. Secondhand chemicals used as antiseptics or detergents must be legal and safe for food contact. It is not advisable to use glass cleaners or other metal disinfectants since they may leave behind hazardous scum. Sterilizer accumulation can concentrate on processed foods, including processed fruits and vegetables, therefore it is important to portion off any remaining chemical residues in food to reassure us. complete eradication There are also several quaternary ammonium compounds like dodecyl trimethyl ammonium chloride and nonionic surfactants like stearyl alcohol ethoxylate. factors affecting its removal in various material contexts, such as the length of the bath or the temperature of the water To evaluate these compounds, liquid chromatography-mass spectrometry is frequently employed. Numerous authors have talked about management strategies and how they affect diets that interact with certain environments [46–50].

## 7.2.2 Contamination from housework courses

Contamination from classes on household tasks: Cleaning and disinfection are essential to reducing diet adulteration because they eliminate the presence of potentially dangerous germs throughout the distribution of nutrition. Secondhand chemicals used as antiseptics or detergents must be legal and safe for food contact. It is not advisable to use glass cleaners or other metal disinfectants since they may leave behind hazardous scum. Sterilizer accumulation can concentrate on processed foods [51–54], including processed fruits and vegetables, therefore it is important to portion off any remaining chemical residues in food to reassure us. complete eradication There are also several quaternary ammonium compounds like dodecyl trimethyl ammonium chloride and nonionic surfactants like stearyl alcohol ethoxylate. factors affecting its removal in various material contexts, such as the length of the bath or the temperature of the water To evaluate these compounds, liquid chromatography-mass spectrometry is frequently employed. Numerous authors have talked about management strategies and how they affect diets that interact with certain environments [55, 56].

## 7.2.3 Contamination by heating

When high cooking temperatures are combined with foreign chemicals, hazardous compounds are formed, which can have a detrimental influence on food quality and safety. Certain toxic compounds can be generated in the food due to processing activities such as boiling, sweltering, dismissing, or fermentation. Boiling is a culinary method that can result in the release of a range of hazardous chemicals into nutriment. Flavour compounds are fashioned by the interaction of oxidized hot fats, proteins, and other sulphur and nitrogen components in the diet. To increase flexibility or flavour, certain compounds are extracted from foods and added to frying oils. Promote pigs found in fried fat as a healthy alternative to fried food. According to a recent EFSA report, the average exposure to 3-monochloro-1,2-propanediol (3MPCD) was 1 mg/kg body weight. The majority of individuals utilize it every day (EFSA, 2011). 3MCPD may be derived from a number of sources, including acid hydrolysis of wheat, soya bean, and other protein products, as well as epichlorohydrin resin used to protect paper from moisture and common cellulosic materials adaptable from used sausage casings. Acrylamide and its precursors are also important polluters of heat transmission. Some solvents, such as nitrosamines, can be produced by combining ordinary nutrition machineries and nutrition flavours. Nitrosodimethylamine is extracted directly from foods by the fire drying or roasting techniques. Nitrosamine formula I is less when evaporated or boiled at low temperatures (100 °C) than frying, roasting, or grilling [41–45]. Chemical nitric oxide concentrations can be measured via colorimetric and spectroscopic methods after gas or liquid chromatography or as a component of that nitroso group. Gas chromatography combined with a specialized thermal sparkle analyser (TEA) indicator is the greatest appropriate, important, and broadly rummage-sale diagnostic technique for detecting flexible nitrosamine. Heat-induced pollutants include polycyclic aromatic hydrocarbons, which are found in roasted and burnt goods, ethyl carbamate, and other items, and furan, which is found in the variability of hot meals, including coffee and canned foods. Furan is a flavouring agent that may be obtained from ascorbic acid, carbohydrate depletion, amino acid depletion, and fatty acid oxidation. Mutagen production is exceedingly low in the absence of fat. In the presence of hard salts and fats, traditional blends, including Maillard forerunners, were burned. As a consequence, it was revealed that substituted imidazoquinoxalines are mutagenic. Tocopherol had minimal influence on the reaction, which was aided by oxidized lipids and iron salt. The unique mutagenesis activity of frying oil is also connected to the breakdown products of nitrogen-free lipid hydrogen peroxide and is not dependent on the fried substrate. Microwave cooking is becoming increasingly popular in households and companies. Food is frequently cooked in its packaging in a microwave oven, which is a common aspect of microwave cooking [46–51].

## 7.3 Causes of food security

Household food insecurity (HFI) is caused by poverty, any of family member's poor health, and subsistence and household management practices. Food security is directly connected to nutritional and health protection, although not always. Individuals acquire nutrition security when their body tissues obtain adequate quantities of nutrients and other vital elements. Nutrition security results from the convergence of household food security, access to healthcare, and other fundamental human requirements such as proper sanitation. Food security is connected to other elements of nutritional security. A family with inadequate food resources, for example, may opt not to seek medical treatment for their children or to purchase prescription medications [50–56]. Families must have unfettered access to good and nutritious food for food security to become a reality. Access to nutritious food, on the other hand, is contingent on having enough economic resources and food easily available in the country, area, and communities where residences are located. National food security is determined by the balance of locally cultivated food versus imported food derived from imported, processed, or animal feed. As a result, a sustainable global food supply system is critical for ensuring both local and global food securities. As a result, understanding and dealing with climate change, agricultural commodity policies, armed conflicts, and ultimately the health of our planet from a home food security perspective in the context of the UN Sustainable Development Goals (SDGs), which seek to eradicate hunger, achieve food security and improved nutrition, and promote sustainable agriculture around the world, is critical [14–16].

## 7.3.1 Effects of food security

HFI indicates powerful biological and psychological forces that may raise the risk of individuals' mental, social, and emotional development in a number of ways throughout their lives. The biological route involves the possibility of a relationship between HFI, malnutrition, dietary habits, and general health. A recent research from the United States, for example, documented the very low food quality of low-income persons at risk of food insecurity. Low-fat cereals, fruits, vegetables, and fish constituted their diet. This is an eating pattern that is strongly associated with an elevated risk of obesity, metabolic syndrome, chronic illnesses including diabetes, and early mortality. Anxiety and concern; emotions of loneliness, deprivation, and isolation; depression; and poor family and social connections among those suffering food insecurity are all part of the mental-emotional approach [41–45].

# 7.4 Carbon-based nanomaterials

## 7.4.1 Ordered mesoporous carbon (OMC)

Mesoporous carbon materials are a novel form of non-silica mesoporous material that has received a lot of interest in recent years. In comparison to mesoporous silicon materials, mesoporous carbon materials offer a few extremely unique qualities, including large surface area and porosity, changeable aperture size, controlled hole shape and wall structure, light generation, and absence of biological toxicity. In addition, by optimizing and controlling integration conditions, high thermal and hydrothermal stability, as well as a large precise zone and aperture volume, can be attained, making this category of property more suited for a wide range of applications, including electrode resources, catalyst ropes, and adsorbent carters. Based on conventional pores, mesoporous carbon can be divided into two types: irregular mesoporous carbon and ordered mesoporous carbon (OMC). Disturbed mesoporous carbon is often produced by the catalytic initiation of metal ions, carbonization of polymers, and carbonitriding or oxidation of silica templates by live aerogels, subsequent in a structure with a low frequency and homogeneity. The synthetic carbon material is therefore a perfect anode material for sodium-ion batteries and other energy-storing systems. OMC supplies are serene of highly anode nanorods and macroporous carbon, which have improved electrochemical constancy and distinct possessions that other supplies do not have, such as an extremely well-ordered hole structure, an effortlessly precise mesoporous structure, the delivery of the size of the narrow holes, as well as the large surface area (2,000 $m^2$ g) and the volume of the specific hole (1.5 $cm^3$ g). OMC is typically synthesized as a solid template utilizing mesoporous silica cell filters as a template, followed by embedding a mesoporous silica template in NaOH or HF solutions. Mesoporous carbon compounds are used in a variety of applications, including material integration, catalyst carriers, adsorption separation, and electrical equipment. Kui and colleagues developed a new APTA sensor based on sulfur and nitrogen-doped OMC (SNOMC), which are applied to determine the values of $Hg^{2+}$ at 1–1000 ppm (Figure 7.1a) as electrochemical sensor. 09 01330 g001a 550 nanomaterials 09 01330 g001b 550 nanomaterials The use of OMC nanoparticles in neural processing. (a) Hg2 + electrochemical sensor construction technique using OMS nanomaterials [45–46].

## 7.4.2 Carbon nanotubes (CNTs)

Iijima discovered one-sided carbon nanotubes (CNTs) and nanomaterials constructed of hexagonal carbon atoms in 1991. CNTs have become one of the most intensively investigated carbon sources due to their unusual three-dimensional assembly, physical and chemical characteristics, and modest training procedures, and great progress has been achieved in numerous study fields. The primary atoms of C in CNTs typically have $sp^2$ hybrid orbitals; nevertheless, when the local topology is established, $sp^3$ hybrid orbitals can be generated. There is some bending between grid structures that are made up of hexagons. The link, which is mainly removed from the outer surface of CNTs owing to the creation of chemical bonds in the mixture and disintegration, becomes the chemical basis of its binding, which is incompatible with specific macromolecules. Their graphene cylinder cats, according to the system, may be split into solo-walled cents and cents (wants). The carbon nanotubes based on graphene are the most interesting materials. The surface of these materials is responsible for the various modifications [15–16]. Gelatine, in wide-ranging, has microscopic imperfections such as holes that may readily move across layers during initial production due to its easy chemical reaction and smooth structure and space, making the chemical structure of mints quite complicated SWCNT electrochemical architectures and features including catalytic activity, stability, electrical conductivity, and biocompatibility are frequently exploited in the edifice of chemical or biological sensors for nutriment security Gelatine is a widely used material. The surface of the gelatine was functionalized with various modifications. As a result, it has many good features, such as good catalytic activity, stability, electrical conductivity, and biocompatibility. In the present times, it is used in analytical chemistry for the chemical or biological sensors of nutrition agents. The goal of this application is to create an electro-chemical sensor based on acetylcholinesterase (pain) for evaluating sensitive and low-cost pesticides in natural and food samples (Figure 7.2a). Mounts are intended to serve two de-velopmental purposes. The first duty is to dramatically increase facilitate electrochemical polymerization. The second function is to minimize Michaelis Menten ($K_m$) solubility by en-zymatic activity. In actual samples, electrochemical sensors based on mounts demonstrated stable, repeatable, and quick responses to several pesticides [1–6].

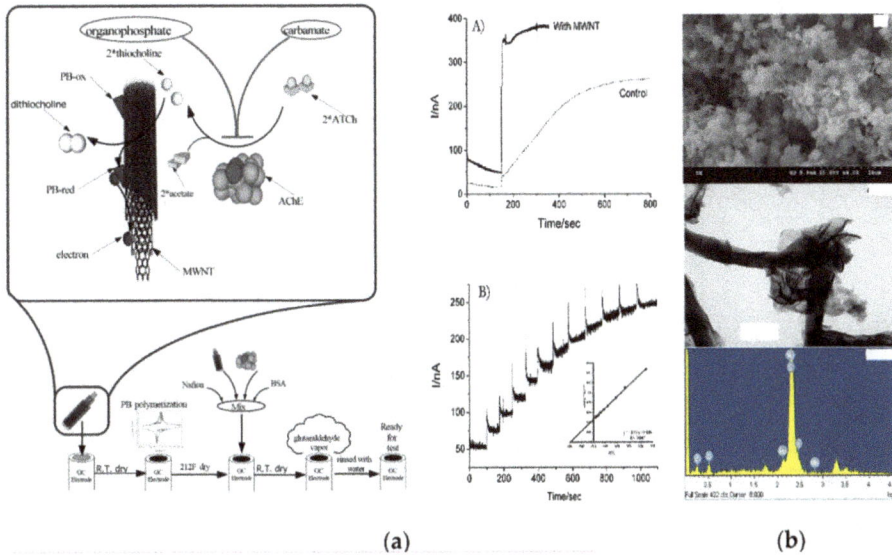

(a)                                                      (b)

**Figure 7.1:** MWCNT material has significant electrochemical sensitivity. (a) An electrochemical sensor based on ache/PB/MWCNT for pesticide detection (index reprinted with permission. Chemical Society, 2008). (b) Exclusion of HAS and EDX from $MoS_2$/desires and complexes (index reprinted with permission. Elsevier, 2017. All rights reserved).

The electrochemical sensitivity of MWNTs was described in Figure 7.2. To dissociate chloramphenicol, a comprehensive-spectrum antibiotic that works by intrusive with bac-terial protein production, a hybrid material featuring adhesion sites between MWCNT nanosheets and molybdenum disulphide ($MoS_2$) was created. 2b). $MoS_2$/month nanocom-posites exhibit exceptional electrochemical characteristics and an amazing capacity to form caps. The modified $MoS_2$/MWNT electrode reacted consistently with CAP con-centrations ranging from 0.08 to 1,392 m, with a detection limit of 0.01502 m. CNT materials with high catalytic activity and conductivity efficiently lower surface strength and expedite electron transport in electrochemical processes. The sensors feature better sensitivity, a wider linear response range, and a quicker reaction time when compared to traditional CNT sensors. Bhardwaj et al. created a simple and low-cost paper-based electrochemical immunosensor that can be built on plates with nonexistent

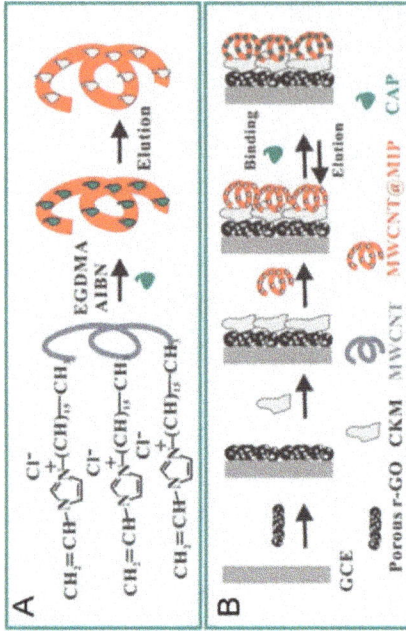

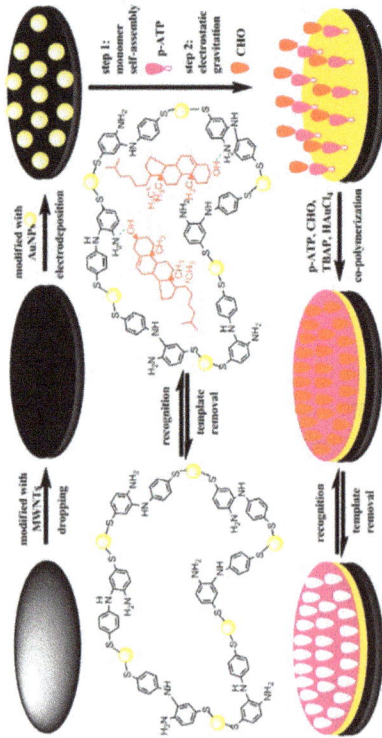

**Figure 7.2:** Function of MWNTs in molecularly embossed biomimetic sensors. (a) The groundwork method of AuNPs/MWNTs/GCE@MIP membrane. (b) Scheme of the construction procedure of a MWCNTs@MIP-CAP-based sensor.

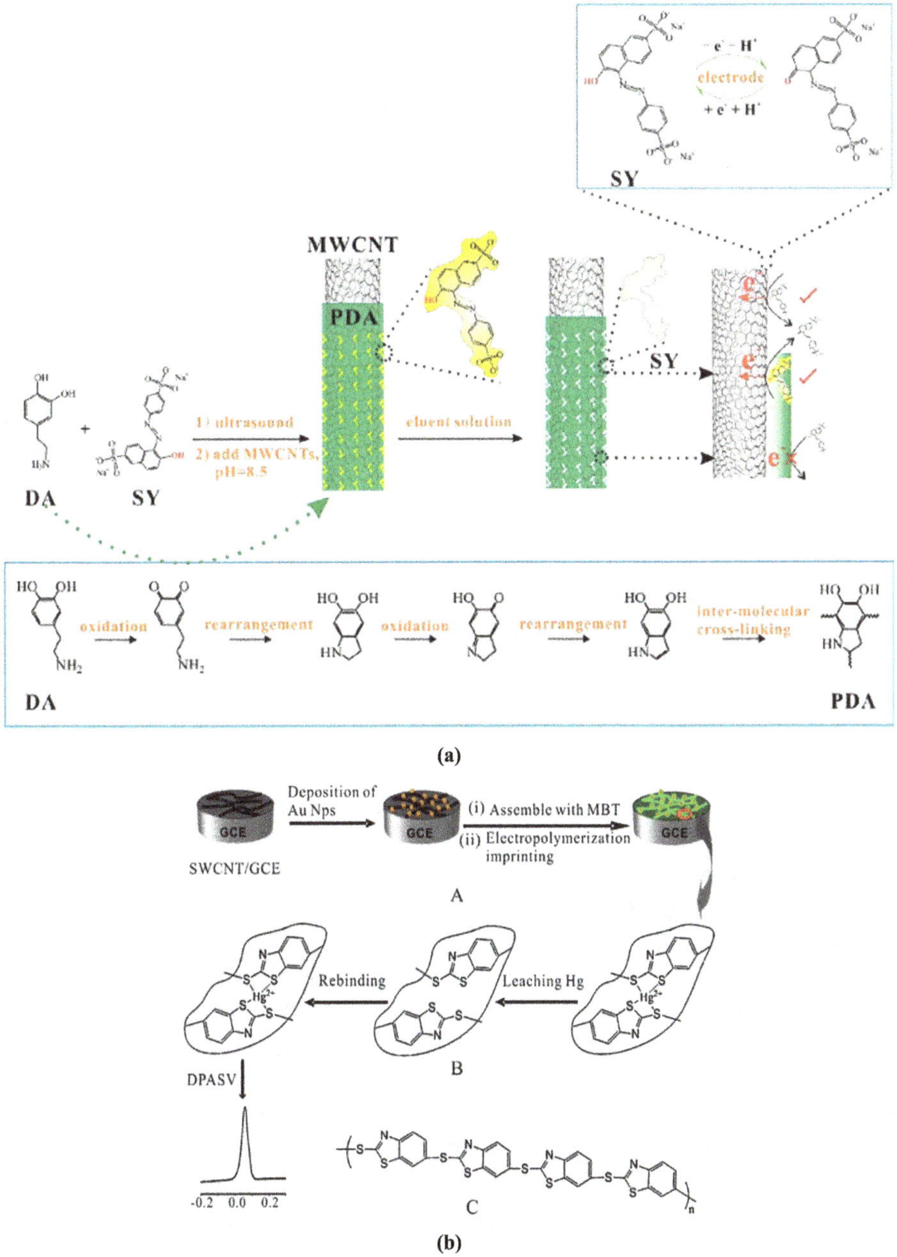

(a)

(b)

**Figure 7.3:** CNT use in MIP-based sensors is revealed in Figure 3. (a) Making a sunset yellow MWCNT@MIP-PDA sensor (Fu, Wu et al. 2012, Yin, Cheng et al. 2018). (b) Diagram showing the Hg(II)-imprinted PMBT, AuNPs, SWCNTs, and GCE (Fu, Wu et al. 2012). (c) Making a gold electrode with a CS-SNP/graphene-MWCNTs composite coating (Lian, Liu et al. 2013).

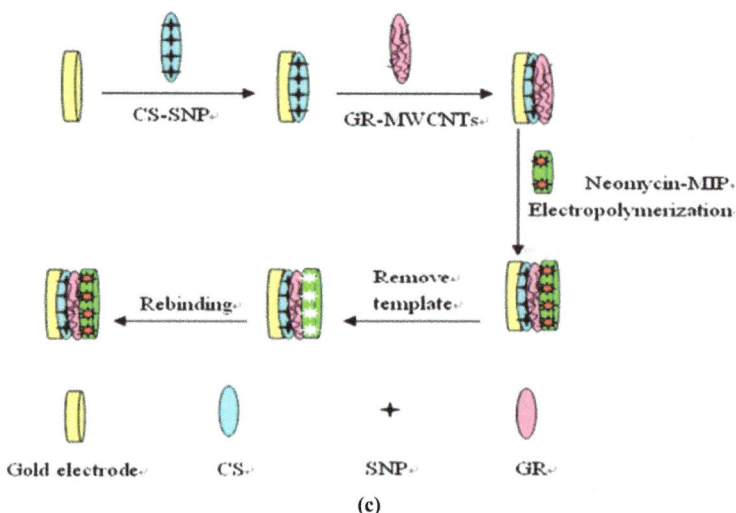

**Figure 7.3** (continued)

bioconjugates. Gold. The anti-aureus antibody smoothly attached to SWNT and migrated to the active electrode region without detecting the analyte, causing the maximum current value to alter. This remarkable sensor revealed a narrow line ($R^2 = 0.976$) between the peak current increase and the log *S. aureus* concentration (10107 CFU), with a concentration limit of 13 CFU in milk with a short period (30 min) and higher concentrations. In the immunological sensor's sensitivity, the multi-junction sensor was developed by Kara et al. The stunts were used to detect the foodborne diseases. Au complexes of polyethyleneimine was coated was used to diagnose the chronic diseases [17–21].

CNT resources with decent catalytic movement and conductivity importantly lessen the surface strength and effectively fasten electron handover in electrochemical responses. When associated with conventional CNT sensors, the sensors generally have higher sensitivity, linear response range, and faster response. Bhardwaj et al. developed a simple and inexpensive paper-based electrochemical immunosensor that can be made on plates using no bioconjugates. Aureus antibody linked to SWNT and moved to the active electrode region, resulting in a change in the maximum current value. This surprising sensor showed a thin line ($R^2 = 0.976$) between the peak current increase and the log *S. aureus* concentration (10,107 CFU), with a concentration limit of 13 CFU in milk with a short duration (30 min) and higher concentrations, resulting in higher concentrations. In sensitivity of the immune sensor, the stunts were used to detect the foodborne diseases. Stunts and polyethylenimine coated with gold tungsten filaments formed a $2 \times 2$ complex line containing streptavidin and biotinylated antibodies. The overview of SWCNTs aimed to diminish upbringing racket and enhance the biorecognition rejoinder amid Ab and Ag. Sol–gel-derivative silica/chitosan nanocomposites have been used to inhibit cholesterol esterase and cholesterol oxidase (cox) in the indium-tin glass. Its nanocomposite reduces the reaction time to 1,002 s, keeps the chest active and stable, and improves sensitivity. Parvin et al. made silver and CNT/copper nanoparticles and nitrate-sensitive nitrate-coated silver optical probe [5–9].

## 7.5 Graphene and its wastes

Graphene is a twofold carbon substantially having a clean-packed honeycomb conductor assembly of solitary-walled carbon atoms. His discoveries, which defied the assumption that two-dimensional crystals did not exist, piqued the scientific community's curiosity. The discovery of graphene sparked a fresh generation of carbon offset research using CNTs. Graphene atoms are composed of $sp^2$-hybridized carbon. To build a full four-way network structure, hybrid orbitals form bit connections with nearby carbon atoms. The specific area of HAS is approximately 2,630 $m^2$ g, and the specific capacity of carbon graphite (GC) for chemical techniques can approach 100,230 F g. The flexible construction of carbon graphite (GC) sheets adds to their large surface area. It enhances electrolyte diffusion by producing nano-sized holes and pores between the layers. As a result, graphene is an excellent electrode for a supercapacitor electrode. With GR serving as electrodes and CNT serving as current collectors, a flexible GR-based thin film supercapacitor was created. The embedded tool achieves high power density (8–14 Wh kg) and specific strength by combining high-performance large-format film with CNT film (250–450 kW kg). Graphene with high electrical conductivity, a sole quantum Hall effect at room malaise, and very rapid electron flow are excellent for nanoelectronic device fabrication. The improved performance of suspended GR-field effect transistors (GR-FETs) in aqueous solutions was reported by Cheng et al. Significantly when the power of low-frequency noise falls by 12 and 6 times for the hole and electron carriers, respectively, the trans-conductance of GR-FETs in the linear operating modes increases by 1.5 and 2 times. The reason for this is that the GR material has a more cohesive structure than the stable material, it was collapses and melts more quickly. The GR layer might be agglomerated. To increase graphene characteristics in data gathering applications, different inorganic and organic compounds or polymers have been utilized. These growth hormone (GH) compounds have numerous characteristics and serve countless protagonists in the development of innovative nutriment protection sensors. Inorganic GH nanocomposites can be created by spreading nanomaterials over the surface of GH sheets. Inanimate NPS can increase the distance amid graphene coats while decreasing the pressures between them, allowing graphene monolayers to retain their structure and characteristics [8–15]. This synergy is critical in the functioning systems. Using a periodic adsorption test, Liu and his colleagues evaluated the influence of two inanimate masses, $SiO_2$ and $Al_2O_3$, on exposure of 17b-estradiol to graphene oxide. The consequences established that the attendance of inanimate silver nanoparticles (NPS) impeded adsorption and lengthened the period obligatory to

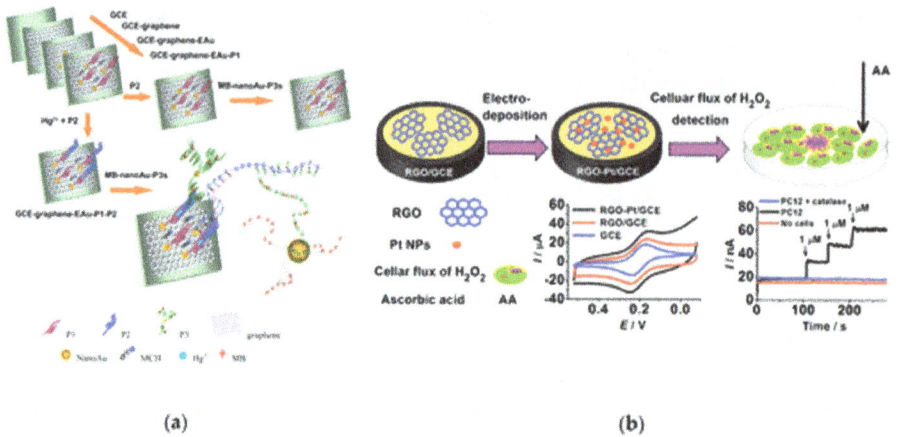

(a)                                             (b)

**Figure 7.4:** CNT use in MIP-based sensors is revealed in Figure 3. (a) Making a sunset yellow MWCNT@MIP-PDA sensor (Fu, Wu et al. 2012, Yin, Cheng et al. 2018). (b) Diagram showing the Hg(II)-imprinted PMBT, AuNPs, SWCNTs, and GCE (Fu, Wu et al. 2012). (c) Making a gold electrode with a CS-SNP/graphene-MWCNTs composite coating (Lian, Liu et al. 2013).

influence adsorption equilibrium in GH. As a result, this training provides novel insights on the endpoints and transfers GH and pollutants in marine ecosystems. Its high graphene rating makes it a superb iron NPS transporter. The fusion of GR and NP sheets can be avoided by depositing metallic NPS on the graphite sheet's surface. Complex structures are often one of a kind of high-rise buildings [5–8].

## 7.6 Conclusions

A variety of nanoscale carbon-based supplies is a unique material due to their unresolved performance. A large figure of theoretical and hands-on studies that explain the grounding, amendment, and tender of carbon-based nanosupplies in the field of food stuff have been conducted. Significant advancement has been made, thus entirely representative of the potential of carbon-built nanosupplies as a new-fangled nerve-building device. The nano technology, and sensory materials will be key to further improvements in the practice of carbon-built nanomaterials in diet investigation.

## References

[1]    Aliaga MA, Chaves-dos-santos SM. Food and nutrition security public initiatives from a human and socioeconomic development perspective: Mapping experiences within the 1996 world food summit signatories. Soc Sci Med. 2014, 104, 74–79.

[2]   Pérez-escamilla R. Food security and the 2015–2030 sustainable development goals: From human to planetary health: Perspectives and opinions. Curr Dev Nutr. 2017, 1(7), e000513.

[3]   Ramakrishnan U. Prevalence of micronutrient malnutrition worldwide. Nutr Rev. 2002, 60(Suppl_5), S46–S52.

[4]   Forouzanfar MH, Afshin A, Alexander LT, Anderson HR, Bhutta ZA, Biryukov S, . . . Carrero JJ. Global, regional, and national comparative risk assessment of 79 behavioural, environmental and occupational, and metabolic risks or clusters of risks, 1990–2015: A systematic analysis for the global burden of disease study 2015. The Lancet. 2016, 388(10053), 1659–1724.

[5]   Smith MD, Rabbitt MP, Coleman-jensen A. Who are the world's food insecure? new evidence from the food and agriculture organization's food insecurity experience scale. World Dev. 2017, 93, 402–412.

[6]   Pérez-escamilla R, Segall-corrêa AM. Food insecurity measurement and indicators. Rev de Nutr. 2008, 21, 15s–26s.

[7]   Pérez-escamilla R. Food security and the 2015–2030 sustainable development goals: From human to planetary health: Perspectives and opinions. Curr Dev Nutr. 2017, 1(7), e000513.

[8]   Lang T, Mason P. Sustainable diet policy development: Implications of multi-criteria and other approaches, 2008–2017. Proc Nutr Soc. 2018, 77(3), 331–346.

[9]   Leung CW, Ding EL, Catalano PJ, Villamor E, Rimm EB, Willett WC. Dietary intake and dietary quality of low-income adults in the supplemental nutrition assistance program. Am J Clin Nutr. 2012, 96(5), 977–988.

[10]  Millen BE, Abrams S, Adams-campbell L, Anderson CA, Brenna JT, Campbell WW, . . . Lichtenstein AH. The 2015 dietary guidelines advisory committee scientific report: Development and major conclusions. Adv Nutr. 2016, 7(3), 438–444.

[11]  Perez-escamilla F, de toledo vianna RP. Food insecurity and the behavioral and intellectual development of children: A review of the evidence. J Appl Res Child. 2012, 3(1), 9.

[12]  Fram MS, Frongillo EA, Jones SJ, Williams RC, Burke MP, DeLoach KP, Blake CE. Children are aware of food insecurity and take responsibility for managing food resources. J Nutr. 2011, 141(6), 1114–1119.

[13]  Bernal J, Frongillo EA, Herrera H, Rivera J. Children live, feel, and respond to experiences of food insecurity that compromise their development and weight status in peri-urban Venezuela. J Nutr. 2012, 142(7), 1343–1349.

[14]  Gubert MB, Spaniol AM, Bortolini GA, Pérez-escamilla R. Household food insecurity, nutritional status and morbidity in Brazilian children. Public Health Nutr. 2016, 19(12), 2240–2245.

[15]  Pérez-escamilla R, Dessalines M, Finnigan M, Pachón H, Hromi-fiedler A, Gupta N. Household food insecurity is associated with childhood malaria in rural Haiti. J Nutr. 2009, 139(11), 2132–2138.

[16]  Rivera JA, Pedraza LS, Martorell R, Gil A. Introduction to the double burden of undernutrition and excess weight in Latin America. Am J Clin Nutr. 2014, 100(6), 1613S–1616S.

[17]  Gubert MB, Spaniol AM, Segall-corrêa AM, Pérez-escamilla R. Understanding the double burden of malnutrition in food insecure households in Brazil. Maternal Child Nutr. 2017, 13(3), e12347.

[18]  Kulkarni VS, Kulkarni VS, Gaiha R. Double burden of malnutrition reexamining the coexistence of undernutrition and overweight among women in India. Int J Health Serv. 2017, 47(1), 108–133.

[19]  Zhang N, Bécares L, Chandola T. Patterns and determinants of double-burden of malnutrition among rural children: Evidence from China. PloS One. 2016, 11(7), e0158119.

[20]  Laraia BA. Food insecurity and chronic disease. Adv Nutr. 2013, 4(2), 203–212.

[21]  Pérez-escamilla R, Villalpando S, Shamah-levy T, Méndez-gómez humarán I. Household food insecurity, diabetes and hypertension among Mexican adults: Results from Ensanut 2012. Salud Publ de Mex. 2014, 56, s62–s70.

[22]  Weigel MM, Armijos RX, Racines M, Cevallos W, Castro NP. Association of household food insecurity with the mental and physical health of low-income urban Ecuadorian women with children. J Environ Public Health. 2016, 2016.

[23] Kollannoor-samuel G, Wagner J, Damio G, Segura-pérez S, Chhabra J, Vega-lópez S, Pérez-escamilla R. Social support modifies the association between household food insecurity and depression among Latinos with uncontrolled type 2 diabetes. J Immigrant Minority Health. 2011, 13(6), 982–989.

[24] Sharpe I, Davison CM. Investigating the role of climate-related disasters in the relationship between food insecurity and mental health for youth aged 15–24 in 142 countries. PLOS Global Public Health. 2022, 2(9), e0000560.

[25] Britto PR, Lye SJ, Proulx K, Yousafzai AK, Matthews SG, Vaivada T, . . . BHUTTA ZA. Early childhood development interventions review group, for the lancet early childhood development series steering committee. nurturing care: Promoting early childhood development. Lancet. 2017, 389 (10064), 91–102.

[26] Bermúdez-millán A, Pérez-escamilla R, Segura-pérez S, Damio G, Chhabra J, Osborn CY, Wagner J. Psychological distress mediates the association between food insecurity and suboptimal sleep quality in Latinos with type 2 diabetes mellitus. J Nutr. 2016, 146(10), 2051–2057.

[27] Jordan ML, Perez-escamilla R, Desai MM, Shamah-levy T. Household food insecurity and sleep patterns among Mexican adults: Results from ENSANUT-2012. J Immigrant Minority Health. 2016, 18(5), 1093–1103.

[28] Brown LR. Could food shortages bring down civilization? Sci Am. 2009, 300(5), 50–57.

[29] Whitmee S, Haines A, Beyrer C, Boltz F, Capon AG, de souza dias BF, . . . Yach D. Safeguarding human health in the Anthropocene epoch: Report of the rockefeller foundation -lancet commission on planetary health. The Lancet. 2015, 386(10007), 1973–2028.

[30] Imamura F, Micha R, Khatibzadeh S, Fahimi S, Shi P, Powles J Global Burden of Diseases Nutrition and Chronic Diseases Expert Group (NutriCoDE). Dietary quality among men and women in 187 countries in 1990 and 2010: A systematic assessment. The Lancet Global Health. 2015, 3(3), e132–e142.

[31] Fagherazzi G, El Fatouhi D, Fournier A, Gusto G, Mancini FR, Balkau B, . . . Bonnet F. Associations between migraine and type 2 diabetes in women: Findings from the E3N cohort study. JAMA Neurol. 2019, 76(3), 257–263.

[32] Pérez-escamilla R, Lutter CK, Rabadan-diehl C, Rubinstein A, Calvillo A, Corvalán C, . . . Rivera JA. Prevention of childhood obesity and food policies in Latin America: From research to practice. Obesity Rev. 2017, 18, 28–38.

[33] Stein D, Weinberger-litman SL, Latzer Y. Psychosocial perspectives and the issue of prevention in childhood obesity. Front Public Health. 2014, 2, 104.

[34] Andrews KR, Silk KS, Eneli IU. Parents as health promoters: A theory of planned behavior perspective on the prevention of childhood obesity. J Health Commun. 2010, 15(1), 95–107.

[35] Daniels LA, Magarey A, Battistutta D, Nicholson JM, Farrell A, Davidson G, Cleghorn G. The NOURISH randomised control trial: Positive feeding practices and food preferences in early childhood-a primary prevention program for childhood obesity. BMC Public Health. 2009, 9(1), 1–10.

[36] Gubert MB, Dos santos SMC, Santos LMP, Perez-escamilla R. A municipal-level analysis of secular trends in severe food insecurity in Brazil between 2004 and 2013. Glob Food Sec. 2017, 14, 61–67.

[37] Pérez-escamilla R. Can experience-based household food security scales help improve food security governance? Glob Food Sec. 2012, 1(2), 120–125.

[38] Gubert MB, Segall-corrêa AM, Spaniol AM, Pedroso J, Coelho SE, Pérez-escamilla R. Household food insecurity in black-slaves descendant communities in Brazil: Has the legacy of slavery truly ended? Public Health Nutri. 2017, 20(8), 1513–1522.

[39] Gubert MB, Segall-corrêa AM, Spaniol AM, Pedroso J, Coelho SE, Pérez-Escamilla R. Household food insecurity in black-slaves descendant communities in Brazil: Has the legacy of slavery truly ended? Public Health Nutri. 2017, 20(8), 1513–1522.

[40] Cook JT, Frank DA. Food security, poverty, and human development in the United States. Ann NY Acad Sci. 2008, 1136(1), 193–209.

[41]  Cafiero C, Viviani S, Nord M. Food security measurement in a global context: The food insecurity experience scale. Measurement. 2018, 116, 146–152.

[42]  Villagomez-ornelas P, Hernandez-lopez P, Carrasco-enriquez B, Barrios-sanchez K, Perez-escamilla R, Melgar-quinonez H. Statistical validity of the Mexican food security scale and the Latin American and Caribbean food security scale. Salud Pública de Méx. 2014, 56, s5–s11.

[43]  Nelson ME, Hamm MW, Hu FB, Abrams SA, Griffin TS. Alignment of healthy dietary patterns and environmental sustainability: A systematic review. Adv Nutr. 2016, 7(6), 1005–1025.

[44]  Tiwari S, Bharadva K, Yadav B, Malik S, Gangal P, Banapurmath CR, . . . Agrawal RK. Infant and young child feeding guidelines, 2016. Indian Pediatrics. 2016, 53(8), 703–713.

[45]  Richter LM, Daelmans B, Lombardi J, Heymann J, Boo FL, Behrman JR Lancet Early Childhood Development Series Steering Committee. Investing in the foundation of sustainable development: Pathways to scale up for early childhood development. The Lancet. 2017, 389(10064), 103–118.

[46]  Vilar-compte M, Teruel GM, Flores-peregrina D, Carroll GJ, Buccini GS, Perez-escamilla R. Costs of maternity leave to support breastfeeding; Brazil, Ghana and Mexico. Bull WHO. 2020, 98(6), 382.

[47]  Jones AD, Ngure FM, Pelto G, Young SL. What are we assessing when we measure food security? A compendium and review of current metrics. Adv Nutr. 2013, 4(5), 481–505.

[48]  Cetthakrikul N, Topothai C, Suphanchaimat R, Tisayaticom K, Limwattananon S, Tangcharoensathien V. Childhood stunting in Thailand: When prolonged breastfeeding interacts with household poverty. BMC Pediatr. 2018, 18(1), 1–9.

[49]  Chen H, Zuo X, Su S, Tang Z, Wu A, Song S, . . . Fan C. An electrochemical sensor for pesticide assays based on carbon nanotube-enhanced acetycholinesterase activity. Analyst. 2008, 133(9), 1182–1186.

[50]  Fu XC, Wu J, Nie L, Xie CG, Liu JH, Huang XJ. Electropolymerized surface ion imprinting films on a gold nanoparticles/single-wall carbon nanotube nanohybrids modified glassy carbon electrode for electrochemical detection of trace mercury (II) in water. Anal Chim Acta. 2012, 720, 29–37.

[51]  Govindasamy M, Chen SM, Mani V, Devasenathipathy R, Umamaheswari R, Santhanaraj KJ, Sathiyan A. Molybdenum disulfide nanosheets coated multiwalled carbon nanotubes composite for highly sensitive determination of chloramphenicol in food samples milk, honey and powdered milk. J Colloid Interface Sci. 2017, 485, 129–136.

[52]  Lian W, Liu S, Yu J, Li J, Cui M, Xu W, Huang J. Electrochemical sensor using neomycin-imprinted film as recognition element based on chitosan-silver nanoparticles/graphene-multiwalled carbon nanotubes composites modified electrode. Biosens Bioelectron. 2013, 44, 70–76.

[53]  Yin ZZ, Cheng SW, Xu LB, Liu HY, Huang K, Li L, . . . Lu YX. Highly sensitive and selective sensor for sunset yellow based on molecularly imprinted polydopamine-coated multi-walled carbon nanotubes. Biosens Bioelectron. 2018, 100, 565–570.

[54]  Yang G, Zhao F. Electrochemical sensor for chloramphenicol based on novel multiwalled carbon nanotubes@ molecularly imprinted polymer. Biosens Bioelectron. 2015, 64, 416–422.

[55]  Zhang Y, Bai X, Wang X, Shiu KK, Zhu Y, Jiang H. Highly sensitive graphene -Pt nanocomposites amperometric biosensor and its application in living cell H2O2 detection. Anal Chem. 2014, 86(19), 9459–9465.

[56]  Zhang Y, Zeng GM, Tang L, Chen J, Zhu Y, He XX, He Y. Electrochemical sensor based on electrodeposited graphene-Au modified electrode and nanoAu carrier amplified signal strategy for attomolar mercury detection. Anal Chem. 2015, 87(2), 989–996.

Sonali Loya and Swati Chandravanshi

# Chapter 8
# Carbon dots in anticancer detection and therapy

**Abstract:** Carbon dots (CDs) are carbon-based nanomaterials. These found many applications in different fields due to their fascinating properties,. The size of CDs ranges from 1 to 10 nm. CDs have been recognized as a promising candidate for various applications such as bioimaging, catalysis, biotherapy, electronics, biosensing, targeted drug delivery, detection of small molecules, treatment of cancer, and other biomedical applications. This chapter mainly highlights the preparation and applications of CDs.

**Keywords:** Carbon dots, nanomaterials, biomedical applications, preparation methods

## 8.1 Introduction

Carbon dots (CDs) are the carbon-based nanomaterials with size ranging from 1 to 10 nm. CDs are mainly incorporated with $sp^2/sp^3$-hybridized carbon core with various surface functional groups. Sun et al. named carbon nanoparticles (CNPs) as "carbon quantum dots" on the basis of their spectral features and properties, which were similar to the silicon quantum dots at that time. As a new member of nanomaterials, CDs exhibit many appreciable advantages such as low cytotoxicity, brilliant fluorescence (FL), good photoluminescence behaviour, good biocompatibility [12], high photostability, and stable chemical inertness. These appreciable features of CDs make them the most promising nanomaterials for multiple applications like biosensors [1–3], bioimaging [4, 5], drug delivery, biotherapy, and solar cells [6, 7].

CDs are broadly classified into three main types: graphene quantum dots (GQDs), carbon nanodots (CNDs), and carbonized polymeric dots (CPDs). GQDs are composed of $sp^2$ crystalline carbons in graphene or graphite manner [14]. Both CPDs and CNDs comprise a spherical core with connected surface groups. CPDs mainly contain aggregated / cross-linked carbon cores and polymeric chain shells.

In today's era, cancer has become one of the major health issues, which affects the world's population because of its high incidence and mortality rate. According to WHO, there were around 10 million deaths in 2020 or we can say nearly 1 in 6 deaths. The correct identification and treatment of cancer is still a big problem in the world. If it is not identified correctly, it will not be able to be treated properly because different techniques

**Sonali Loya, Swati Chandravanshi,** Govt. Nehru P.G. college, Dongargarh, Chhattisgarh, India

https://doi.org/10.1515/9783110799958-008

are employed in different tumour cells. Various types of cancer treatment have some advantages and also some disadvantages, affecting the quality of life of the patient. There are many chemotherapy drugs that affect not only the cancer cells but also the healthy cells, due to which there are some side effects such as vomiting and hair loss in patients [8]. For the new era of cancer research, CDs could be a promising candidate for the diagnosis and drug delivery [9].

The properties of CDs are governed by quantum confinement and surface state [13], which can also be changed by using different precursors or methods [15, 16]. CD can be made in such a way that they can exhibit different functional groups like amine, carbonyl, hydroxyl, carboxyl, and ether. Due to this versatile design capability, we can obtain CDs of different size and surface functional groups. This helps us to modulate the physical and chemical properties of CDs. These features of CDs make them the promising materials in cancer treatment by using them for bioimaging to drug delivery and also good agents for photodynamic therapy (PDT) [10] and photothermal therapy (PTT) [11].

# 8.2 Preparation of CDs

There are various methods to synthesize nanoparticles. All these methods are divided into two basic approaches: top-down approach and bottom-up approach. Top-down approach involves the exfoliation and cutting down of macroscopic carbon structures to the carbon particle in a nanoscale range. Chemical exfoliation, laser ablation, and ultrasonic-assisted treatment are some examples of this method. Bottom-up approach involves the building up of a nanomaterial from bottom by combining atom to atom, molecule to molecule, and cluster to cluster. Solvothermal method, pyrolysis, and chemical vapour deposition (CVD) are few examples of this approach.

## 8.2.1 Top-down approach

### 8.2.1.1 Chemical exfoliation

Chemical exfoliation is a very simple method to produce high-quality CDs in large scale without using any complex technique. In this method, various precursors like carbon nanotubes, graphene oxide (GO), and carbon fibre are cleaved to a desired nanostructure size by using strong acids or oxidizing agents. From the past years, many researchers have prepared different types of CDs by this method which has been described as follows:

In 2007, Mao and co-workers prepared multicolour fluorescent CNPs from the combustion soot of candles by means of oxidation using strong acids. This was then purified using polyacrylamide gel electrophoresis. The CNPs thus formed were small, having size less than 2 nm and were soluble in $H_2O$ [17].

In 2011, Peng et al. synthesized GQDs by chemical oxidation and breaking of micro-meter-sized carbon fibres using $H_2SO_4$ and $HNO_3$ as in Figure 8.1a [18].

In 2016, Zhao et al. prepared GQDs by chemical exfoliation of petroleum asphaltene by using mild oxidizing agents, that is, the mixture of concentrated $HNO_3$ and $H_2SO_4$. They were then neutralized by aqueous $NH_3$. GQDs comprise two layers of graphene nanosheets, and their surface is composed of mainly O- and N-based functional groups [19]. They showed strong FL properties.

In 2019, Gunjal et al. prepared CDs by chemical oxidation of waste tea residue using 0.1 M $HNO_3$ [20]. The solution is then cooled and centrifuged to remove large particles. It was then neutralized with $Na_2CO_3$. After a process of dialysis, a clear yellow suspension of CDs was obtained.

In 2014, Sun et al. prepared fluorinated GQDs by chemical exfoliation of fluorinated graphene oxide with concentrated $HNO_3$ and concentrated $H_2SO_4$ [21]. The solution was refluxed in the presence of microwave radiation for 6 h. After cooling the solution mild ultrasonication was performed. It was then treated with Sodium Carbonate. The solution was filtered. Here they obtained yellow solution, which after dialysis produces blue fluorescent GQD-F.

In 2017–2018, Soni et al. prepared N-, S-co-doped carbon quantum dots (CQDs) by chemical oxidative cleavage of palm shell powder using triflic acid as shown in Figure 8.1b [22]. The CQDs thus formed showed strong photoluminescence and good dispersibility. It was also shown that their size could be changed by changing the length of amino acid chains.

Kailasa and co-workers synthesized three (blue, green, and yellow) fluorescent colour CDs by acidic ($H_2SO_4$) oxidation of tomato [23]. CDs thus produced showed good water dispersibility and high quantum yield (QY).

In 2019, Desai et al. prepared *Cucumis melo* CDs by the acidic oxidation of muskmelon (*C. melo*) fruit [24]. The acids used were $H_2SO_4$ and $H_3PO_4$.

In 2017, Nair et al. synthesized high-quality GQDs by the oxidative cleavage of GO by using potassium permanganate in 30 min [25]. The QY of GQDs was up to 23.8%, and also the product yield was high around 75–81%.

In 2015, Zhu et al. prepared GQDs by the oxidative cleavage of GO using hydroxyl radicals. This was obtained by the decomposition of $H_2O_2$ in the presence of a catalyst tungsten oxide nanowire ($W_{18}O_{49}$). This method does not produce any by-product [26].

### 8.2.1.2 Laser ablation method

In laser ablation method, a part of the material is evaporated from the surface by means of laser irradiation. The surface of the target material absorbs high-powered laser beam incident on it which makes the temperature of the absorbing material to

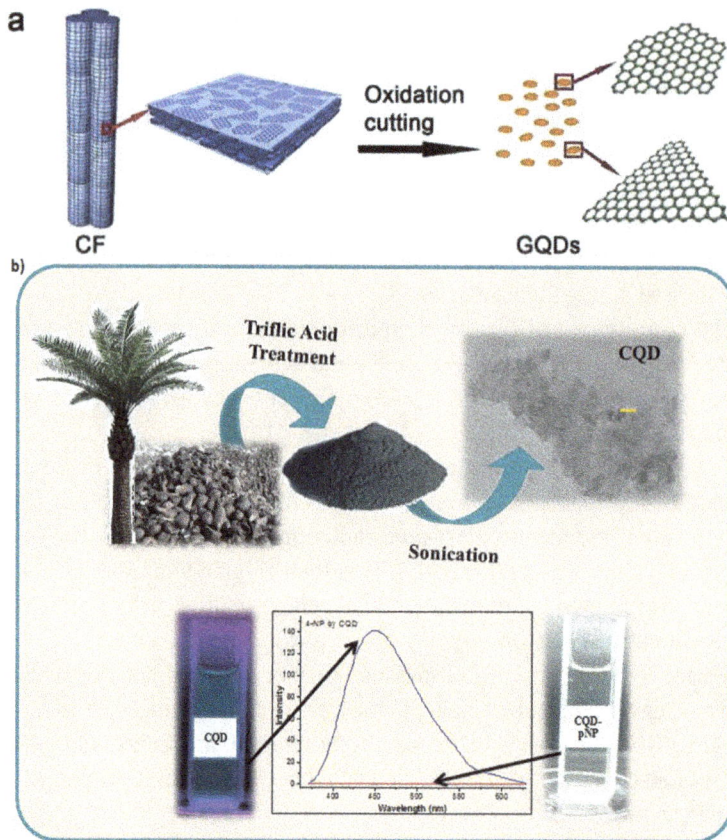

**Figure 8.1:** (a) Oxidative cutting of carbon fibre into GQDs [18] (Copyright 2012, American Chemical Society) and (b) chemical exfoliation process of palm shell powder [22] (Copyright 2018, Elsevier).

increase rapidly. This makes the material of the surface to vaporize into laser plume. Sometimes, the vaporized material condensates into cluster of particles, which are then either deposited on the substrate or collected through the filter system. Various attempts were made to synthesize CDs from laser ablation method, which is described as follows:

In 2006, Sun et al. prepared CDs by laser ablation of a mixture of graphite powder and cement in the presence of water vapour with argon as a carrier gas. A Q-switched Nd-YAG laser was used for ablation, as shown in Figure 8.2a [27].

In 2016, Kang et al. prepared GQDs from multi-walled carbon nanotube by using pulsed laser ablation (PLA) technique, as shown in Figure 8.2b [29].

In 2019, Ren et al. prepared N-doped micropore CQDs from sustainable and waste *Platanus* biomass using PLA technique as shown in Figure 8.2c [28]. The QY was 32.4%.

In 2019–2020, Cui et al. prepared CQDs from low-cost carbon cloth by using dual-beam PLA system. In this process, a single laser beam was used, which was divided into two laser beams in order to reduce the laser ablation time, as shown in Figure 8.2d [30]. The QY of CQDs was 35.4%.

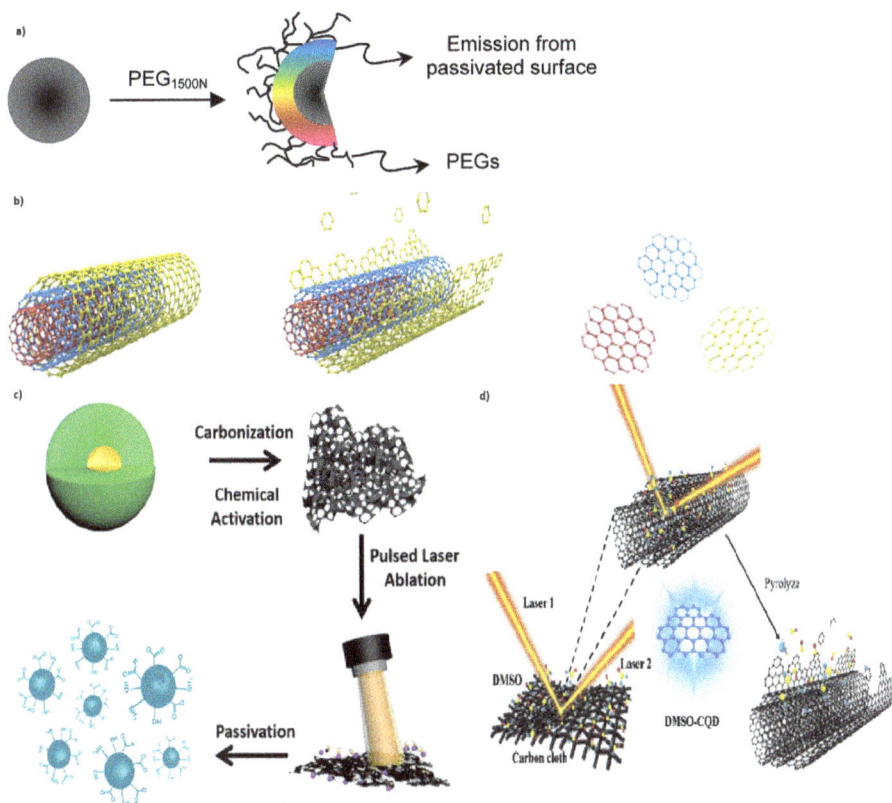

**Figure 8.2:** (a) Representation of carbon dots with PEG1500N species attached to the surface [27] (Copyright 2006, American Chemical Society); (b) figure showing the exfoliation of multi-walled carbon nanotube to GQDs (Scientific reports, Open access); (c) synthesis of N-doped micropore carbon quantum dots from waste *Platanus* biomass; (MDPI open access); and (d) mechanism of DMSO-CQD domains passivated by DMSO molecules [30] (Copyright 2020, Elsevier).

### 8.2.1.3 Ultrasonic–assisted treatment

Here, alternate high-pressure and low-pressure waves are created, resulting in the formation and disruption of small bubbles in solution. From the cavitation of small bubbles, a strong hydrodynamic shear force is generated which cuts the macroscopic carbon materials into nanoscale CDs.

In 2011–2012, Zhuo et al. synthesized GQDs from graphene by ultrasonic method. First, the graphene was oxidized using concentrated $H_2SO_4$ and concentrated $HNO_3$, which was then treated ultrasonically with an ultrasonic instrument. Then the concentrated $H_2SO_4$ and $HNO_3$ were removed. After re-dispersion in $H_2O$ filtration and dialysis, they obtained GQDs. Since then, bulk carbon materials such as GO, carbon nanofibres, MWCNTs, and graphite were investigated as starting materials for the synthesis of GQDs using ultrasound in aqueous solution or organic solvents [31].

In 2014, Song et al. prepared high-quality GQDs from graphene intercalation compounds (GICs) with controlled oxidation [32]. First of all, potassium–sodium tartrate was grinded and mixed with graphite which then reacted to the autoclave vessel for 24 h at 250 °C. The GICs thus formed are exfoliated in water under ultrasonic-assisted method, resulting in the formation of GQDs.

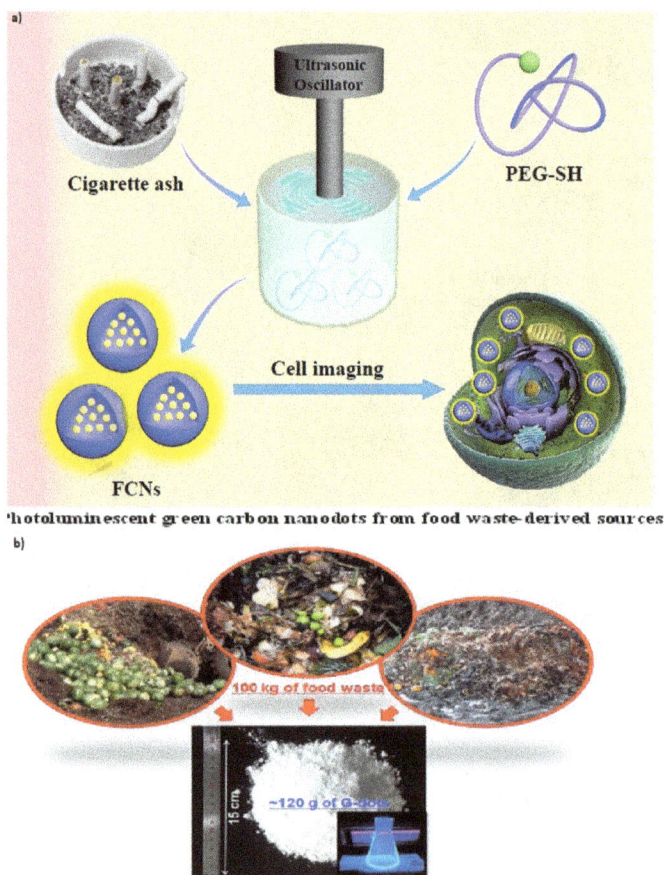

Photoluminescent green carbon nanodots from food waste-derived sources

**Figure 8.3:** (a) Formation of polymer-functionalized CQDs by ultrasonic-assisted treatment [33] (Copyright 2018, Elsevier). (b) Large-scale synthesis of G-dots from large food waste [34] (Copyright 2014, American Chemical Society).

In 2018, Huang et al. synthesized high-quality methoxy polyethylene glycol (PEG)-functionalized fluorescent CNPs using cigarette ash through a one-pot ultrasonic irradiation treatment using thiol group-terminated PEG as precursor as shown in Figure 8.3a [33].

In 2013–2014, Park et al. prepared CNDs from waste food materials by simple ultrasonic irradiation treatment, as shown in Figure 8.3b [34].

## 8.2.2 Bottom-up approach

### 8.2.2.1 Microwave synthesis

This method involves the formation of nanoparticles by microwave irradiation of solution. Microwave irradiation has good penetration effect, which homogeneously heats up the reaction solution. This results in uniform nucleation and rapid crystal growth. This method has various advantages like cost-effectiveness, provide uniform heat, and short time of reaction.

In 2012, Li et al. prepared stabilizer-free greenish yellow luminescent GQDs from graphene oxide nanosheets under acid conditions through microwave-assisted treatment [35].

In 2017, Yao et al. synthesized fluorescent CQDs by the use of waste crab shell using microwave-assisted approach as shown in Figure 8.4a [36]. They used transition metal ions such as $Gd^{3+}$, $Mn^{2+}$, and $Eu^{3+}$ to incorporate into carbon matrix.

In 2016, Kumawat and co-workers prepared GQDs through microwave-assisted green synthesis route using *Mangifera indica* (mango) leaves as a carbon source as shown in Figure 8.4b [37]. The size of GQDs from 2 to 8 nm shows bright red luminescence. The obtained GQDs showed excellent biocompatibility and photostability, making them suitable for magnetic resonance (MR) imaging and thermal sensing of live cells.

In 2014–2015, Pires et al. synthesized CQDs using microwave-assisted technique from an aqueous solution of raw cashew gum [38]. No passivation reagent was used.

In 2019, Ren et al. synthesized nitrogen-doped GQDs using sodium citrate and triethanol amine as raw materials through microwave-assisted approach [39]. No other harsh chemical was used. The QY was 8%.

In 2019, Ricardo et al. prepared CDs from the aqueous solution of citric acid and urea placed in a glass beaker through microwave-assisted treatment [40].

### 8.2.2.2 Hydrothermal method

In hydrothermal method, the materials are dissolved in water under high temperature and high pressure. The dissolved substance is then crystallized to obtain the desired end products. The process is usually performed in a steel pressure vessel also

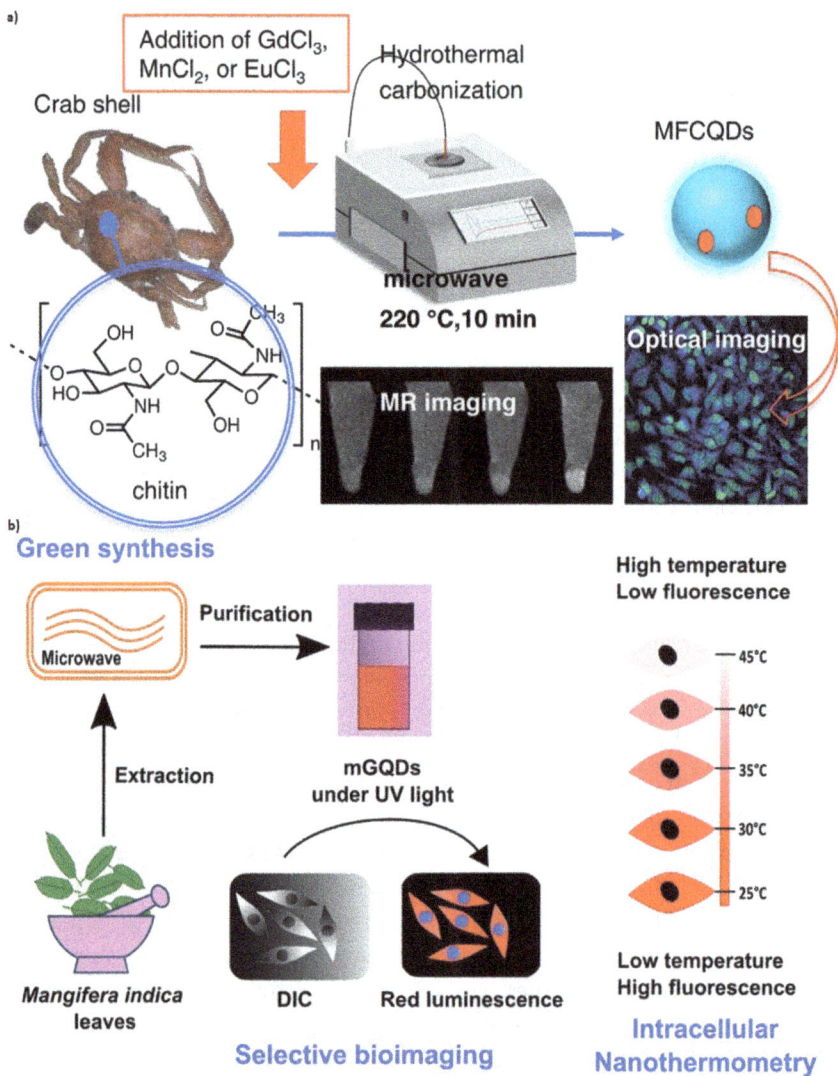

a)

Addition of GdCl$_3$, MnCl$_2$, or EuCl$_3$

Crab shell

Hydrothermal carbonization

MFCQDs

microwave
220 °C, 10 min

MR imaging

Optical imaging

OH

CH$_3$

NH

O

HO

NH

OH

CH$_3$

chitin

b)
**Green synthesis**

Microwave

Purification

Extraction

mGQDs under UV light

High temperature
Low fluorescence

45°C
40°C
35°C
30°C
25°C

Low temperature
High fluorescence

*Mangifera indica* leaves

DIC

Red luminescence

**Selective bioimaging**

**Intracellular Nanothermometry**

**Figure 8.4:** (a) Preparation of MFCQDs by a microwave-assisted hydrothermal method [36] (Copyright 2017, American Chemical Society). (b) Preparation of soluble CDs from mango leaves and their various applications [37] (Copyright 2017, American Chemical Society).

known as autoclave, which may or may not be coated with protective Teflon coatings. This process offers an advantage of low-cost, non-toxic, and simple approach.

In 2009, Pan et al. synthesized ultrafine GQDs from preoxidized graphene sheets (GSs) by using hydrothermal method. The cutting of GSs involves the complete breaking of mixed epoxy chains present on the surface, which consists of less epoxy groups and more carbonyl groups into CQDs [41].

In 2016–2017, Zhao et al. prepared GQDs from GO using hydrothermal method with assistance of $KO_2$, as shown in Figure 8.5a [42]. The QY was 8.9%.

In 2018, Halder et al. synthesized GQDs from pre-synthesized GO as a precursor and low amount of $H_2O_2$ as an oxidant by simple hydrothermal method (Figure 8.5b) [43]. The synthesized GQDs were uniformly small-sized of approximately 5 nm.

In 2013–2014, Mehta et al. synthesized water-dispersible fluorescent CDs from *Saccharum officinarum* juice (Figure 8.5c) [44].

In 2012, Lu et al. prepared fluorescent CNPs from wastes of pomelo peel as a carbon source through hydrothermal method, as shown in Figure 8.5d [45]. The prepared CNPs were water soluble and have QY of 6.9%.

In 2013–2014, Liu et al. synthesized CQDs using bamboo leaves via green hydrothermal method. Branched polyethylenimine (BPEI)-capped CQDs were prepared by coating CQDs with BPEI by electrostatic adsorption, as shown in Figure 8.5d [46]. The average size of CQDs prepared was 3.6 nm and the QY was 7.1%.

In 2017–2018, Essner et al. prepared a series of CDs from citric acid using hydrothermal and microwave routes followed by dialysis or ultrafiltration purification steps. They showed that impurities which are produced during CD synthesis should be removed to get good results [47].

### 8.2.2.3 Chemical vapour deposition

CVD method is a well-known approach used to fabricate CQDs. In this method, the carbon source is taken in gaseous phase, and the source of energy such as plasma or resistively heated coil is used to transfer energy to a gaseous carbon molecule. In general, hydrocarbons such as methane and CO are made to flow through the quartz tube placed in an oven at high temperature around 720 °C. At such high temperatures, the hydrocarbons are broken down to produce pure carbon molecules, which then diffuse towards the substrate that is heated and coated with a catalyst. Here, the carbon molecules bind with the substrate. There undergo some reactions producing CNPs of desired size. The size of the final product could be determined by modulating some parameters like source of carbon, flow rate, growth time, and temperature of the substrate.

Fan et al. prepared CQDs by the CVD method. They used methane gas as a carbon source. The copper foil was rinsed with HCl and alcohol to remove the oxidized surface. It was then heated to 1,000 °C in the presence of $H_2$ (10 mL $min^{-1}$) and argon (200 mL $min^{-1}$) for 40 min. Then, hydrogen was turned off and argon was kept for another 10 min to remove residual hydrogen. Methane (2 mL $min^{-1}$) was introduced in the reaction tube for 3 s. The synthesized CQDs had size in the range of 5–15 nm [48].

In 2015, Huang et al. prepared GQDs fabricated on silicon wafer by CVD [49]. This method was simple, cost-effective, eco-friendly, and absence of chemical functional

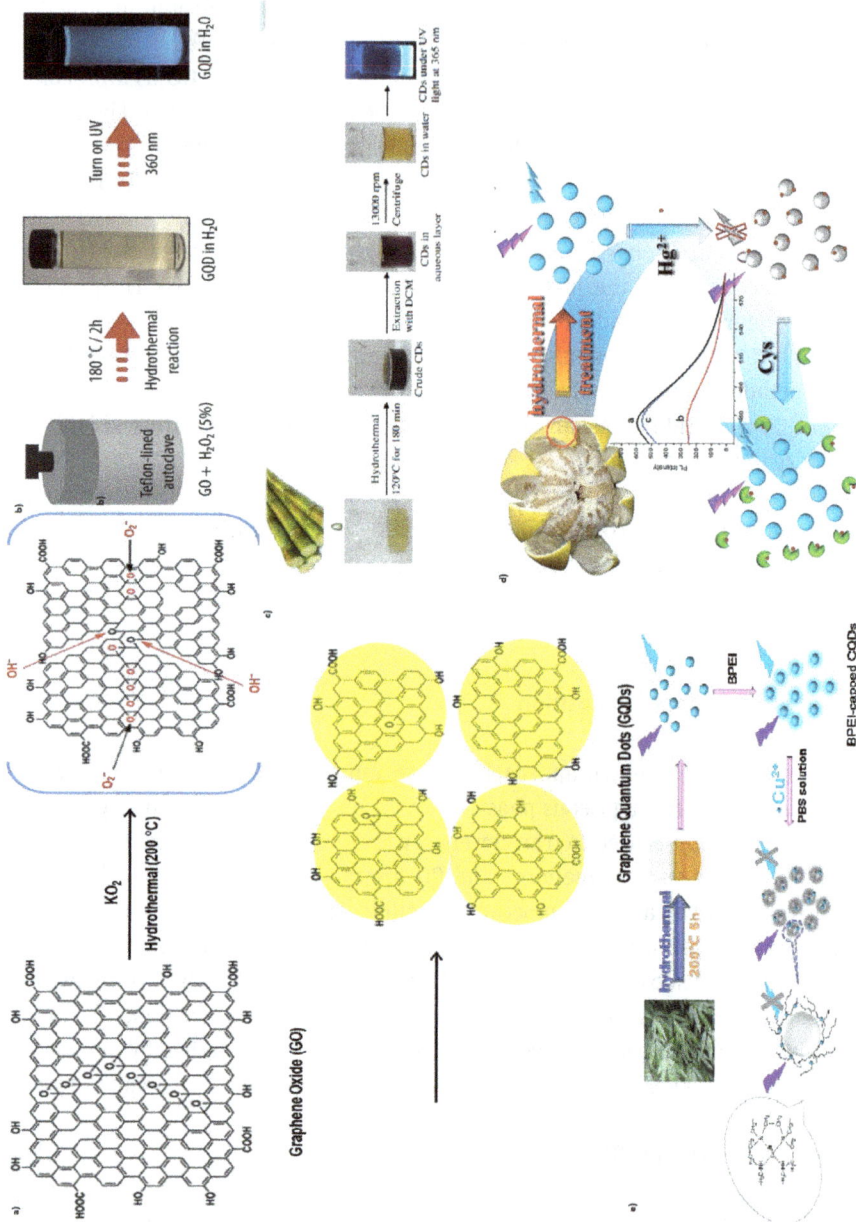

**Figure 8.5:** (a) Hydrothermal cutting procedure for GO with the assistance of KO$_2$ [42] (Copyright 2017, Elsevier); (b) synthesis of GQDs from GO and the picture with and without 360 nm wavelength UV excitation [43] (Copyright 2018, American Chemical Society); (c) synthesis of carbon dots from *Saccharum officinarum* juice by hydrothermal method [44]; (Copyright 2014, Elsevier); (d) formation of CDs from pomelo peel [45] (Copyright 2012, American Chemical Society); and (e) synthesis and application of CQDs (Copyright 2014, Elsevier).

groups. The prepared GQDs were single crystalline and highly pure with an average thickness of 1.2 nm and an average diameter of 7.5 nm.

The N-GQDs were also prepared by using chitosan as a carbon source (Figure 8.6) [50]. The synthesized N-GQDs had an average diameter of 12 nm and thickness of 3 nm.

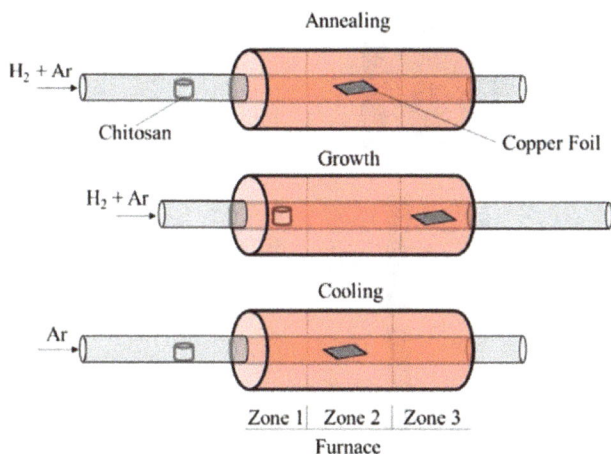

**Figure 8.6:** Synthesis of N-GQDs by CVD [50] (Copyright 2018, American Chemical Society).

### 8.2.2.4 Pyrolysis

Pyrolysis is a very influential technique to form fluorescent CDs using macroscopic carbon structures as a starting material. This involves four main steps, that is, heating, dehydration, degradation, and carbonization, which act as important factors for converting the organic carbon containing substance into CQDs under high temperature. The carbon precursors are broken down into CNPs by using highly concentrated alkali or acid.

In 2011, Zhou et al. synthesized water-soluble fluorescent CDs from the low-temperature carbonization and simple filtration of watermelon peel as a carbon source as shown in Figure 8.7a [51]. The process involves two steps. First, the carbonization of watermelon peel at 220 °C for 2 h in ambient air condition is followed by the ultrasonic treatment for 30 min, filtration, and centrifugation.

In 2019, Praneerad et al. prepared CDs by the pyrolysis of durian peel waste [52]. The formed CDs had a QY of 11% and an average size of about 10 nm.

In 2013, Sun et al. synthesized sulphur- and nitrogen-co-doped CDs (S–N-C-dots) by using sulphuric acid, carbonization, and etching of hair fibre, as shown in Figure 8.7b [53].

In 2013, Wee et al. prepared CDs through one-pot carbonization of bovine serum albumin (BSA) protein. About 1 mL of BSA was mixed with 3 mL of concentrated $H_2SO_4$.

This mixture was then transferred into water bath at 50 °C for at least 2 h. NaOH solution was used to neutralize it, and dialysis was done to obtain pure CDs [54].

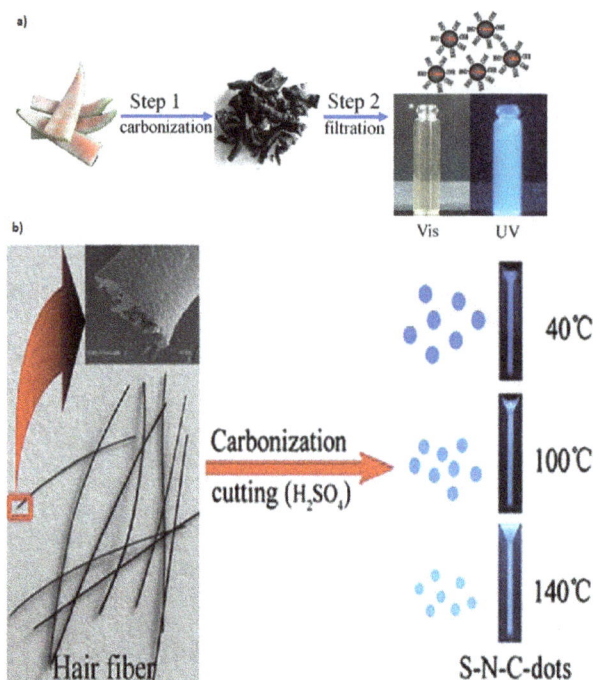

**Figure 8.7:** (a) Synthesis of water-soluble fluorescent C-dots from watermelon peel [51] (Copyright 2012, Elsevier) and (b) cutting of hair fibre into S–N–C-dots [53] (Copyright 2013, Elsevier).

### 8.2.2.5 Solvothermal method

Solvothermal method is similar to the hydrothermal approach. The only difference is that in solvothermal method, the water solution is replaced by one or other several solvents sealed with Teflon equipped with a steel autoclave. The mixture of solvent and the raw carbon source is made to react at high temperature and high pressure.

In 2016, Tian et al. prepared GQDs by the application of $H_2O_2$ in $N,N$-dimethylformamide environment by solvothermal method, as shown in Figure 8.8a [55]. Concentrated sulphuric acid and nitric acid were completely avoided to treat the raw material. The prepared GQDs show strong blue emission, and the QY was 15%.

In 2016, Liu et al. synthesized fluorescent CDs using one-pot hydrothermal treatment of rose-heart radish as shown in Figure 8.8b [56]. About 2 g of freshly chopped rose-heart radish was added into 10 mL of ultrapure water, which was then transferred into 25 mL Teflon-lined autoclave and heated at 180 °C for 3 h in an oven. The QY was 13.6%.

Qian et al. prepared N-doped CQDs by simple solvothermal process using $CCl_4$ and diamine mixture at 200 °C [57].

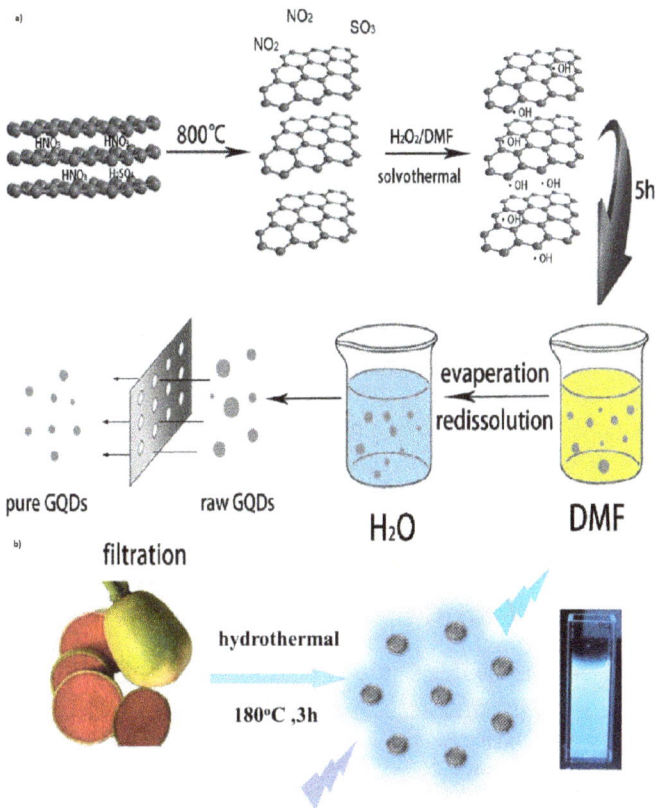

**Figure 8.8:** (a) Preparation of GQDs by solvothermal method [55] (Copyright 2018, Elsevier) and (b) synthesis of N-CDs from rose-heart radish, along with photograph of the sample under 365 nm UV lamp excitation [56] (Copyright 2017, Elsevier).

## 8.3 Applications of carbon dots

### 8.3.1 Carbon dots for cancer diagnosis

CDs have several properties, that is, less harmful, biocompatibility, photostability, and chemical inertness in medical fields. Due to these properties, CDs are considered for cancer diagnosis [58].

### 8.3.1.1 Fluorescence imaging probe carbon dots

FL CDs have excellent multicolour emission, hydrophilicity, biocompatibility, lower toxicity, low photodamage, and less auto FL disruption of biological samples, and their ease in preparation. Because of their properties, FL CDs have been used as imaging probe for cancer [59, 60]. Many research groups reported that cancer cells could be specifically identified using CDs by improving the FL properties using different precursors (Table 8.1). For example, solid-state fluorescent CDs were synthesized by using boric acid and ethylenediamine as an initial material by hydrothermal method. This became boron-doped CDs, which showed favourable solubility and robust FL in each aqueous and solid medium [61]. Sun et al. prepared highly efficient pure red emission CDs (R-CDs) with the high QY (22.9%). Citric acid and formamide were used in the form of precursors for the formation of R-CDs with 43.9% (high) photothermal conversion efficiency (PCE) under irradiation of 671 nm laser light. R-CDs were proved to be excellent in cancer diagnosis because of their high PCE [62]. Similarly, R-CDs were prepared by using pulp-free lemon juice as a precursor. R-CDs were of low cost, eco-friendly, have high QY (28%), and were monodispersed (diameter was 4.6 nm) by using this method. R-CDs were used as a luminescent probe for cancer diagnosis through these properties [63]. Highly fluorescent near-infrared (NIR)-emitting CDs were synthesized by using lemon juice and formamide as precursors with QY (31%) via a solvothermal method [64].

**Table 8.1:** Different precursors and preparation method for FL-CDs.

| S. no. | Precursors | Preparation method | Quantum yield (%) | Remarks | Reference |
|---|---|---|---|---|---|
| 1 | Boric acid and ethylenediamine | Hydrothermal | 22 | B-doped | [61] |
| 2 | Citric acid and formamide | Microwave | 22.9 | N-doped | [62] |
| 3 | Pulp-free lemon juice | Solvothermal | 28 | N-doped | [63] |
| 4 | Pulp-free lemon juice and formamide | Solvothermal | 31 | N-doped | [64] |
| 5 | Gelatin | Hydrothermal | 31.6 | – | [65] |
| 6 | Chitosan and acetic acid | Hydrothermal | 43 | – | [66] |
| 7 | Citric acid and urea | Thermal pyrolysis | 12.9, 35.1, and 52.6 | | [67] |
| 8 | k-Carrageenan and folic acid | Hydrothermal | 76.12 | N- and S-doped | [68] |
| 9 | N,N-Dimethyl, N,N-diethyl, and N,N-dipropyl-p-phenylenediamine | Solvothermal | 86 | N-doped | [69] |

Liang's group synthesized highly fluorescent CDs from gelatin by hydrothermal method with the QY of 31.6%. CDs were prepared by a simple approach for cancer diagnosis due to their long emission lifetime, low toxicity, steady emission, good dispersibility, and good compatibility with cells [65]. Functionalization increases the fluorescent properties. So, amino-functionalized fluorescent CDs were prepared by using chitosan and acetic acid with QY of 43%, showing low cytotoxicity and excellent biocompatibility [66]. Multiple colours from blue to red emissive CDs were reported by Miao's group in 2018, in which CDs were synthesized by thermal pyrolysis of citric acid and urea by managing the surface functionalization and graphitization. The QYs for blue, green, and red emission were up to 52.6%, 35.1%, and 12.9%, respectively [67]. Das group synthesized photoluminescent CDs by using k-carrageenan and folic acid (FA) as precursors with nitrogen and sulphur doped by hydrothermal method. CDs were simple and efficient for cancer diagnosis with high QY of 76.12%, and have good water solubility, excellent photostability, and biocompatibility [68]. Efficient red bandgap emission CDs were synthesized by using *N,N*-dimethyl, *N,N*-diethyl, and *N,N*-dipropyl-*p*-phenylenediamine as initial materials by solvothermal method. The QY was up to 86.0% in ethanol [69]. Wang's group reported that trichrome–tryptophan–sorbitol CDs (TC-WS-CDs) were prepared from natural biocompatible tryptophan and sorbitol by one-pot hydrothermal method for diagnosis of hepatocellular carcinoma (Figure 8.9) [70].

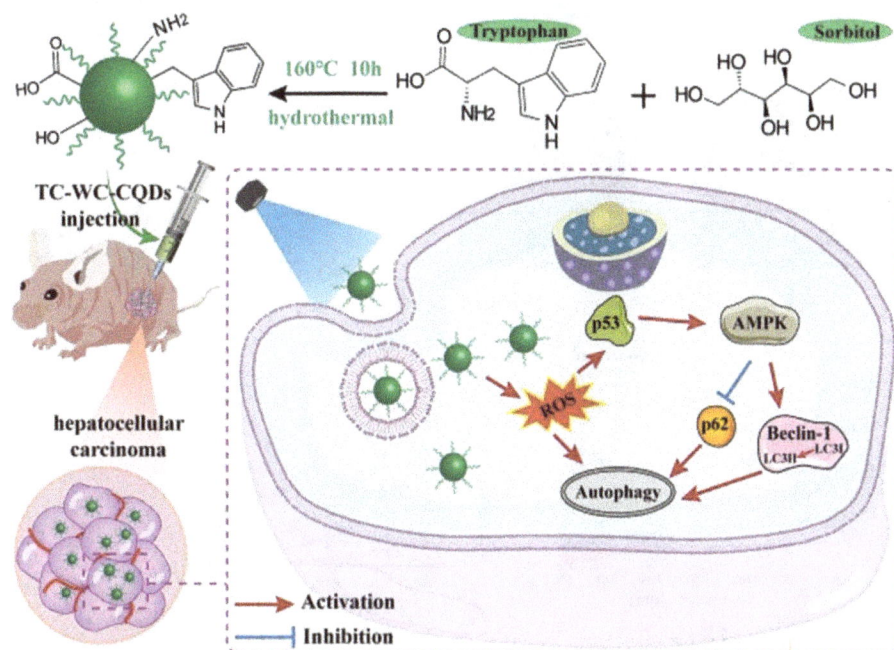

**Figure 8.9:** Synthesis and light-induced antitumor mechanism of trichrome–tryptophan–sorbitol carbon quantum dots [70] (Copyright 2022, Open access).

### 8.3.1.2 Photoacoustic (PA) imaging probe carbon dots

FL imaging probes have various properties such as biocompatibility, ease to preparation, lower toxicity, excellent multicolour emission, hydrophilicity, and low autofluorescence interference to biological samples, but have intrinsic limitation like limited
penetration depth [71]. Therefore, photoacoustic (PA) imaging probe is used as a new
emerging bioimaging method for the diagnosis of cancer with deep tissue penetration
and great resolution [72]. A non-invasive biomedical imaging technique is PA imaging.
When pulsed laser is irradiated to a material, ultrasonic wave (acoustic wave) is
formed, which reconstructs the image of the light when the pulsed laser is irradiated to
a material, an ultrasonic wave (acoustic wave) is formed, which reconstructs the image
of the light [73]. Lee's group was synthesized NIR-absorbing N-doped CDs (N-CDs) for PA
imaging for liver cancer by employing nitric acid as a source of nitrogen and citric acid
as a source of carbon. NIR-N-CDs showed good photostability and absorbance in the
NIR region. Figure 8.10 shows the synthesis of N-CDs and PA imaging [74]. Moreover,
CDs were prepared by using natural biomass parasitic fungus *Hypocrella bambusae* (HB)

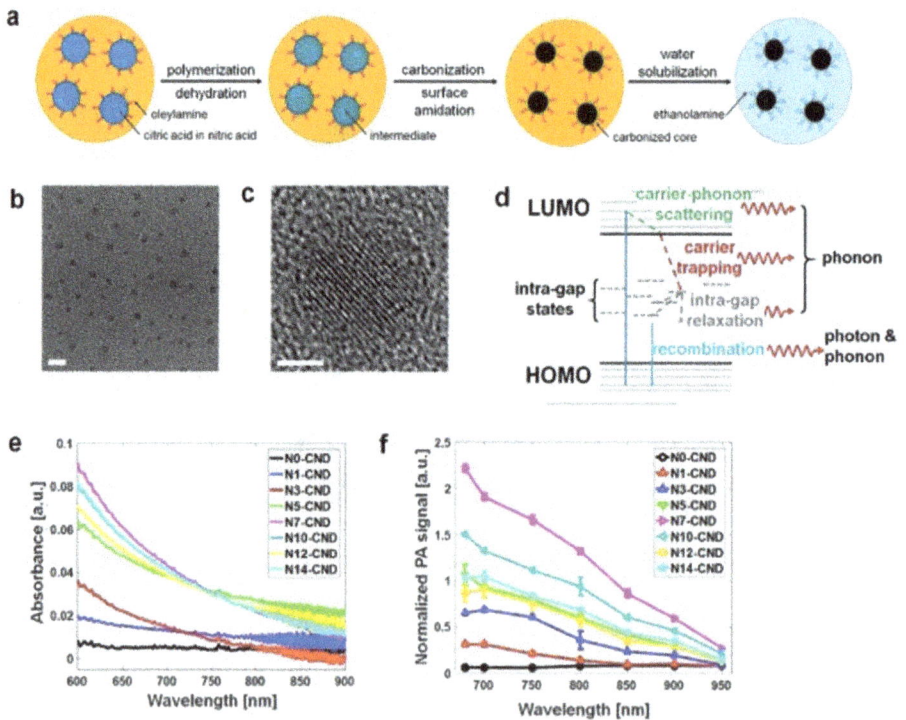

**Figure 8.10:** (a) Synthesis of N-CNDs; (b) TEM image of N-CNDs; (c) partial graphitic structure in the core
of N-CNDs; (d) electronic structure; (e) optical absorption spectrum of CDs; and (f) PA amplitude
spectrum [74] (Copyright 2016 Ivspring International, Open access).

in bamboo as a precursor by solvothermal method. HB-CDs were applied for bimodal FL/PA imaging and PDT/PTT to cancer diagnosis because of their properties such as wide absorption (350–800 nm), red emission (at 610 nm), low toxicity, and good water solubility [75]. Citric acid and urea were used as precursors by Xu and his co-workers for preparing supra-CDs. Supra-CDs were used as contrast agents for PA imaging with good irradiating power densities to 1 W cm$^{-2}$ and photothermal agent for PTT realized under 655 nm laser irradiation [76]. Wu's group also synthesized CDs for PA imaging by using porphyrin for breast cancer ablation [77].

### 8.3.1.3 Magnetic resonance imaging probe carbon dots

The most potent imaging techniques, such as MR imaging, have been primarily utilized in diagnostic imaging. They have also been used to offer morphological, physiological, and even molecular information about the body because of their non-invasive characteristics and high spatial resolution. Hence, many MR imaging tools described cellular and molecular changes in cancer [78]. MR imaging is used to provide good contrast image of soft tissues and verify that tumours have been removed surgically [79]. Many CDs were used as MR imaging probes with incorporation of metal ions and metal-free CDs for the diagnosis of cancer. For example, Du et al. formed gadolinium-doped CDs (Gd@CDs) by one-step hydrothermal method for MR imaging of tumours, in which gadopentetic acid (Gd-DTPA) is used as Gd source and glycine as the surface passivation agent. With a longitudinal relaxivity rate ($r_1$) of 6.45 mM$^{-1}$ s$^{-1}$ and great biocompatibility, Gd-CDs show excellent performance. Gd-CDs have excellent T1 contrast agents and good for radiotherapy of tumours [80]. Similarly, Gd@CDs have been synthesized by using 3,4-dihydroxyhydrocinnamic acid, 2,2′-(ethylenedioxy)bis(ethylamine), and gadolinium chloride via hydrothermal method, as shown in Figure 8.11. DOX@IR825@Gd@CDs have been formed by using doxorubicin (DOX) hydrochloride drug and NIR photothermal agent, IR825, MR imaging for triple-negative breast cancer [81].

Ramos's team also prepared CDs that were combined with nitrogen and lanthanides (such as Gd and Yb) using a microwave-assisted hydrothermal technique for multimodal contrast agent for imaging of cancer with excellent QY (66%) [82]. Gd$^{3+}$ ion has toxicity which is produced by Gd-CDs so for bio safety concerns Wang's group sythesized fluorine and nitrogen co-doped carbon dots with Fe(III) complex. Fe$^{3+}$@F N-CDs were synthesized by using glucose and levofloxacin via microwave-assisted thermal decomposition method. The longitudinal relaxivities ($r_1$) of free Fe$^{3+}$, Fe$^{3+}$@ CD complex, and Fe$^{3+}$@F, N-CD complex were 1.59, 4.23, and 5.79 mM$^{-1}$ s$^{-1}$, respectively. Fe$^{3+}$@F, N-CDs have large coordination constant (1.06 × 10$^7$), high relaxivity rate, low toxicity, good photoluminescence, and less synthesis cost. Therefore, Fe$^{3+}$@F, N-CDs were used for MR imaging probe for cancer diagnosis [83]. Similarly, non-toxic magnetofluorescence Mn-CDs were synthesized by manganese(II) phthalocyanine as a precursor at 180 °C with simultaneous bimodal FL/MR imaging characteristic via solvothermal method.

With a maximal peak at 745 nm and a T1-weighted magnetic resonance relaxivity rate ($r_1$) of 6.97 mM$^{-1}$ s$^{-1}$ for a multifunctional nanotheranostic method, Mn-CDs can be employed as a smart contrast agent for FL [84]. Nimi's group reported zerovalent iron (ZVI)-CDs for MR imaging, which was citrate-stabilized (C@ZVI@CDs). C@ZVI@CDs are of 10 nm size, show paramagnetic properties, and have longitudinal magnetic relaxivity rate of 4.93 mM$^{-1}$ s$^{-1}$. ZVI@CDs were used as multifunctional CDs, C@ZVI@CDs as MR angiogram in vivo, and paramagnetic ZVI-CDs (P@ZVI@CDs) in optical imaging [85]. All the above-mentioned CDs were metal-incorporated CDs. Novel metal-free CDs have been developed to act as safe contrast agents for T1-weighted MR imaging. Metal-free boron-doped CDs were reported for MR imaging of cancer diagnosis by Wang's group. It was formed by applying 4-vinylphenylboronic acid and boric acid as initial materials. B-CDs have a high $r_1$ value of 18.27 mM$^{-1}$ s$^{-1}$, which improves the contrast in in vivo T1-weighted imaging [86].

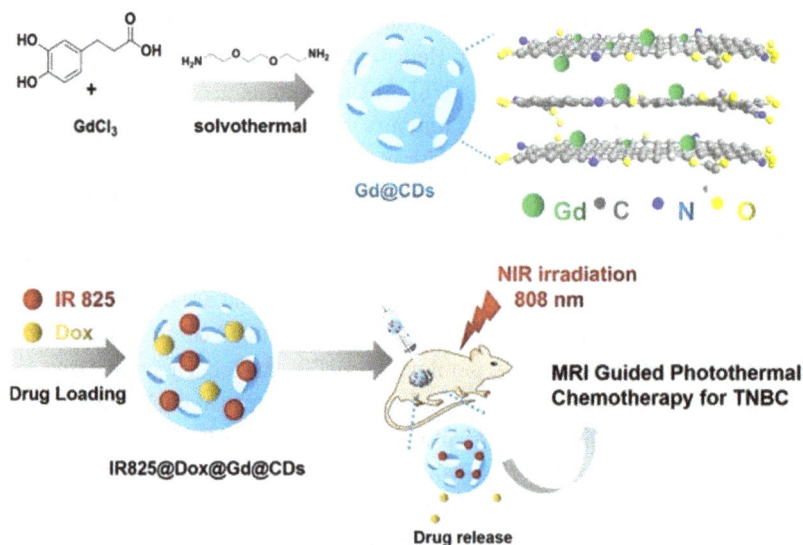

**Figure 8.11:** Gd@CD-based multifunctional carbon nanoplatform design method for triple-negative breast cancer MRI-guided photothermal chemotherapy [81] (Copyright 2021 Hindawi, Open access).

## 8.4 Carbon dots for cancer therapy

Several uses of CDs in cancer therapy were reported. CDs have been used in drug delivery, PDT, PTT, and multimodal cancer therapy.

## 8.4.1 CDs for drug delivery

Chemotherapy drugs have some drawbacks such as poorly soluble in water, some side effects, potential in drug delivery as a result of luminescence and adaptable surface chemistry, high biocompatibility, simple internalization by cells, increased drug solubility, and bioavailability. Wang and co-workers synthesized CDs by using citric acid monohydrate as carbon source and grafting FA on CDs (FA-CDs) with active targeted drug delivery ability. DOX is an anticancer drug, which was loaded in FA-CDs and then formed FA-CD-DOX. It provides outstanding FL imaging capabilities for liver cancer cells (Figure 8.12). FA-CD-DOX has 97% (high) FL QY and high targeting ability than free DOX [87].

Similar to this, Yang's team described nuclear localization signal (NLS) peptide CDs loaded with DOX via an acid-labile hydrogen bond, which were demonstrated to have a higher ability to prevent tumour growth than free DOX. Consequently, NLS-CDs containing DOX serve as promising drug delivery systems for the treatment of cancer [88]. Green fluorescent HP-CDs were prepared by using hyaluronic acid (HA) and polyetherimide by hydrothermal method. Ferrocenylseleno-dopamine (FcDA) is an anticancer drug which is assembled on the surface of HP-CDs, and then formed CDs@FcDA nanoprobe. It has been used for redox-gated cancer cell imaging and drug delivery [89]. Yang's group developed CDs conjugated with β-cyclodextrin (β-CD/CDs), which act as a nanocarrier for DOX (anticancer drug). According to the host–guest chemistry, DOX was bound into the cavity of β-CD with maximum loading ratio of 27.3% at pH 7.4 and released drug at pH 5.0 [90]. Neodymium-doped CDs were synthesized and fabricated with poly-β-CD, which show photoluminescence and magnetic behaviour. CDs were used as a nanocarrier for camptothecin (anticancer drug). The host–guest chemistry is pH dependent [91]. Mathad's group synthesized CDs from neem (*Azadirachta indica*), which was anchored with β-CD for anticancer drug delivery by the host–guest inclusion chemistry. β-CD/CD glassy carbon electrode (β-CD@CDs/GCE) was prepared for the simple, eco-friendly, sensitive, cost-effective determination of anticancer drug (lapatinib). β-CD@CD/GCE is a good efficient electrochemical sensor for cancer treatment [92].

## 8.4.2 CDs for photodynamic therapy

A highly promising and newly developed non-invasive approach of treating cancer that is activated by light is called PDT. PDT can be used alone or in conjunction with ionizing radiation, chemotherapy, or surgery. In PDT, photosensitive species irradiated with specific wavelength light forms reactive oxygen species, such as $^1O_2$, $O_2^{\cdot-}$, $H_2O_2$, and OH radicals, when it comes into contact with oxygen, which can induce cancer cell lysis and death [93, 94]. Some photosensitizing drugs like porphyrin-related drugs are used in PDT and for cancer diagnosis (FL diagnosis). Photosensitizer has some drawbacks such as low selectivity, reduced water solubility, photostability, and photosensitivity. Therefore,

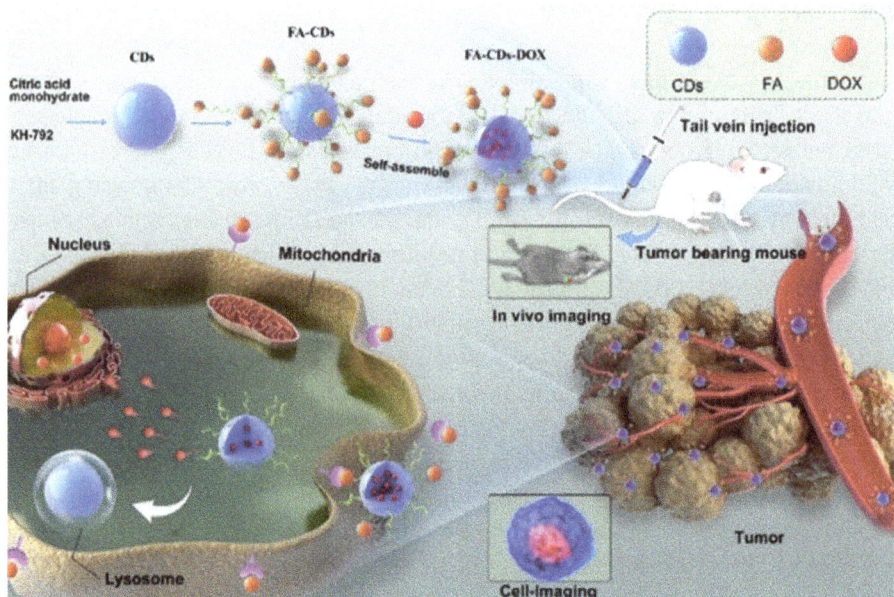

**Figure 8.12:** Imaging and targeted therapy of liver cancer using FA-CD-DOX [87] (Copyright 2020, Elsevier).

various techniques have been tried to blend photosensitizing medications with other carriers, including CDs, gold nanoparticles, and carbon nanotubes or fluorescent CDs used as a photosensitizer [95]. A unique green fluorescent CQD was recently prepared by Yue's team using the natural vitamin riboflavin (VB2) as a photosensitizer. Some characteristics of VB2-CDs include their good water solubility, biocompatibility, and strong singlet oxygen generation capacity, in which VB2-CDs showed bright green FL for PDT, and CDs inhibited the growth of tumours [96].

Similarly, Wu's group reported F-, N-CDs for PDT of hypoxia tumour. F-, N-CDs were irradiated with LED light (400–500 nm, 15 mW cm$^{-2}$), which emit bright green FL and produce hydroxyl radical and superoxide anions ($O_2^-$). F-, N-CDs have shown as bioimaging agents and photosensitizers with excellent water solubility and low cytotoxicity [97]. In 2021, Xu's group developed new R-CDs by using phosphate and methylene blue by hydrothermal method. R-CDs have biocompatibility, photostability, and good singlet oxygen yield of 0.91. So, R-CDs were used in PDT materials [98]. Huang and co-workers synthesized a chlorine e6-conjugated CDs (C-dots-Ce6) with good water solubility, low cytotoxicity, good biocompatibility, good photosensitizer FL detection (PFD), and good photostability. Excellent imaging and tumour homing capabilities for PFD and PDT of cancer in vivo were demonstrated by C-dots-Ce6 [99]. Similarly, CD-chlorin e6-hyaluronate (C-dots-Ce6-HA) was prepared by the interaction between diaminohexane-modified HA and Ce6 carboxylic group, which produced more singlet oxygen as compared to free Ce6 for facile PDT of melanoma skin cancer (Figure 8.13) [100]. Moreover, natural biomass

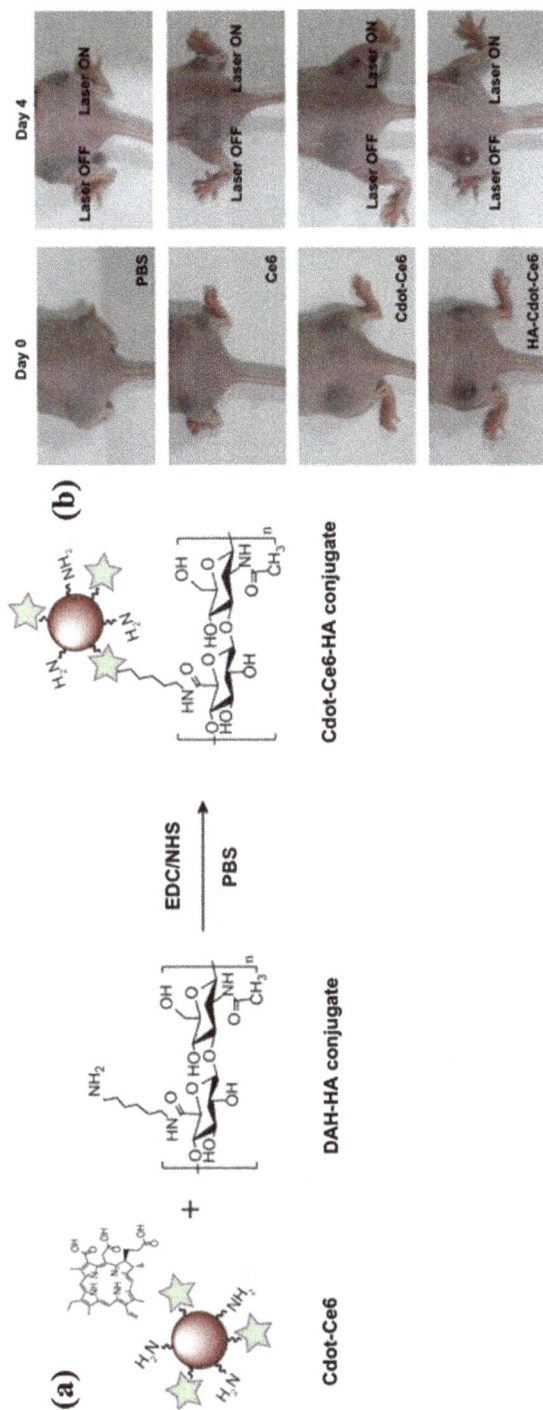

**Figure 8.13:** (a) The synthesis of C-dots-Ce6-HA conjugate using the EDC/NHS chemistry and (b) photo images showing the therapeutic effect of photodynamic therapy [100] (Copyright 2015, Elsevier).

(pheophytin powder) was applied as carbon source for the synthesis of NIR-light-emitting (680 nm) CDs by microwave method. The QY was 0.62 with high singlet oxygen generation. To improve the water solubility, DSPE-mPEG2000 was used with CDs. The CDs were shown as FL/PDT imaging agent and FL images of mice after injection of CD assembly in vivo. These hydrophobic CDs were good for cancer treatment [101].

## 8.4.3 CDs for photothermal therapy

In the last decade, the traditional cancer therapies like radiotherapy and chemotherapy have side effects; for reducing side effects, PTT has been acknowledged as an effective and non-invasive approach for treating cancer. PTT has a photothermal agent with photothermal effect to convert light into heat. This elevated temperature (heat) kills cancer cells by avoiding significant side effects on normal cells [102]. Many nanoparticles have been used as PTT because of their photothermal effects in the NIR region, biocompatibility, and photostability. The commonly used PTT nanoparticles are gold nanorod, nanoshells, nanocage, nanotriangle, nanoflowers, and metallic nanoparticles. Nowadays, CDs have also been used as photothermal agents (Table 8.2) [103]. Sun's group synthesized highly efficient R-CDs by the hydrothermal process, in which the precursor was citric acid. It has unique properties like red emission ($\lambda_{max}$ = 640 nm), respectable QY (22.9%), and high PCE of 43.9% irradiated with 671 nm laser light. Due to their high PCE, R-CDs convert laser energy into heat and are an ideal

Table 8.2: Different precursors, preparation methods and PCE (%) value for light-induced CDs.

| S. no. | Precursors | Preparation method | Wavelength of irradiation (nm) | Photothermal conversion efficiency (%) | References |
|---|---|---|---|---|---|
| 1 | Citric acid | Hydrothermal | 671 | 43.9 | [62] |
| 2 | Polythiophene benzoic acid | Hydrothermal | 635 | 36.2 | [104] |
| 3 | Citric acid and urea | Hydrothermal | 655 | 59.19 | [76] |
| 4 | 1,3,6-Trinitropyrene and polyethylenimine | Molecular fusion | 808 | 38.3 | [105] |
| 5 | Cyanine dye (CyOH) and polyethylene glycol | Solvothermal | 808 | 38.7 | [106] |
| 6 | Nigrosin and manganese acetate | Solvothermal | 808 | 7.6 | [107] |
| 7 | Carbon nanopowder, graphite, grapheme, and carbon nanotube | Hydrothermal | 808 | 39.9, 45.0, 45.7, and 24.3 | [108] |

photothermal cancer therapy agent [62]. CDs with intrinsic theranostic properties were prepared from polythiophene benzoic acid as a precursor. They exhibited that the newly prepared CDs show photodynamic and photothermal effects in 635 nm laser ablation. It has good PCE of up to 36.2% and singlet oxygen generation efficiency of 27%; hence, as a PTT agent for cancer treatment, CDs significantly inhibit tumour development when exposed to NIR laser irradiation [104]. Supra-CNDs were prepared by citric acid and urea as precursors via hydrothermal process. This work exhibits that the prepared supra-CDs are highly efficient in PTT for cancer, due to their high PCE of 59.19% with 655 nm laser, and also visualized by bimodal FL/PA imaging. In vivo PA imaging shows that supra-CDs are injected and aggregated in tumour tissues, and 655 nm laser irradiation inhibits tumour growth [76].

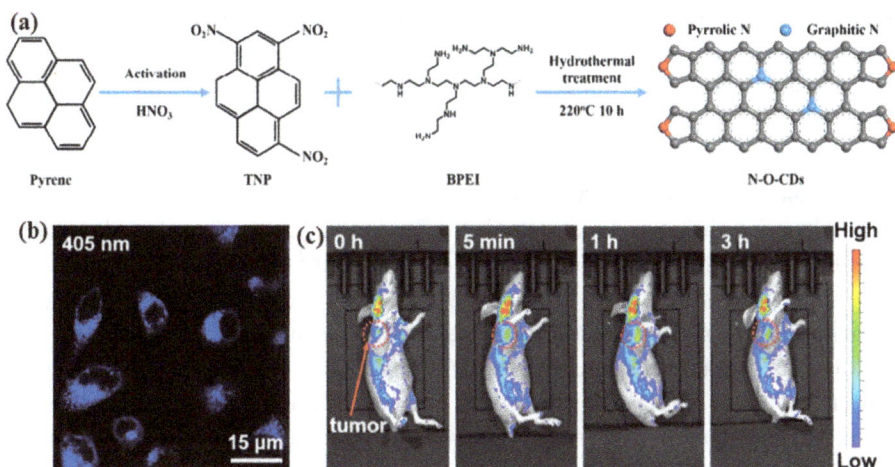

**Figure 8.14:** (a) Synthetic method of N-O-CDs; (b) confocal laser scanning microscopy (CLSM) images; and (c) in vivo photothermal imaging of the tumour site from a mouse intratumour injection with N-O-CDs at 0 h, 5 min, 1 h, and 3 h post-treatment [105] (Copyright 2018, Elsevier).

1,3,6-Trinitropyrene and polyethylenimine were used as precursors by Geng's team to prepare nitrogen- and oxygen-co-doped CDs (N-O-CDs) (Figure 8.14) by the molecular fusion route with high PCE of 38.3% (irradiated with 808 nm laser) and exhibit a good optical absorbance in the NIR region [105]. Zheng et al. formed new CD (CyCDs) PTT agent by using hydrophobic cyanine dye (CyOH) and poly(ethylene glycol) by solvothermal method. CyCDs act as ideal nanotheranostic agents with high PCE of 38.7% (600– 900 nm) for NIR FL imaging and PTT in vitro and in vivo [106]. In addition, non-metals are added to CDs to improve their photophysical and photochemical characteristics. One of the crucial trace elements is Mn and doped into CDs. Manganese-doped nigrosin CDs (Mn-NCDs) have long emission wavelength (653 nm), excellent photothermal effect, and wide absorbance with NIR light. Mn-NCDs show multidimensional theranostic agents such as FL/PA and PTT in vivo with PCE of 7.6%. In PTT, species are irradiated to 808 nm laser light, and the temperature of tumour tissues with Mn-NCDs accumulate and inhibit tumour growth [107]. Recently, Shi et al. prepared CDs from carbon nanopowder, graphite, grapheme, and carbon nanotubes as carbon sources. The PCEs of carbon nanopowder-derived CDs, graphite CDs, graphene CDs, and carbon nanotube-derived CDs were 39.9%, 45.0%, 45.7%, and 24.3%, respectively (irradiated with 808 nm laser light). Due to their ideal PCE, all CDs exhibited high biocompatibility and as photothermal therapy agents [108].

## 8.4.4 CDs for multimodal cancer therapy

Multimodal therapy combines more than one method of treatment. It was created to satisfy stringent therapeutic standards and get over each phototherapeutic technique's inherent drawbacks by combining different therapeutic modalities into a single platform and increased anticancer ability. Gai's group reported organic and inorganic photosensitizers for PDT, four kinds of photothermal materials for PTT, and photolabile groups for chemotherapy. It discussed about PTT/chemo-co-therapy, PTT/PDT co-therapy, radiotherapy-composed co-therapy, and PDT/chemo-co-therapy [109]. Jiao's group designed a multifunctional nanoplatform of AS1411-Gd-CD as a phototheranostic agent. Gd-CDs were formed by using citric acid, urea, ammonium fluoride, and gadolinium chloride hexahydrate by the solvothermal method. For improving tumour targetability, AS1411 aptamers were conjugated with Gd-CDs. AS1411-GD-CDs have multiple properties such as good biocompatibility, physicochemical properties, optical performance, MR contrast agent, and high PCE. Because of these properties, AS1411-Gd-CDs were used as outstanding FL/MR dual-modal imaging nanoplatform and excellent PTT agent in cancer theranostics. Figure 8.15 illustrates the preparation of AS1411-Gd-CDs and FL/MR-guided PTT of tumour [110]. Similarly, multifunctional gold nanorod @ silica-CDs (GNR@$SiO_2$-CDs) act as a phototheranostic agent. It was incorporated with GNR applying $SiO_2$ as a scaffold. CDs act as FL and PDT (635 nm laser light), and GNRs were used as PA imaging and PTT (808 nm laser light) agents. They have higher efficacy in treating cancer than PDT and PTT alone because of their high sensitivity, good spatial resolution of FL/PA imaging, low toxicity, and superior biocompatibility [111].

PTT required high-power laser irradiation. Here, Sun's group developed Ce6-modified R-CDs, in which photosensitizer chlorin-e6 (0.56% mass) was anchored onto amino-rich R-CDs under 671 nm NIR laser. It acts as multimodal FL/PDT/PTT synergistic cancer therapy by reduced irradiation power with PEC of 46% [112]. Scialabba et al. designed an NIR CD PTT agent (CDs-PEG-BT@IT) with biotin and the anticancer drug irinotecan (16–28%) to recognize tumour cells. Using CD-PEG-BT@IT as nanoheaters, tumours can experience localized hyperthermia and a significant release of chemotherapeutic drugs [113]. R-CDs were synthesized by using polythiophene benzoic acid with red light emission at 640–680 nm. It showed dual-multimodal PDT (with singlet oxygen QY of 27%) and PTT (with PCE of 36.2%) effects. By using a single laser, CDs' singlet oxygen QY and

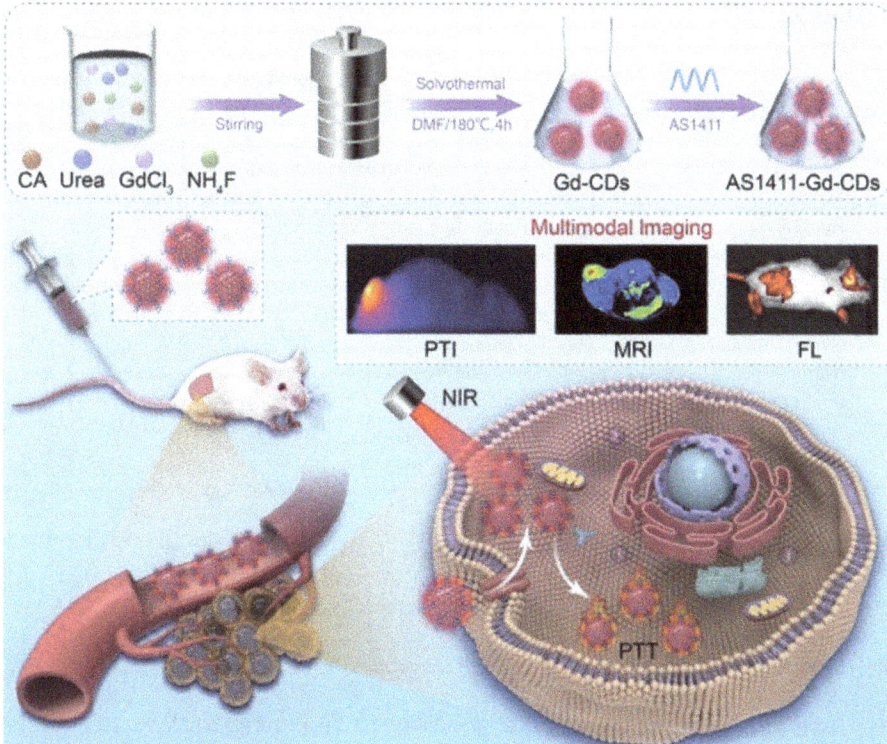

**Figure 8.15:** Schematic representation of the preparation of AS1411-Gd-CDs and FL/MR-guided PTT of tumour [110] (Copyright 2022, Elsevier).

strong PCE qualities allow them to be employed as effective PDT/PTT treatment agents in vivo in addition to being red FL imaging probes [104].

# 8.5 Conclusion

CDs have shown their applications in many biomedical fields. In this chapter, we provided a summary of recent uses of CDs as crucial tools for cancer detection and treatment. We reported their biomedical application for cancer diagnosis (FL, PA, and MR imaging probes) and therapy (drug delivery, PDT, PTT, and multimodal cancer therapy). With the benefit of CD's low cost, high water solubility, biocompatibility, low cytotoxicity, excellent cell permeability, excellent FL imaging sensitivity, photostability, and anticancer impact, we presented in vitro and in vivo research of CDs. Therefore, CDs might be a useful biomedical tool for treating and diagnosing cancer.

# References

[1]     Wang J, Li RS, Zhang HZ, Wang N, Zhang Z, Huang CZ. Highly fluorescent carbon dots as selective and visual probes for sensing copper ions in living cells via an electron transfer process. Biosens Bioelectron. 2017, 97, 157–163.

[2]     Jana J, Lee HJ, Chung JS, Kim MH, Hur SH. Blue emitting nitrogen-doped carbon dots as a fluorescent probe for nitrite ion sensing and cell-imaging. Anal Chim Acta. 2019, 1079, 212–219.

[3]     Hu J, Tang F, Jiang Y, Liu C. Rapid screening and quantitative detection of Salmonella using a quantum dot nanobead-based biosensor. Analyst. 2020, 145, 2184–2190.

[4]     Liu ML, Chen BB, Li CM, Huang CZ. Carbon dots: Synthesis, formation mechanism, fluorescence origin and sensing applications. Green Chem. 2019, 21, 449–471.

[5]     Qin KH, Zhang DF, Ding YF, Zheng XD, Xiang YY, Hua JH, Zhang Q, Ji XL, Li B, Wei YL. Applications of hydrothermal synthesis of Escherichia coli derived carbon dots in in vitro and in vivo imaging and p-nitrophenol detection. Analyst. 2020, 145, 177–183.

[6]     Hu C, Li MY, Qiu JS, Sun YP. Design and fabrication of carbon dots for energy conversion and storage. Chem Soc Rev. 2019, 48, 2315–2337.

[7]     Cao Y, Cheng Y, Sun MT. Graphene-based SERS for sensor and catalysis. Appl Spectrosc Rev. 2021, 58, 1–38.

[8]     Safdie FM, Dorff T, Quinn D, Fontana L, Wei M, Lee C, Cohen P, Longo VD. Fasting and cancer treatment in humans: A case series report. Aging. 2009, 1, 988–1007.

[9]     Couvreur P. Nanoparticles in drug delivery: Past, present and future. Adv Drug Deliv Rev. 2013, 65, 21–23.

[10]    Lucky SS, Soo KC, Zhang Y. Nanoparticles in Photodynamic Therapy. Chem Rev. 2015, 115, 1990–2042.

[11]    Jaque D, Martinez Maestro L, Del Rosal B, Haro-Gonzalez P, Benayas A, Plaza JL, Martin Rodriguez E, Garcia Sole J. Nanoparticles for photothermal therapies. Nanoscale. 2014, 6, 9494–9530.

[12]    Namdari P, Negahdari B, Eatemadi A. Synthesis, properties and biomedical applications of carbon-based quantum dots: An updated review. Biomed Pharmacother. 2017, 87, 209–222.

[13]    Zhu S, Song Y, Wang J, Wan H, Zhang Y, Ning Y, Yang B. Photoluminescence mechanism in graphene quantum dots: Quantum confinement effect and surface/edge state. Nano Today. 2017, 13, 10–14.

[14]    Yuan F, Li S, Fan Z, Meng X, Fan L, Yang S. Shining carbon dots: Synthesis and biomedical and optoelectronic applications. Nano Today. 2016, 11, 565–586.

[15]    Sagbas S, Sahiner N. Carbon dots: Preparation, properties, and application. Nanocarbon Compos. 2019, 651–676.

[16]    Miao S, Liang K, Zhu J, Yang B, Zhao D, Kong B. Hetero-atom-doped carbon dots: Doping strategies, properties and applications. Nano Today. 2020, 33, 100879.

[17]    Liu H, Ye T, Mao C. Fluorescent Carbon Nanoparticles Derived from Candle Soot. Angew Chem Int Ed. 2007, 46, 6473–6475.

[18]    Peng J, Gao W, Gupta BK, Liu Z, Romero-Aburto R, Ge LH, Song LH, Alemany LB, Zhan XB, Gao GH et al. Graphene quantum dots derived from carbon fibers. Nano Lett. 2012, 12, 844–849.

[19]    Zhao P, Yang M, Fan W, Wang X, Tang F, Yang C, Dou X, Li S, Wang Y, Cao Y. Facile one-pot conversion of petroleum asphaltene to high quality green fluorescent graphene quantum dots and their application in cell imaging. Part Syst Charact. 2016, 33, 635–644.

[20]    Gunjal DB, Gurav YM, Gore AH, Naik VM, Waghmare RD, Patil CS, Sohn D, Anbhule PV, Shejwal RV, Kolekar GB. Nitrogen doped waste tea residue derived carbon dots for selective quantification of tetracycline in urine and pharmaceutical samples and yeast cell imaging application. Opt Mater. 2019, 98, 109484.

[21] Sun HJ, Ji HW, Ju EG, Guan YJ, Ren JS, Qu XG. Synthesis of fluorinated and nonfluorinated graphene quantum dots through a new top-down strategy for long-time cellular imaging. Chem Eur J. 2015, 21, 3791–3797.

[22] Soni H, Pamidimukkala PS. Green synthesis of N, S co-doped carbon quantum dots from triflic acid treated palm shell waste and their application in nitrophenol sensing. Mater Res Bull. 2018, 108, 250–254.

[23] Kailasa SK, Ha S, Baek SH, Phan LM, Kim SJ, Kwak K, Park TJ. Tuning of carbon dots emission color for sensing of Fe3+ion and bioimaging applications. Mater Sci Eng C. 2019, 98, 834–842.

[24] Desai SL, Jha SJ, Basu H, Singhal RK, Kailasa SK. Acid oxidation of muskmelon fruit for the fabrication of carbon dots with specific emission colors for recognition of $Hg^{2+}$ ions and cell imaging. ACS Omega. 2019, 4, 19332–19340.

[25] Nair R, Thomas R, Sankar V, Muhammad H, Dong M, Pillai S. Rapid, acid-free synthesis of high-quality graphene quantum dots for aggregation induced sensing of metal ions and bioimaging. ACS Omega. 2017, 2, 8051–8061.

[26] Zhu C, Yang S, Wang G, Mo R, He P, Sun J, Di Z, Kang Z, Yuan N, Ding J et al. A new mild, clean and highly efficient method for the preparation of grapheme quantum dots without by-products. J Mater Chem B. 2015, 3, 6871–6876.

[27] Sun YP, Zhou B, Lin Y, Wang W, Fernando KAS, Pathak P, Meziani MJ, Harruff BA, Wang X, Wang H et al. Quantum-sized carbon dots for bright and colorful photoluminescence. J Am Chem Soc. 2006, 128, 7756.

[28] Ren X, Zhang F, Guo BP, Gao N, Zhang XL. Synthesis of N-doped micropore carbon quantum dots with high quantum yield and dual-wavelength photoluminescence emission from biomass for cellular imaging. Nanomaterials. 2019, 9, 495.

[29] Kang S, Mhin S, Han H, Kim K, Jones JL, Ryu JH, Kang JS, Kim SH, Shim KB. Ultrafast method for selective design of graphene quantum dots with highly efficient blue emission. Sci Rep. 2016, 6, 38423.

[30] Cui L, Ren X, Wang JG, Sun MT. Synthesis of homogeneous carbon quantum dots by ultrafast dual-beam pulsed laser ablation for bioimaging. Mater Today Nano. 2020, 12, 10009.

[31] Zhuo S, Shao M, Lee ST. Upconversion and downconversion fluorescent graphene quantum dots: Ultrasonic preparation and photocatalysis. ACS Nano. 2012, 6, 1059–1064.

[32] Song S, Jang M, Chung J, Jin S, Kim B, Hur SH, Yoo S, Cho YH, Jeon S. Highly efficient light-emitting diode of grapheme quantum dots fabricated from graphite intercalation compounds. Adv Opt Mater. 2014, 2, 1016–1023.

[33] Huang HY, Cui Y, Liu MY, Chen JY, Wan Q, Wen YQ, Wei Y. A one-step ultrasonic irradiation assisted strategy for the preparation of polymer-functionalized carbon quantum dots and their biological imaging. J Colloid Interface Sci. 2018, 532, 767–773.

[34] Park SY, Lee HU, Park ES, Lee SC, Lee JW, Jeong SW, Chi HK, Lee YC, Yun SH, Lee J. Photoluminescent green carbon nanodots from food-waste-derived sources: Large-scale synthesis, properties, and biomedical applications. ACS Appl Mater Interfaces. 2014, 6, 3365–3370.

[35] Li L, Ji J, Fei R, Wang C, Lu Q, Zhang J, Jiang LP, Zhu JJ. A facile microwave avenue to electrochemiluminescent two-color graphene quantum dots. Adv Funct Mater. 2012, 22, 2971–2979.

[36] Yao YY, Gedda G, Girma WM, Yen CL, Ling YC, Chang JY. Magnetofluorescent carbon dots derived from crab shell for targeted dual-modality bioimaging and drug delivery. ACS Appl Mat Interf. 2017, 9, 13887–13899.

[37] Kumawat MK, Thakur M, Gurung RB, Srivastava R. Graphene quantum dots from Mangifera indica: Application in near-infrared bioimaging and intracellular nanothermometry. ACS Sustain Chem Eng. 2017, 5, 1382–1391.

[38] Pires NR, Santos CMW, Sousa RR, De Paula RCM, Cunha PLR, Feitosa JPA. Novel and fast microwave-assisted synthesis of carbon quantum dots from raw cashew gum. J Braz Chem Soc. 2015, 26, 1274–1282.

[39] Ren Q, Ga L, Ai J. Rapid synthesis of highly fluorescent nitrogen-doped graphene quantum dots for effective detection of ferric ions and as fluorescent ink. ACS Omega. 2019, 4, 15842–15848.

[40] Sendao RMS, CristaJ DMA, Afonso ACP, Yuso MVM, Algarra M, Silva JCE, Silva LP. Insight into the hybrid luminescence showed by carbon dots and molecular fluorophores in solution. Phys Chem Chem Phys. 2019, 21, 20919.

[41] Pan D, Zhang J, Li Z, Wu M. Hydrothermal route for cutting graphene sheets into blue-luminescent graphene quantum dots. Adv Mater. 2010, 22, 734–738.

[42] Zhao Y, Wu X, Sun S, Ma L, Zhang L, Lin H. A facile and high-efficient approach to yellow emissive graphene quantum dots from grapheme oxide. Carbon. 2017, 124, 342–347.

[43] Halder A, Godoy-Gallardo M, Ashley J, Feng X, Zhou T, Hosta-Rigau L, Sun Y. One-pot green synthesis of biocompatible grapheme quantum dots and their cell uptake studies. ACS Appl Bio Mater. 2018, 2, 452–461.

[44] Mehta VN, Jha S, Kailasa SK. One-pot green synthesis of carbon dots by using Saccharum officinarum juice for fluorescent imaging of bacteria (Escherichia coli) and yeast (Saccharomyces cerevisiae) cells. Mater Sci Eng C. 2014, 38, 20–27.

[45] Lu WB, Qin XY, Liu S, Chang GH, Zhang YW, Luo YL, Asiri AM, Al-Youbi AO, Sun XP. Economical, green synthesis of fluorescent carbon nanoparticles and their use as probes for sensitive and selective detection of mercury(II). Ions Anal Chem. 2012, 84, 5351–5357.

[46] Liu Y, Zhao Y, Zhang Y. One-step green synthesized fluorescent carbon nanodots from bamboo leaves for copper(II) ion detection. Sens Actuators B Chem. 2014, 196, 647–652.

[47] Essner JB, Kist JA, Polo-Parada L, Baker GA. Artifacts and errors associated with the ubiquitous presence of fluorescent impurities in carbon nanodots. Chem Mater. 2018, 30.

[48] Fan L, Zhu M, Lee X, Zhang R, Wang K, Wei J, Zhong M, Wu D, Zhu H. Direct synthesis of graphene quantum dots by chemical vapor deposition. Part Syst Charact. 2013, 30, 764–769.

[49] Huang K, Lu W, Yu X, Jin C, Yang D. Highly pure and luminescent graphene quantum dots on silicon directly grown by chemical vapor deposition. Part Part Syst Charact. 2016, 33, 8–14.

[50] Kumar S, Aziz S, Girshevitz O, Nessim G. One-step synthesis of N-doped graphene quantum dots from chitosan as a sole precursor using chemical vapor deposition. J Phys Chem C. 2018, 122, 2343–2349.

[51] Zhou J, Sheng Z, Han H, Zou M, Li C. Facile synthesis of fluorescent carbon dots using watermelon peel as a carbon source. Mater Lett. 2012, 66, 222–224.

[52] Praneerad J, Neungnoraj K, In I, Paoprasert P. Environmentally friendly supercapacitor based on carbon dots from durian peel as an electrode. Key Eng Mater. 2019, 803, 115–119.

[53] Sun D, Ban R, Zhang PH, Wu GH, Zhang JR, Zhu JJ. Hair fiber as a precursor for synthesizing of sulfur- and nitrogen-codoped carbon dots with tunable luminescence properties. Carbon. 2013, 64, 424–443.

[54] Wee SS, Ng YH, Ng SM. Synthesis of fluorescent carbon dots via simple acid hydrolysis of bovine serum albumin and its potential as sensitive sensing probe for lead (II) ions. Talanta. 2013, 116, 71–76.

[55] Tian RB, Zhong ST, Wu J, Jiang W, Shen Y, Wang T. Solvothermal method to prepare graphene quantum dots by hydrogen peroxide. Opt Mater. 2016, 60, 204–208.

[56] Liu W, Diao H, Chang H, Wang H, Li T, Wei W. Green synthesis of carbon dots from rose-heart radish and application forFe3+ detection and cell imaging. Sens Actuators B Chem. 2017, 241, 190–198.

[57]   Qian Z, Ma J, Shan X, Feng H, Shao L, Chen J. Highly luminescent N-doped carbon quantum dots as an effective multifunctional fluorescence sensing platform. Chem Eur J. 2014, 20, 2254–2263.
[58]   Lim SY, Shen W, Gao Z. Carbon quantum dots and their applications. Chem Soc Rev. 2015, 44, 362–381.
[59]   Jia Q, Zhao Z, Liang K, Nan F, Li Y, Wang J, Ge J, Wang P. Recent advances and prospects of carbon dots in cancer nanotheranostics. Mater Chem Front. 2020, 4, 449–471.
[60]   Singh G, Kaur H, Sharma A, Singh J, Alajangi HK, Kumar S, Singla N, Kaur IP, Barnwal RP. Carbon based nanodots in early diagnosis of cancer. Front Chem. 2021, 69, 1–12.
[61]   Shen C, Wang J, Cao Y, Lu Y. Facile access to B-doped solid-state fluorescent carbon dots toward light emitting devices and cell imaging agents. J Mater Chem C. 2015, 3, 6668–6675.
[62]   Sun S, Zhang L, Jiang K, Wu A, Lin H. Toward high-efficient red emissive carbon dots: Facile preparation, unique properties, and applications as multifunctional theranostic agents. Chem Mater. 2016, 28, 8659–8668.
[63]   Ding H, Ji Y, Wei JS, Gao QY, Zhou ZY, Xiong HM. Facile synthesis of red-emitting carbon dots from pulp-free lemon juice for bioimaging. J Mater Chem B. 2017, 5, 5272–5527.
[64]   Ding H, Zhou X, Qin B, Zhou Z, Zhao Y. Highly fluorescent near-infrared emitting carbon dots derived from lemon juice and its bioimaging application. J Lumin. 2019, 211, 298–304.
[65]   Liang Q, Ma W, Shi Y, Li Z, Yang X. Easy synthesis of highly fluorescent carbon quantum dots from gelatin and their luminescent properties and applications. Carbon. 2013, 60, 421–428.
[66]   Yang Y, Cui J, Zheng M, Hu C, Tan S, Xiao Y, Yang Q, Liu Y. One-step synthesis of amino-functionalized fluorescent carbon nanoparticles by hydrothermal carbonization of chitosan. Chem Comm. 2012, 48, 380–382.
[67]   Miao X, Qu D, Yang D, Nie B, Zhao Y, Fan H, Sun Z. Synthesis of carbon dots with multiple color emission by controlled graphitization and surface functionalization. Adv Mater. 2018, 30, 1704740.
[68]   Das P, Ganguly S, Agarwal T, Maity P, Ghosh S, Choudhary S, Gangopadhyay S, Maiti TK, Dhara S, Banerjee S, Das NC. Heteroatom doped blue luminescent carbon dots as a nano-probe for targeted cell labeling and anticancer drug delivery vehicle. Mater Chem Phys. 2019, 237, 1–27.
[69]   Jia H, Wang Z, Yuan T, Yuan F, Li X, Li Y, Tan Z, Fan L, Yang S. Electroluminescent warm white light-emitting diodes based on passivation enabled bright red bandgap emission carbon quantum dots. Adv Sci. 2019, 6, 1900397.
[70]   Wang Y, Chen J, Tian J, Wang G, Luo W, Huang Z, Huang Y, Li N, Guo M, Fan X. Tryptophan-sorbitol based carbon quantum dots for theranostics against hepatocellular carcinoma. J Nanobiotechnol. 2022, 20, 1–16.
[71]   Bu L, Shen B, Cheng Z. Fluorescent imaging of cancerous tissues for targeted surgery. Adv Drug Deliv Rev. 2014, 76, 21–38.
[72]   Weber J, Beard PC, Bohndiek SE. Contrast agents for molecular photoacoustic imaging. Nat Methods. 2016, 13, 639–650.
[73]   Zhao Z, Swartchick CB, Chan J. Targeted contrast agents and activatable probes for photoacoustic imaging of cancer. Chem Soc Rev. 2022, 51, 829–868.
[74]   Lee C, Kwon W, Beack S, Lee D, Park Y, Kim H, Hahn SK, Rhee SW, Kim C. Biodegradable nitrogen-doped carbon nanodots for non-invasive photoacoustic imaging and photothermal therapy. Theranostics. 2016, 6, 2196–2208.
[75]   Jia Q, Zheng X, Ge J, Liu W, Ren H, Chen S, Wen Y, Zhang H, Wu J, Wang P. Synthesis of carbon dots from Hypocrella bambusae for bimodal fluorescence/ photoacoustic imaging-guided synergistic photodynamic/ photothermal therapy of cancer. J Colloid Interface Sci. 2018, 526, 302–311.
[76]   Xu G, Bao X, Chen J, Zhang B, Li D, Zhou D, Wang X, Liu C, Wang Y, Qu S. In vivo tumor photoacoustic imaging and photothermal therapy based on supra-(carbon nanodots). Adv Healthc Mater. 2018, 8, 1800995–1801001.

[77]   Wu F, Su H, Cai Y, Wong WK, Jiang W, Zhu X. Porphyrin-implanted carbon nanodots for photoacoustic imaging and in vivo breast cancer ablation. Appl Bio Mater. 2018, 1, 110–117.

[78]   Haris M, Yadav SK, Rizwan A, Singh A, Wang E, Hariharan H, Reddy R, Marincola FM. Molecular magnetic resonance imaging in cancer. J Transl Med. 2015, 13, 1–16.

[79]   Hu Y, Mignani S, Majoral JP, Shen M, Shi X. Nanoscale metal–organic frameworks for therapeutic, imaging, and sensing applications. Chem Soc Rev. 2018, 47, 1874–1900.

[80]   Du F, Zhang L, Zhang L, Zhang M, Gong A, Tan Y, Miao J, Gong Y, Sun M, Ju H, Wu C, Zou S. Engineered gadolinium-doped carbon dots for magnetic resonance imaging-guided radiotherapy of tumors. Biomaterials. 2017, 121, 109–120.

[81]   Jiang Q, Liu L, Li Q, Cao Y, Chen D, Du Q, Yang X, Huang D, Pei R, Chen X, Huang G. NIR-laser-triggered gadolinium-doped carbon dots for magnetic resonance imaging, drug delivery and combined photothermal chemotherapy for triple negative breast cancer. J Nanobiotechnol. 2021, 19, 1–15.

[82]   Ramos DB, Canga JC, Mayo JC, Sainz RM, Encinar JR, Costa-Fernandez JM. Carbon quantum dots codoped with nitrogen and lanthanides for multimodal imaging. Adv Funct Mater. 2019, 29, 1903884–1903894.

[83]   Wang J, Hu X, Ding H, Huang X, Xu M, Li Z, Wang D, Yan X, Lu Y, Xu Y, Chen Y, Morais PC, Tian Y, Zhang RQ, Bi H. Fluorine and nitrogen co-doped carbon dots complexation with Fe(III) as a T1 contrast agent for magnetic resonance imaging. ACS Appl Mater Interfaces. 2019, 11, 1–14.

[84]   Jia Q, Ge J, Liu W, Zheng X, Chen S, Wen Y, Zhang H, Wang P. A magnetofluorescent carbon dot assembly as an acidic H2O2-driven oxygenerator to regulate tumor hypoxia for simultaneous bimodal imaging and enhanced photodynamic therapy. Adv Mater. 2018, 30, 1706090–1706100.

[85]   Nimi N, Saraswathy A, Nazeer SS, Francis N, Shenoy SJ, Jayasree RS. Multifunctional hybrid nanoconstruct of zerovalent iron and carbon dots for magnetic resonance angiography and optical imaging: An in vivo study. Biomaterials. 2018, 171, 46–56.

[86]   Wang H, Revia R, Wang K, Kant RJ, Mu Q, Gai Z, Hong K, Zhang M. Paramagnetic properties of metal-free boron-doped graphene quantum dots and their application for safe magnetic resonance imaging. Adv Mater. 2017, 29, 1605416–1605422.

[87]   Wang S, Chen L, Wang J, Du J, Li Q, Gao Q, Gao Y, Yu S, Yang Y. Enhanced-fluorescent imaging and targeted therapy of liver cancer using highly luminescent carbon dots-conjugated foliate. Mat Sci Eng C. 2020, 116, 111233–111244.

[88]   Yang L, Wang Z, Wang J, Jiang W, Jiang X, Bai Z, H Y, Jiang J, Wang D, Yang L. Doxorubicin conjugated functionalizable carbon dots for nucleus targeted delivery and enhanced therapeutic efficacy. Nanoscale. 2016, 8, 6801–6809.

[89]   Lu X, Wang X, Li A, Zhou T, Zhang L, Qu J, Mao Z, Gu X, Zhang X, Jing S. Ferrocenylseleno-dopamine functionalized carbon dots for redox-gated imaging and drug delivery in cancer cells. Dyes Pigm. 2022, 205, 110586–110598.

[90]   Yang T, Huang JL, Wang YT, Zheng AQ, Shu Y, Wang JH. β-Cyclodextrin-decorated carbon dots serve as nanocarriers for targeted drug delivery and controlled release. Chem Nano Mat. 2019, 5, 479–487.

[91]   Alexander A, Pillai AS, Manikantan V, Varalakshmi GS, Akash BA, Enoch IVMV. Magnetic and luminescent neodymium-doped carbon dot–cyclodextrin polymer nanocomposite as an anticancer drug-carrier. Mater Lett. 2022, 313, 131830–131846.

[92]   Mathad AS, Seetharamappa J, Kalanur SS. β-Cyclodextrin anchored neem carbon dots for enhanced electrochemical sensing performance of an anticancer drug, lapatinib via host-guest inclusion. J Mol Liq. 2022, 350, 1–12.

[93]   Li Q, Li Y, Min T, Gong J, Du L, Phillips DL, Liu J, Lam JWY, Sung HHY, Williams ID, Kwok RTK, Ho CL, Li K, Wang J, Tang BZ. Time-dependent photodynamic therapy for multiple targets: A highly efficient

AIE-active photosensitizer for selective bacterial elimination and cancer cell ablation. Angew Chem Int Ed. 2019, 58, 2–10.

[94] Robertson CA, Hawkins ED, Abrahamse H. Photodynamic therapy (PDT): A short review on cellular mechanisms and cancer research applications for PDT. J Photochem Photobiol B Biol. 2009, 96, 1–8.

[95] Berg K, Selbo PK, Weyergang A, Dietze A, Prasmickaite L, Bonsted A, Engesaeter B, Angell-Petersen E, Warloe T, Frandsen N. Porphyrin-related photosensitizers for cancer imaging and therapeutic applications. J Microsc. 2005, 218, 133–147.

[96] Yue J, Li L, Jiang C, Mei Q, Dong WF, Yan R. Riboflavin-based carbon dots with high singlet oxygen generation for photodynamic therapy. J Mater Chem B. 2021, 9, 7972–7978.

[97] Wu X, Xu M, Wang S, Abbas K, Huang X, Zhang R, Tedesco AC, Bi H. F,N-Doped carbon dots as efficient Type I photosensitizers for photodynamic therapy. Dalton Trans. 2022, 51, 2296–2303.

[98] Xu Y, Wang C, Ran G, Chen D, Pang Q, Song Q. Phosphate-assisted transformation of methylene blue to red-emissive carbon dots with enhanced singlet oxygen generation for photodynamic therapy. Appl Nano Mater. 2021, 5, 4820–4828.

[99] Huang P, Lin J, Wang X, Wang Z, Zhang C, He M, Wang K, Chen F, Li Z, Shen G. Light-triggered theranostics based on photosensitizer-conjugated carbon dots for simultaneous enhanced fluorescence imaging and photodynamic therapy. Adv Mater. 2012, 24, 5104–5110.

[100] Beack S, Kong WH, Jung HS, Do IH, Han S, Kim H, Kim KS, Yun SH, Hahn SK. Photodynamic therapy of melanoma skin cancer using carbon dot-chlorin e6-hyaluronate conjugate. Acta Biomater. 2015, 26, 295–305.

[101] Wen Y, Jia Q, Nan F, Zheng X, Liu W, Wu J, Ren H, Ge J, Wang P. Pheophytin derived near-infrared-light responsive carbon dot assembly as a new phototheranotic agent for bioimaging and photodynamic therapy. J Chem Asian. 2019, 14, 2162–2168.

[102] Chen J, Ning C, Zhou Z, Yu P, Zhu Y, Tan G, Mao C. Nanomaterials as photothermal therapeutic agents. Prog Mater Sci. 2019, 99, 1–26.

[103] Nocito G, Petralia S, Malanga M, Béni S, Calabrese G, Parenti R, Conoci S, Sortino S. Biofriendly route to near-infrared active gold nanotriangles and nanoflowers through nitric oxide photorelease for photothermal applications. ACS Appl Nano Mater. 2019, 2, 7916–7923.

[104] Ge J, Jia Q, Liu W, Lan M, Zhou B, Guo L, Zhou H, Zhang H, Wang Y, Gu Y. Carbon dots with intrinsic theranostic properties for bioimaging, red-light-triggered photodynamic/photothermal simultaneous therapy in vitro and in Vivo. Adv Healthc Mater. 2016, 5, 665–675.

[105] Geng B, Yang D, Pan D, Wang L, Zheng F, Shen W, Zhang C, Li X. NIR-responsive carbon dots for efficient photothermal cancer therapy at low power densities. Carbon. 2018, 134, 153–162.

[106] Zheng M, Li Y, Liu S, Wang W, Xie Z, Jing X. One pot to synthesize multifunctional carbon dots for near infrared fluorescence imaging and photothermal cancer therapy. ACS Appl Mater Interfaces. 2016, 8, 23533–23541.

[107] Yang C, Li Y, Yang Y, Tong R, He L, Long E, Liang L, Cai L. Multidimensional theranostics for tumor fluorescence imaging, photoacoustic imaging and photothermal treatment based on manganese doped carbon dots. J Biomed Nanotechnol. 2018, 14, 1590–1600.

[108] Shi W, Han Q, Wu J, Ji C, Zhou Y, Li S, Gao L, Leblanc RM, Peng Z. Synthesis mechanisms, structural models, and photothermal therapy applications of top-down carbon dots from carbon powder, graphite, graphene, and carbon nanotubes. Int J Mol Sci. 2022, 23, 1456–1472.

[109] Gai S, Yang G, Yang P, He F, Lin J, Jin D, Xing B. Recent advances in functional nanomaterials for light triggered cancer therapy. Nano Today. 2018, 19, 146–187.

[110] Jiao M, Wang Y, Wang W, Zhou X, Xu J, Xing Y, Chen L, Zhang Y, Chen M, Xu K, Zheng S. Gadolinium doped red-emissive carbon dots as targeted theranostic agents for fluorescence & MR imaging guided cancer phototherapy. Chem Eng J. 2022, 440, 135965–135977.

[111] Jia Q, Ge J, Liu W, Liu S, Niu G, Guo L, Zhang H, Wang P. Gold nanorod@silica-carbon dots as multifunctional phototheranostics for fluorescence and photoacoustic imaging guided synergistic photodynamic/photothermal therapy. Nanoscale. 2016, 8, 13067–13077.

[112] Sun S, Chen J, Jiang K, Tang Z, Wang Y, Li Z, Liu C, Wu A, Lin H. Ce6-modified carbon dots for multimodal-imaging-guided and single-NIR-laser-triggered photothermal/photodynamic synergistic cancer therapy by reduced irradiation power. ACS Appl Mater Interfaces. 2019, 11, 5791–5803.

[113] Scialabba C, Sciortino A, Messina F, Buscarino G, Cannas M, Roscigno G, Condorelli G, Cavallaro G, Giammona G, Mauro N. Highly homogeneous biotinylated carbon nanodots: Red-emitting nanoheaters as theranostic agents toward precision cancer medicine. ACS Appl Mater Interfaces. 2019, 11, 19854–19866.

Mahdie Matin, Mahtab Mirhoseinian, Alireza Alikhanian,
Golnar Bayatani, Mohammad Nazari Montazer, Mohammad Mahdavi,
Burak Tüzün*and Parham Taslimi

# Chapter 9
# Carbon dots in photodynamic therapy

**Abstract:** Carbon dots (CDs) as a subclass of carbon-based nanoparticles were first discovered in 2004 and attracted the attention of scientists worldwide to their extinguish physiochemical properties; nowadays CDs have a broad range of applications including bioimaging, photocatalysis, fluorescent ink, sensors, lasers, and LED. In this chapter we will have a look into the synthesis, structure, and applications of CDs on biomedical science such as bioimaging, biosensors, drug and gene delivery, and photodynamic therapy. Photodynamic therapy is the method to transfer light energy to its surrounding using photosensitizer structure trials, which showed meaningful response to this approach in many types of malignancies, CDs could be a suitable option to use as photosensitizers.

**Keywords:** Carbon dots, Photodynamic therapy, Cancer, Nanoparticles, Biomedical sciences

## 9.1 Introduction

Carbon dots (CDs) are a subclass of carbon-based nanoparticles that were first discovered from arc-discharge soot during the purification of single-wall nanotubes in 2004. CDs immediately attracted significant attention due to their nontoxic (or less toxic), water-soluble, highly fluorescent, photoluminescent properties [1–3]. Carbon-based materials have unique optical properties, including acceptable biocompatibility and low toxicity. Carbon nanotube, fullerene, and CDs are the most famous carbon-based materials [35]. In this chapter, we focus on CDs and explain some researches that

---
*Corresponding author: Burak Tüzün, Plant and Animal Production Department, Technical Sciences Vocational School of Sivas, Sivas Cumhuriyet University, Sivas, Turkey,
e-mail: theburaktuzun@yahoo.com, http://orcid.org/0000-0002-0420-2043
Mahdie Matin, Mahtab Mirhoseinian, Alireza Alikhanian, Golnar Bayatani,
Mohammad Nazari Montazer, Mohammad Mahdavi, Endocrinology and Metabolism Research Center, Endocrinology and Metabolism Clinical Sciences Institute, Tehran University of Medical Sciences, Tehran, Iran
Parham Taslimi, Department of Biotechnology, Faculty of Science, Bartin University, 74100 Bartin, Turkey

https://doi.org/10.1515/9783110799958-009

have been reported in the past years. The therapeutic method with CDs and other carbon-based materials is helpful in the cancer therapy as treatments for skin cancer, bladder cancer [36], lung cancer [37], and other diseases like covid_19 [38] and atherosclerosis [37]. As mentioned, CDs can produce reactive oxygen species (ROS) which is an essential factor in photodynamic therapy. CDs have little use in photobleaching, unlike fluorescent intensity which is more practical. The CDs have excellent $^1O_2$ quantum yield, dazzling fluorescence, minimal toxicity, and good dispersity. The longer the irradiation time, the greater the concentration-dependent cytotoxicity of CDs. There are drawbacks employing CDs in photodynamic treatment, including limited exclusivity and a short-effective lifespan for ROS synthesis.

## 9.1.1 Structure and application of carbon dot

### 9.1.1.1 Structure

Their size is reported to be less than 10 nm, mainly in the range of 2 to 8 nm, which facilitates their use in biological applications. However, the size value of CDs can be adjusted by different nanocomposites; for example, the diameter of CDs coated by hydroxyl groups was reported to be 3.1 ± 0.5 nm [2, 3].

CDs consist of clusters of carbon (53.93%) with various other atoms such as nitrogen (1.30%), oxygen (40.33%), hydrogen (2.56%), and some doped elements such as sulphur and phosphorus [3, 5]. The position of heteroatom doping has been studied in CDs, not just for the surface structure but also hypothesized in the core. Some organic functioning groups are also believed to be in the core, such as carbonyls or different kinds of nitrogen-containing groups [1].

Nuclear magnetic resonance (NMR) measures showed that carbon atoms in C-dots were an amorphous core of pure $sp^3$ carbon or $sp^2$ hybridized with unsaturated $sp^3$ carbon atoms, implying that C-dots are conjugated systems [1, 5].

Most CDs are hydrophilic with good solubility because the oxygen-containing functioning groups emanated from precursors or are produced during synthesis. Nevertheless, hydrophobic CDs have been prepared through partially carbonizing hydrophobic precursors and surface modification by hydrophobic molecules. Lately, it has been shown that hydrophilic CDs can be made hydrophobic following the covalent attachment of dodecyl amine, as revealed by the spontaneous transferring of CDs from water into a toluene phase [2].

Due to the diverse structure of CDs and their varied cores, we cannot assign a single structure to them, and each CD has a different structure, so here we discuss the structure of some of the CDs on which research has been completed. A quasi-spherical morphology defines CDs. A pure carbon core for CDs would be obtained from different structures may be either a collection of 2D structures (e.g., graphitic) or a structure with 3D bonding (e.g., diamond-like) or an amorphous structure [1–3]. Furthermore, based on what groups are used as CD starting materials, it is possible to have different morphologies, for example:

- If a banana is the starting material, CDs will have a mean size of 3 nm; are spherical-shape; and include carbon, oxygen, and potassium. Also, they contain $sp^3$ carbon groups and hydroxyl connected carbon, which made the interlayer spacing of CDs to be 0.42 nm, which is higher than the interlayer spacing of graphite (0.33); thus, these CDs are crystalline in nature.
- If food caramels and orange peel wastes are the starting material, CDs will be amorphous and has no crystalline nature as the XRD displayed a broad amorphous peak.
- If graphite is the starting material, CDs will have a diamond-like structure.
- If the CDs derive from multiwalled carbon nanotubes by electrochemical oxidation, will show graphitic nature [3].

The CDs have well-defined chemical structures and surface morphology, additionally chiral performances. These properties are mainly responsible for the various chemical and physical properties.

The three substantial and beneficial CDs in biomedical applications are Y-CD (yellow CDs), B-CD (black CDs), and CND (carbon nitride dots). For this reason, we also review their structural properties. CND and Y-CD are very close in size but have different optical properties, which lead to different structural models. Based on the results of AFM (Atom force microscope) and TEM (Transmission microscope), it seem that these three CDs have a spherical shape like the rest of the CDs, and the results of Raman spectroscopy, ESR (Erythrocyte sedimentation rate), and XRD (X-Ray diffraction analysis) suggested an amorphous structure for B-CD and CND and a crystalline structure for Y-CD [1].

## 9.1.2 Application

C-dots have various applications in broad fields like bioimaging, photocatalysis, fluorescent ink, sensors, lasers, and LED, due to their advantages such as low cost, ease in preparation, fluorescence with photobleaching, and biocompatibility [5, 8]. Here we discuss their biological applications.

### 9.1.2.1 Bioimaging

Bioimaging refers to capturing images of biological material by treating it with a fluorescent molecule. The use of quantum dots (QDs) such as CdSe in optical imaging and particular sensing of toxic metals in vivo and in vitro conditions has been reported, but it is a less explored area of optical sensing, and they have imposed severe health and environmental risks. Highly fluorescent CDs that are synthesized from natural sources with good properties such as solubility, high chemical, physical and optical stability, low toxicity, and the presence of many hydrophilic surface functional groups can overcome the limitations of organic dyes and other synthetic fluorophores; They can serve as a biocompatible alternative to QDs in bioimaging applications (both in vitro and in vivo) [4] [5, 9].

### 9.1.2.1.1 In vitro imaging

C-dots have been used for taking fluorescence images of mammalian cells, plant cells, and microorganisms, and for long-term cell tracking due to their fluorescent properties and excellent optical stability [8].

Ya-Ping Sun et al. imaged *Escherichia coli* ATCC 25 922 cells incubated with PEG1500N-passivated C-dots using a confocal microscope, showing cyan, green, and red colours in the whole-cell region each using 475, 530, and 560 nm long-pass filters. Cell bioimaging results of mammalian Caco-2 cells, Ehrlich ascites carcinoma cells (EAC) from mice, pig kidney cell line (LLC-PK1 cells), murine P19 progenitor cells, and MG-63 cells showed that when the CDs entered cells, they were mainly localized in the cytoplasm. Using water-dispersible CDs that use green tea as starting material, images of MCF-10A, MCF-7, and MDA-MB-231 were taken, which CDs localized on the membrane and cytoplasm. Although CDs are mostly placed in the cytoplasm in biological imaging, carbon dots prepared by electrochemical method from ascorbic acid, *m*-phenylenediamine, and L-cysteine are suitable for imaging nuclear areas in fixed and living cells [5, 8].

### 9.1.2.1.2 In vivo imaging

There are many reports on the use of CDs for fluorescence imaging in vivo, which indicate that this method has no or little toxicity to cells and can effectively penetrate the skin and tissue of mice. Bioimaging of mice has obtained after injection of red fluorescent C-dots with a wavelength greater than 600 nm when excited at a wavelength of 535 nm, indicating their penetration. Also, CDs functionalized with oligomeric poly ethylene glycol (PEG) inside the mouse body was evaluated, and the results showed that these fluorescent dots are not toxic to the selected cell lines [5, 8].

### 9.1.2.2 Biosensor

Biosensors are devices used to analyse organic and inorganic molecules of living organisms. Optical biosensors work by converting intangible information about target analytes into detectable optical signals. Compared with traditional nanomaterials, CDs have various biological advantages for detecting metal ions, biological pH, sugars, proteins, enzymes, and nucleic acid, due to their excellent properties, biocompatibility, and luminescence properties [5, 10].

When C-dots interact with metal ions, their fluorescence can be quenched through electron transfer or the inner filter effect. This property detects $Hg^{2+}$, $Ag^+$, $Fe^{3+}$, and $Cu^{2+}$ ions. However, it should be noted that the detection of each ion was achieved by conjugating the C-spots with corresponding ligands [5, 8]. Moreover, C-dots can detect the physiological pH of living cells and tissues. C-dots-TPY probe, which emits fluorescent light in the visible region with the excitation of two photons at a wavelength of 800 nm, monitors the pH gradient in the range of 6.0–8.5. It was successfully employed in living cells and tumour tissues [5].

Gaikwad et al. presented a practical solution for directly detecting fungal spores from the environment. For the first time, they used CD thin film to detect fungal spores in the environment using fluorescence spectroscopy, which shows high sensitivity for detecting fungal spores [11].

C-dots-based fluorescence resonance energy transfer (FRET) ratio metric fluorescent sensor was developed for detecting $H_2S$ in aqueous media and serum, as well as inside living cells. They used C-dots anchored with naphthalimide-azide served as an energy donor. The C-dots-naphthalimide-azide sensor showed a detection limit of 10 nm, the lowest among fluorescent $H_2S$ sensors. Furthermore, it showed a significant shift (190 nm) between donor excitation and acceptor emission, which could exclude any influence of excitation backscattering effects on fluorescence detection [5].

### 9.1.2.3 Disease-detection system

The importance of fluorescent nanoprobes in biomedical research and practice has been rapidly increasing due to recent advances in fluorescence microscopy, laser technologies, and nanotechnology. C-dots are particularly suitable for biological sample analysis [5]. For example, to detect diseases related to the abnormal expression and distribution of SA (surface area) on the cell surface or in body fluids, such as various cases of cancer and disease, as well as heart and neurological diseases, the fluorescence sensor caused by C-dot functionalized with boronic acid can be used as an alternative analytical tool to investigate diseases without the need for an instrument [12]. Also, for the detection and quantification of dopamine hydrochloride, serotonin, and glutamic acid as potential molecular biomarkers for Alzheimer's disease and other neuro-related diseases, they used highly fluorescent and spherical (water-soluble) C-dots, formed using citric acid, as a MALDI-MS matrix [13].

### 9.1.2.4 Gene and drug delivery system

Gene therapy, involving the delivery of appropriate exogenous genes (DNA or RNA) into cells and their expression for specific enhancing or suppressing effects, is a good cure for life-threatening diseases. Moreover, imaging is indispensable in tracing, diagnosis, treatment, and confirmation of therapy, promoting the development of bio-imaging and simultaneous disease diagnosis and therapy, in both gene and drug delivery systems. As mentioned before, C-dots are becoming a competent candidate for imaging-concomitant gene and drug delivery due to their excellent fluorescence performance, trim sizes, low cytotoxicity, and good transmembrane ability [5].

Yu-Fen Wu et al. synthesized a multifunctional theranostic nano agent, fc-rPEI-Cdots/siRNA, which absorbs at 360 nm and emits at 460 nm, the wavelength of blue light. Their size is 143 nm, and they disperse well in an aqueous solution. Also, their positive surface charge can be complexed with negatively charged siRNAs; therefore, it acts as a siRNA carrier and releases siRNA in a reducing environment. Enhanced accumulation of fc-rPEI-Cdots was observed in lung cancer cells compared to normal fibroblasts. The viability of H460 treated with fc-rPEI-Cdots/ pooled siRNA complex for 3 days is significantly reduced to nearly 30%. In conclusion, our novel theranostic fc-rPEI-Cdots/ siRNA nanoagents have the potential for lung cancer targeting and treatment [14]. In another report, C-dots were used as a visual tool to monitor the association and dissociation of polymeric carrier/plasmid DNA (pDNA) complex during transfection in a non-labelled manner, providing an efficient strategy to study the mechanism of polymer-mediated pDNA delivery [5].

Most of the recent pharmaceutical agents suffer from severe secondary side effects due to systemic delivery, which could potentially be avoided by using tissue-specific drug delivery systems. Various drug delivery systems, such as lipid- or polymer-based nanoparticles, have been designed to improve drug pharmacological and therapeutic properties. For example, we expect efforts to develop novel bone-targeting C-dots biomaterials to benefit the treatment of skeletal diseases such as osteoporosis potentially [5, 15].

## 9.2 Carbon dots synthesis

In general, carbon dots synthesis strategies are categorized into "top-down" and "bottom-up" [16]. Top-down methods rely on breaking down macro-scale materials for producing carbon nanoparticles [17–19]. On the other hand, bottom-up methods include polymerization and carbonization of molecular precursors [20].

The top-down strategies consist of ultrasonic synthesis, chemical exfoliation, electrochemical oxidation, arc-discharge, and laser ablation. The bottom-up methods include microwave synthesis, hydrothermal method, solvothermal method, pyrolysis/carbonization, and chemical vapour deposition (Figure 9.1) [16].

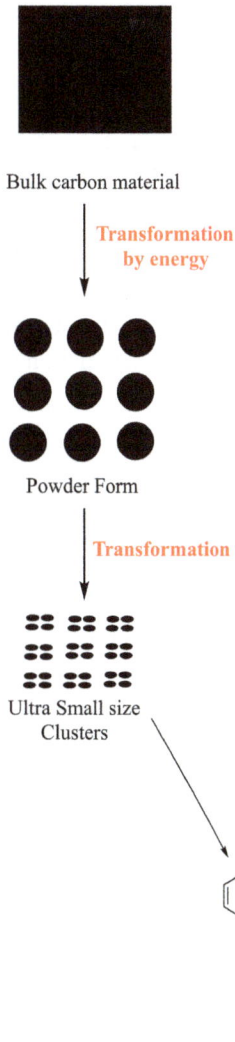

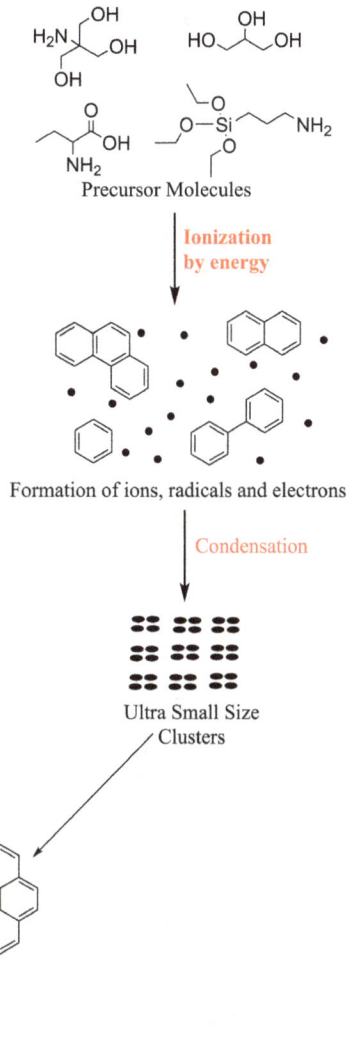

**Figure 9.1:** Main synthesis strategies of CDs. Bottom up technic: CDs are synthesized from smaller carbon precursor via applying energy (electrochemical/chemical, thermal, and laser). The source molecules make CDs by ionization, dissociation, evaporation, and then condensation. Top-down technic: larger carbon structures transform to ultra-small fragments and then CDs, via applying energy (thermal, mechanical, chemical, and ultrasonic) [20].

## 9.2.1 Top-down approaches

### 9.2.1.1 Chemical exfoliation

Chemical exfoliation is an easy way for large production of high-quality CDs without complex equipment. Precursors include carbon fibers, graphene oxide, and carbon nanotubes that separate by strong acids or oxidizing agents [16].

### 9.2.1.2 Laser ablation

Laser ablation is a short period and a simple operation [16]. Researchers prepared CDs from toluene and graphite powders by this technique. The size of the CDs and photoluminescence properties can be controlled by laser furnace, spot size, and irradiation time; for example, smaller CDs can be made by expanding the irradiation time [21, 22].

### 9.2.1.3 Ultrasonic treatment

In this convenient technique, large carbon materials collapse through high energy ultrasonic sound waves. Researchers fabricated CDs by ultrasonic treatment using oligomer-polyamide resin, ascorbic acid, and ammonia as carbon source [23, 24].

## 9.2.2 Bottom-up approaches

### 9.2.2.1 Microwave synthesis

This is a green and cost-effective strategy that is widely used to synthesize CDs in less time and can provide uniform heat to form CDs.

### 9.2.2.2 Pyrolysis/carbonization

Pyrolysis is a great method for making fluorescent CDs via using macroscopic carbon structures as precursors. Advantages of this strategy are short reaction time, low cost, easy operation, solvent-free approaches, and scalable production. Four critical processes consist of heating, dehydration, degradation, and carbonization at high temperature. Carbon precursors are converted into carbon nanoparticles using high concentration alkali or acid in the pyrolysis process [16].

### 9.2.2.3 Hydrothermal

The advantages of hydrothermal method are low cost and non-toxicity and in comparison, to other synthetic strategies, this method is a simple approach for making CQDs (Carbon quantum dots); generally, the water solution of mixtures surrounded with Teflon in an oven and reacted hydrothermally at high pressure and high temperature [16].

### 9.2.2.4 Solvothermal

This fabrication strategy has the advantage of low cost and requires simple equipment. This method is different from the hydrothermal method and the water solution is replaced with one or more solvents enclosed with Teflon equipped with a steel autoclave. The reaction of solvent and raw carbon source mixture occurs at high pressure and high temperature [16, 25].

### 9.2.2.5 Chemical vapour deposition

Chemical vapour deposition (CVD) method has been widely investigated in recent years. In the CVD method, the final product size can be determined by adjusting these parameters including carbon source, growth time, hydrogen ($H_2$) flow rate, and substrate temperature in Table 9.1 [16].

**Table 9.1:** Various methods of carbon dots synthesis [16].

| Synthetic methods | | Merits | Demerits |
|---|---|---|---|
| Top-down | Chemical exfoliation | Most accessible, various sources | Harsh conditions, drastic processes, multiple-steps, poor control over sizes |
| | Laser ablation | Fast, effective, highly tunable | Low quantum yield, poor control over sizes, modification is necessary |
| | Ultrasonic-assisted treatment | Easy operation | Instrumental wastage, high energy cost |
| Bottom-up | Microwave synthesis | Fast, scalable, inexpensive, eco-friendly | Poor control over sizes |
| | Hydrothermal | Inexpensive, eco-friendly, non-toxic | Poor control over sizes |
| | Solvothermal | Inexpensive, eco-friendly, non-toxic | Poor control over sizes |

**Table 9.1** (continued)

| Synthetic methods | Merits | Demerits |
|---|---|---|
| Pyrolysis/carbonization | Solvent-free, low-cost, large-scale production | Non-uniform size distribution |
| Chemical vapour deposition | Controllable morphology and size, high yield | Complicated operation, high cost |

# 9.3 Photodynamic therapy

Photodynamic therapy also known as PDT is an attractive, generally safe, alternative treatment for both oncological and non-oncological diseases, which is based on selective sensitization of various tissues to light [26–28]. PDT operates on a two-stepped procedure of locally or systematically applied, tumour-localizing photosensitizing agent (PS), which then is activated by a specific wavelength and intensity of light, illuminated through different fibers and endoscopes [26, 28]. Accumulation of light-absorbed photosensitizers in pathological tissue causes a chain of photochemical and photobiological reactions also known as photodynamic reaction (PDR), which results in photodamage an eventual death of target cells [26, 29]. Various substances and techniques can optimize PDT's efficiency such as electroporation and nanocarriers application which can increase photosensitizer's tissue selectivity [29]. Photodynamic therapy's selective process makes it an optimal method in dermatology, oncology, gynaecology, and urology, furthermore PDT's good safety profile offers elderly and immune-suppressed patient who aren't able to go through with surgery as another option [29, 30].

## 9.3.1 History

Light has been used as a mean to treat diseases like psoriasis and vitiligo in Egyptian, Indian, and Chinese culture for more than 3,000 years but it wasn't until 1900 when a medical student, Oscaar Raab accidentally discovered the cytotoxic effects of light on microorganisms. He was studying the interactions of fluorescent dyes on *Paramecium* when he realized that intense light applied to acridine can kill the surrounding cells [31, 32]. In 1901, Neil Finsen used red and UV lights in order to prevent the formation of small pox pustules and battle for cutaneous tuberculosis, respectfully, and he received the Nobel Prize in 1903 for these endeavors on phototherapy [31]. On the same year, VonTappier and Jesionek used topically applied 5% eosin solution in combination with white light illumination to treat basal cell carcinoma. Later in 1907, VonTappier and Jodibauer defined the term "Photodynamic therapy" as interactions between dynamic light, a PS agent, and oxygen resulting in tissue destruction [28, 31]. Hausman reported the phototoxicity of haematoporphyrin on mice skin in 1911 and two years later first experiment of PDT was done on human by Meyer-Betz who tried 200 mg of haematoporphyrin on his own skin [31]. For a few decades, PDT and its successes were forgotten about until in 1955, Schwartz developed haematoporphyrin derivative (HPD) from acetylation and reduction of crude haematoporphyrin with sulphuric and acetic acid [31]. HPD was found to be more tissue specific, efficient in smaller doses, and twice as phototoxic as crude haematoporphyrin [33]. In 1960, Lipson and Baldes realized that Schwartz's HPD can be used for photodetection purposes as HPD accumulation in neoplastic lesions gave them a visual advantage during surgery [26, 31]. Diamond work on phototoxicity properties of HDP against gliomas both in vivo and in vitro in 1972 [31].

A big break came through in 1975 when Doughetry and co-workers rediscovered the value of PDT in oncology. He described a commercially suitable PS agent (HPD) in junction with a reliable red light source, causing mammary tumour growth eradication in mice [28, 32]. J.F. Kelly and co-workers reported the same result in mice bladder carcinoma in the same year; therefore, the clinical trials begun for both skin and bladder carcinoma patients in 1976 [28, 31]:

- Doughetry–Skin Carcinoma; 25 patients participated in this trial and 98 out of 113 skin tumours completely responded to PDT.
- Kelly–Bladder Carcinoma; 5 patients were diagnosed with PDT photodetection and there was tumour necrosis observed in 1 patient with recurrent bladder carcinoma.
- Hayate–Lung and Gastric Carcinoma.
- Balchum–Lung Carcinoma.
- McCaughen–Esophageal Carcinoma.

Numerous successful studies were done over the years on various types of cancers in their early stages until in 1993, photodynamic therapy (photofrin application against bladder cancer) was FDA approved in Canada [28]. Today, there are other approved PS agents used in PDT in order to treat various types of cancers and a number of photosensitizers are in clinical trials, optimizing the efficiency of PDT [31].

Nowadays, photodynamic therapy is employed for breast, gynaecological, intraocular, brain, head and neck, intraperitoneal, pancreatic tumours as well as cutaneous malignancies [31].

## 9.3.2 Principle of PDT

### 9.3.2.1 Photosensitizing agent

A photosensitizer is a natural or synthetic structure which is able to transfer light energy to its surrounding [32]. There are more than 3,000 PS agents (e.g., Chlorophyll) but only a few numbers are suited for photodynamic therapy. After HPD's approval in

1993, several PS agents have been developed. Although PSs differ in photochemical and pharmacological properties, they must have certain characteristics in common [34]:

-   Non-toxic until activated by clinically useful light wavelength.
-   Pain-free treatment.
-   Chemically pure and stable.
-   Easy to synthesis and commercially available.
-   Highly selective and with tumour affinity.
-   Appropriate absorption (400–800 nm); if the wavelength is lower than 400 nm, it will cause excessive photosensitivity and if it's higher than 800 nm, it won't have enough energy to excite PS molecules.
-   Reliable generation of PDR.
-   Hydrophilic or soluble in body tissue for easy systemic application.
-   Minimal dark toxicity in the dark.
-   Rapidly eliminated from patient's body and with no toxic degradation product after treatment [29, 32].

Currently, various appropriate PS agents are being employed for their clinical application. HPD (photofrin, sodium porfimer) is the first approved and the first generation of PSs, thus it holds the most clinical experience [32]. Photofrin can be employed for lung, bladder, and esophageal cancer treatment [34]. The treatment time is 20 min per lesion and is reported to be pain-free, easy, and reliable [32]. Photofrin is a mixture of 60 molecules containing various monomers and oligomers of haematoporphyrin; consequently, it has a low chemical purity [29, 32]. It has a poor tissue penetration due to maximum light absorption of only 630 nm [29]. HPD's long half-life and high tissue accumulation causes skin hypersensitivity, and no exposure to direct sunlight is instructed to the patient as sever burn can occur [29, 32]. Photofrin's disadvantages instigated the development of second-generation photosensitizers.

Second-generation PS agents have a higher light absorption of 650–800 nm, higher chemical purity, and a higher yield of singlet oxygen, nevertheless their water solubility is a meagre [29]. M-tetrahydroxophenyl chlorine (Foscan, mTHPC) is synthetically purified plant-based chlorine that can be employed to treat diverse stages of head and neck cancer [32, 34]. Foscan is very active and produces a rapid PDR, so treatment is extremely effective and is completed in only few seconds. However, due to Foscan's high energy, the patient must be kept in a dark room for 24 h after drug application in order to avoid severe burn or dark toxicity and the treatment is reported to be painful [32].

Mono-L-aspartyl chlorine e6 (MACE, Fotolon, NPe6, LS11) is also a plant-based chlorine which is highly effective but with no dark toxicity. Moreover, treatment can be done only after few hours in the same day and thus is convenient for the patient. Fotosens, a dye-based photosensitizer, is also activated shortly after application and the treatment takes 20 min to finish [32].

Aminolevulinic (ALA) is a prodrug altering to IV protoporphyrin via enzyme intermediation and is used generally for cutaneous carcinomas. ALA can be applied topically, orally, and intravenously due to its low toxicity but it is reported to be severely painful during illumination [30, 32]. There are other second PS agent that aren't engaged as much, for instance, texaphyrin, thiopurine derivative, and benzoporphyrin derivative [29].

Today, third PS generation is being developed with the attention is on PS-substrate affinity increase. To achieve this goal, PDT's different modifications are being investigated. These modifications include but not limited to combination of PS with target receptor-focussed molecules, nanocarriers application, PS and LDL (Low density lipoprotein) combined to answer tumour cell's need of cholesterol, PS and cancerous monoclonal antibody conjugation, and more [29].

### 9.3.2.2 Light

The wavelength of light employed in photodynamic therapy is photosensitizer specific and unique. ALA, photofrin, MACE, and Foscan can be activated by a wide range of visible light from red to blue, thus the wavelength can be selected based on clinical application. Clinical red light (630 nm) has a higher wavelength and therefor a deeper penetration of 0.5 cm, which can be used for both surface and deep tumours, while clinical blue light (400 nm) with lower wavelength can be employed only for superficial tumours via its 1 mm penetration [32].

Various light sources have been tested and used over the years such as gold vapour laser, Nd/YAG-pumped dye laser, argon-pumped laser, and copper-pumped laser, but these light systems are expensive and hard to handle. Today, the preferred light source is different types of light-emitting diodes (LEDs) since they are not only small, portable, cost-efficient, and easy to use, but also with same results in PDT as other systems [34].

### 9.3.2.3 Mechanism

Appropriate wavelength of light transfers light energy to PS molecules via electron transmission producing a PDR [32]. After illumination, photons imitate PS molecules from a ground energy state (singlet state) to a short-lived excited singlet state and later to a rather long-lived electronically excited state called triple state [31]. The excited triplet state can then source three outcomes (Figure 9.2):

a) Light and heat release: PDR is a fluorescent process and can be detected through high fluorescent regions (tumour cells) and low fluorescent regions (normal healthy cells). Loss of fluorescent post-PDT can also indicate that the tumour has been ablated [32].

b) Type I photochemical reaction: Triplet state PS interacts with the substrate (cell membrane and molecules) directly and transfers a hydrogen atom or electron, generating free radicals. Oxygenated products are formed from free radicals and oxygen reactions [31]. The resulting oxidative stress leads to cell destruction [29].

c) Type II photochemical reaction: In this assumingly more important pathway, triplet state PS transfers its energy directly to oxygen molecules [29, 31], and then these new-fangled singlet oxygen molecules form reactive oxygen species also known as ROSs [31].

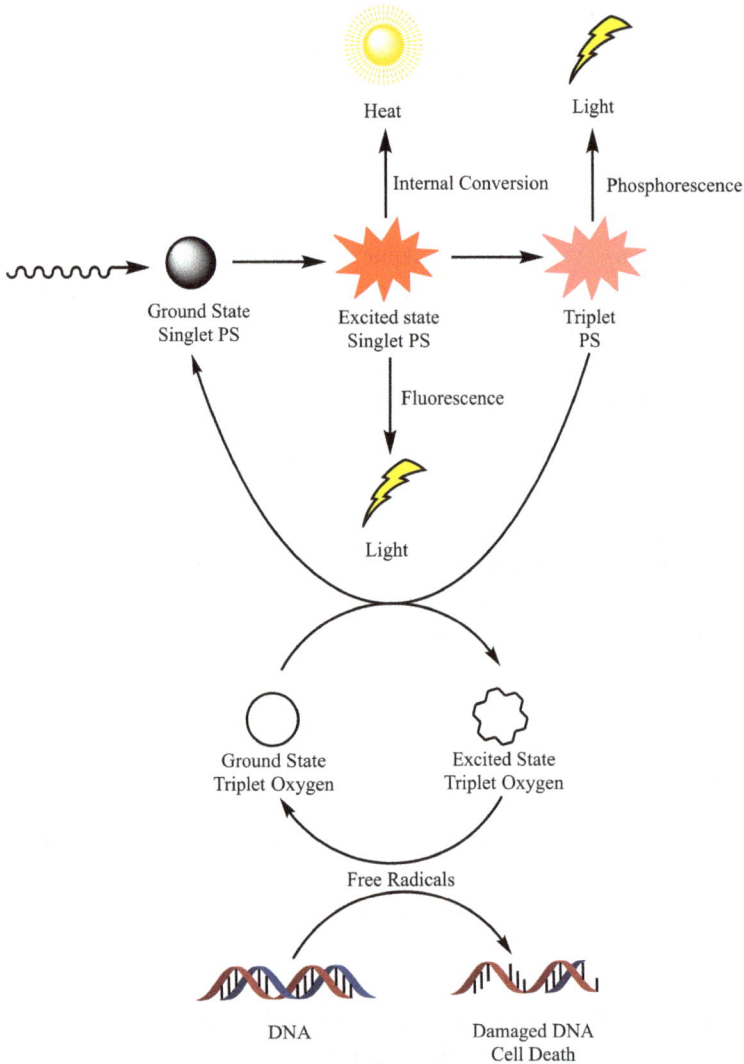

**Figure 9.2:** Photodynamic reaction mechanism [29].

Type I and type II reactions can only occur under non-anoxic areas and can both transpire simultaneously; the ratio however, depends on PS type and affinity, substrate concentration, pH, and oxygen concentration [29, 31]. ROS short half-life ($< 0.04 \mu s$) and extreme reactivity affects cells in destruction radius ($< 0.02 \mu m$) through three different but linked mechanisms; direct tumour photodamage, tumour-associated vascular damage, and immune response against tumour [31].

ROS can directly cause tumour cell death by both apoptosis and necrosis [31, 32]. Different PSs have diverse ways of arrival mechanisms to tumour site, such as receptor-mediated phagocytosis/endocytosis, low-density lipoprotein receptor binding, uptake by tyrosine kinase/epidermal growth factor receptor, diffusion, biodistribution, and other experimental ways. Photofrin and MACE take multiple pathways to tumour location and concentrate in cell and organelle membrane like mitochondria, while Foscan takes Golgi and endoplasmic reticulum to accumulate [32].

Once light is illuminated, cellular and sub-cellular destruction and eventual tumour cell necrosis follows. Calcium and metabolic byproducts release cause cells to malfunction and overwhelm repair functions. Cytokines and toxic chemicals leakage from organelles has a lethal bystander effect on nearby cells, furthermore, they can induce a regional and systemic immune reaction [32]. Tumour apoptosis can also be initiated by photodynamic therapy when low light dose is employed. As a result, cells cease to function but there is no bystander effect nor an immune respond [32]. Non-homogenous drug distribution of PS and oxygen unavailability within the tumour tissue don't allow this mechanism to eradicate the tumour unaided [31].

Necrosis and apoptosis of vasculature contribute both directly and indirectly to tumour destruction. Tumour cell's viability relies on nutrition supplies carried through blood vessels. Then again, blood vessels maintenance depends on growth factors provided by tumour cells [31]. Endothelial vascular cells can concentrate photosensitizers due to its comparable receptors and pathways. When the accumulated PS is activated by illumination, PDR sources vascular wall disruption. This disruption can distress the blood flow resulting in microvascular collapse, hypoxia, and anoxia. Lack of oxygen and other supplies causes vascular shut-down and necrosis of both nanovasculature and tumour [31, 32]. Toxic chemicals and calcium release result in collapse of microvasculature as well, blocking feeding of tumour [32]. Vasoconstriction, thrombus formation, and platelets aggregation have also been reported [31]. Vasculature apoptosis can also occur, resulting in tumour hypoxia and destruction [32].

Photodynamic therapy has an immunomodulatory effect and can suppress the immune system. Cell necrosis cascades an immune response through inflammatory mediators (e.g., cytokines, growth factors, and proteins) released from pathological tissue. These mediators stimulate several white blood cells' (including neutrophils and macrophages) activation that can eventually result in tumour cell death. Macrophages phagocytize PDR's damaged cells and debris upon arrival and also activate CD8 cytotoxic T lymphocytes through CD4 helper T lymphocytes. When photodynamic therapy is completed cytotoxic T-cells can induce cancerous cell apoptosis [32].

### 9.3.2.4 Clinical procedure

PS drug application varies in clinical use. While in most PDTs, PS is applied systemically via intravenous injections, few superficial skin cancer, premalignant skin lesions, and oral mucosa can be treated with topically applied PS (usually ALA) [34]. Moreover, maximized PDR is achieved with clinically determined but minimal photosensitizer dosage for the reason that PS molecules have the opportunity to concentrate in tumour tissue only [32].

Drug-light interval or DLI is the incubation period for photosensitizer drug to accumulate in target tissue before light illumination [34]. A short interval (15 min) provides PS with enough time to concentrate in vascular compartment, causing vascular statis and thrombus followed by direct tumour kill, but then in a longer DLI (4 h), PS accumulates in extravascular compartment. In conclusion, intervals differ in time based on their clinical application and tumour destruction but they mostly take 24–72 h [31].

After an appropriate DLI passes, light is illuminated by different techniques:
–  Superficial PDT is employed with thin and accessible tumours. Light is delivered by a fibre with a microlens (applicator) on its tip.
–  Interstitial PDT is used to treat tumours thicker than 1 cm. Laser fibre and applicator are implanted into the tumour tissue through needles.
–  Intra-active PDT is applied in hollow organ carcinomas like uterus and bronchus.
–  Intra-operative PDT is photodynamic therapy and surgery combined and is employed with anatomically complex areas [34].

Light dosage can also be altered. High light doses result in necrotic pathways and immune response, while low light doses induce apoptosis [32].

Altering drug dose and application, light dose and technique and drug-light interval allows clinicians to optimize a proper PDR providing necrotic pathways, apoptotic pathways, selective vascular destruction, and selective tumour destruction [32].

## 9.3.3 Benefits and disadvantages

To better realize PDT's advantages, we must compare it with other forms of cancer therapy. Photodynamic therapy process generally requires one single PS injection (it can be repetitive if needed), followed by a single clinical illumination. Both steps are expected to be through in a short amount of time (usually one day), while chemotherapy, radiotherapy, and surgery can take weeks to month to be completed. Furthermore, PDT is cost-efficient, on an out-patient basis and can increase life expectancy. Most importantly, it protects normal healthy tissues, creating a safety margin from photodamaging. Minimal long-standing fibrosis results in a functional recovery without scaring due to a matrix provided by PDR for tissue regeneration since subepithelial collagen and elastin aren't damaged [28].

Although PDT is a valuable alternative cancer treatment, it is not without limitation. Photodynamic therapy is a localized treatment; hence it is limited to local, small,
and accessible tumours. However, with fibre-optic technology advancing, appropriate
light can be transmitted to virtually everywhere in the body [28]. Furthermore, PDT
side effects have been reported in different clinical applications. Early onset side effect of photodynamic therapy in cutaneous diseases treatment involves:

– Pain; it is the most frequent (58%) and limiting side effect. Pain starts early during
illumination, picks in the first minute, and decreases from then. It can induce
other symptoms like hypertension and can negatively influence patient's life
quality. Many PDR mechanisms can mediate pain, for example, produced ROSs
stimulate sensory nerves conducting pain also, hypoxia triggers pain signals due
to the low oxygen level around mitochondria-rich nerve endings. Intensity of
pain depends on the depth of singlet oxygen production and therefore differs in
PSs. Pain management measures can be taken via air analgesia, topical anesthesia, nerve blocks, and opioid ingestion (Figure 9.3).
– Photosensitivity; systemic administrated PS leaves residues in skin and as soon as
these accumulated molecules are activated by normal sunlight, it can cause severe tissue morbidity and clinical first or second degree burns. Photosensitivity
differs from one photosensitizer to the other but light precautions must always
be taken.
– LSRs; there is erythema and oedema with 89% prevalence, scaling with 80%, pustules with 6%, and erosions with 1.2%.
– Urticaria; histamine release from mast cells in dermis causing skin to itch during
treatment but it has only been reported in ALA and MAL (Methyl aminolevulinate)
photodynamic therapies.
– Innate and adaptive immune suppression.
– Other trivial side effects like miscellanea, liver parameters changes, allergic reactions, and swelling [30, 34]

Late onset side effects have been reported as well, nonetheless they aren't as common. They consist of pigmentary changes (1%), mild to moderate scaring (0.8%), bul-
lous pemphigoid, and most concerning, carcinogenicity [30].

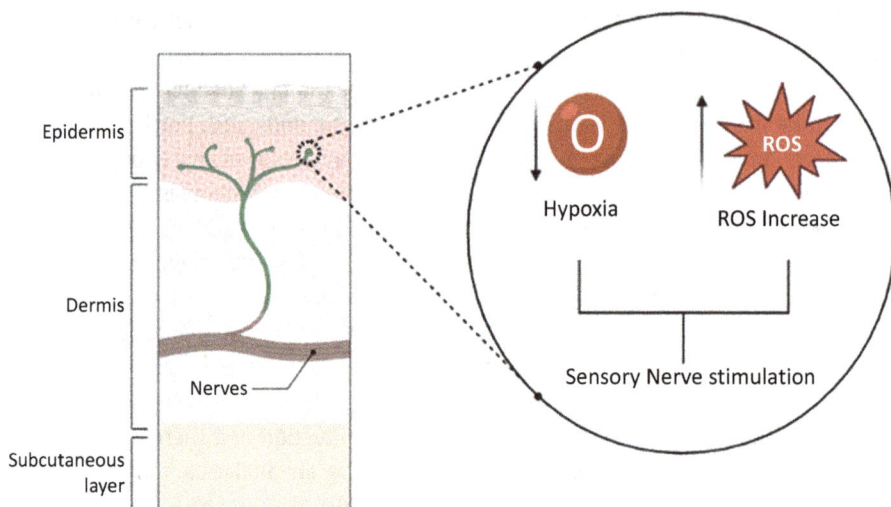

**Figure 9.3:** Pain-inducing mechanisms of PDR.

## 9.4 Some studies about carbon dots in photodynamic therapy

Many CDs-based structures have been synthesized and their effects on the cell and animal model have been studied. We will explain their results and methods as following.

### 9.4.1 CCOF-1 and CCOF-2

We will briefly discuss a valuable study by Shuang Chen et al. about covalent organic frameworks and published in advanced functional materials journal [39]. CCOF-1 and CCOF-2 are useful structures in photodynamic therapy made of CDs and P-phenylenediamine/BODIP [39]. These structures are modified with PEG and became CCOF-1PEG and CCOF-2PEG (Figure 9.4). CCOF-1PEG has absorption of 300–900 nm. This information was obtained from UV–vis absorption and PL spectra. TEM image illustrates average diameter of CCOF-1PEG to be 430 nm. DFT methods show that the size of CCOF-1PEG is 2.6 nm. Moreover, CCOF-2PEG has an absorption of 200–900 nm. The average diameter of CCOF-2PEG is 430 nm, while alone CDs have absorption of 200–800 nm, and the average size is 5.37 nm [39].

CCOF-2PEG demonstrates high capability in photodynamic therapy because of inducing hydroxyl radicals ($\cdot$OH); in order to reveal this, HeLa cell line is employed. Cells incubate with CCOF-2PEG for 0.5 to 2 h, after cell irradiation by laser, utilizing

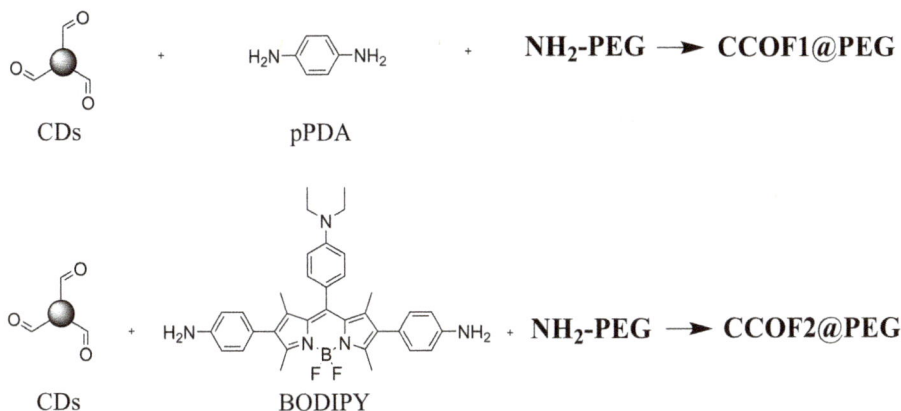

**Figure 9.4:** Schematic indication of CCOF-1@PEG and CCOF-2PEG synthesis [39].

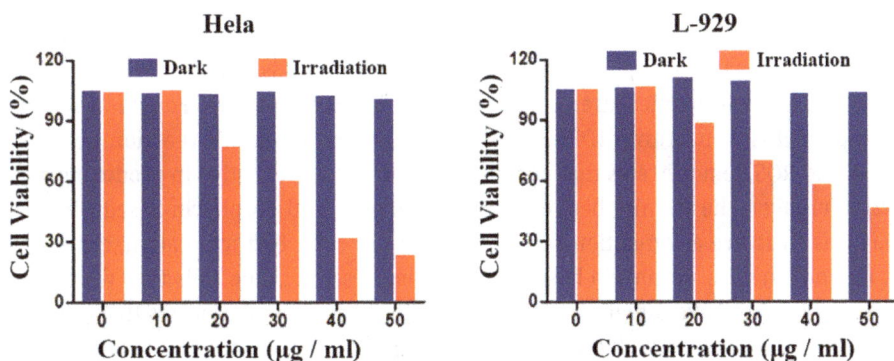

**Figure 9.5:** These charts illustrate the low cytotoxicity of CCOF-2PEG in two cell lines, irradiation and without irradiation [39].

2,7-dichlorofluorescein di acetate (DCFH-DA). In photodynamic therapy, low cytotoxicity is essential. In order to check the cytotoxicity of CCOF-2PEG tumour cells (cancer cell lines like HeLa cells) and normal cells (e.g., L929 cells), incubate with CCOF-2PEG without laser irradiating. In this situation, the methyl thiazolyltetrazolium (MTT) assay on these cells indicates that cell death is shallow (Figure 9.5) [39]. In contrast, CCOF-2PEG has an excellent ability to induce ROS and doesn't have cytotoxicity [39].

The performance of CCOF-2PEG in anti-tumour activity was studied and results show high anti-tumour activity. This study used U14 tumour-bearing mice as an animal model. Mice were divided into four groups; the first and second group were injected with PBS and the third and fourth group were injected with CCOF-2PEG; however, groups two and four were illuminated by laser and tumours were weighted. The third group was also irradiated by laser and measured but tumour's weight was reduced [39].

Another essential feature of structure that is used in photodynamic therapy is biocompatibility. In this study, biocompatibility of CCOF-2PEG was investigated by staining important organs such as heart, liver, lung, and kidney with haematoxylin and eosin (H&E). Result confirmed the biocompatibility of CCOF-2PEG [39].

## 9.4.2 Porphyrin-based carbon dots

A valuable study was done by Yang Li et al. and published in advanced healthcare materials Journal about porphyrin-based CDs. In future, we will explain the results and methods of this study momentarily.

Photoluminescent CDs have excellent properties that make them suitable for photodynamic therapy [40, 41]; therefore, other CDs would not be considered. Although, porphyrin-based CDs can be helpful as they are synthesized from 3-phenyl porphyrin and chitosan. Electron microscopy studies reveal a diameter of porphyrin-based CDs about 2.9 nm and spectroscopy showed high absorbance of 410 nm. As mentioned, the important mechanism of photosensitizers in producing ROSs, thus the ability of produced ROS should be measured in porphyrin-based CDs structure. For this purpose, 1,3-diphenyl-isobenzofuran (DPBF) was used. The extraordinary reaction between DPBF and ROS demonstrates that porphyrin-based CDs are effective in producing ROS [42]. Moreover, the vicinity between lysosome's prob and porphyrin-based CDs display this structure as an option for specific place in cells. Post irradiation, cell death in incubation with porphyrin-based CDs was impressive, while cells that were only incubated didn't have a high death rate [42]. This observation indicates that porphyrin-based CDs although with acceptable cell death, don't have cytotoxicity. For in vitro studies, H22 tumour-bearing KM mice were divided into 4 groups in Table 9.2 [42].

**Table 9.2:** Four groups and their conditions (information displayed is based on article [42]).

| Group 1 | TPP CDs + 650 nm radiation | Tumour size decreased to 56 nm |
| Group 2 | TPP + 650 nm radiation | Tumour size increased to 400 nm |
| Group 3 | TPP CDs | Tumour size increased to 900 nm |
| Group 4 | saline | Tumour size increased to 1,000 nm. |

## 9.4.3 Carbons dots that interact white the nucleus

Nucleus is a crucial part of cells. Tumour cells divide at high speed and they need ribosomes in order to accomplish this. Nucleolus is fundamental for ribosome synthesis, thus two CDs have been introduced. The first one is red emissive two-photon CDs [43] and the second one is Se/N-doped CDs [44]. In future, we will explain these structures.

### 9.4.3.1  Red emissive two-photon carbon dots

A comprehensive research was done by Shangzhao Yi et al. and published in a carbon journal. In future, we will explain that briefly.

For synthesizing this structure, solvothermal synthesis method was used (Figure 9.6) [43]. Employing electron microscopy, X-ray method, and spectrophotometry, two-photon carbon dots (TP-CDs) were characterized: average size is about 4 nm; two-photon absorption is between 680 to 1,000 nm, and emission is approximately 600 nm [43].

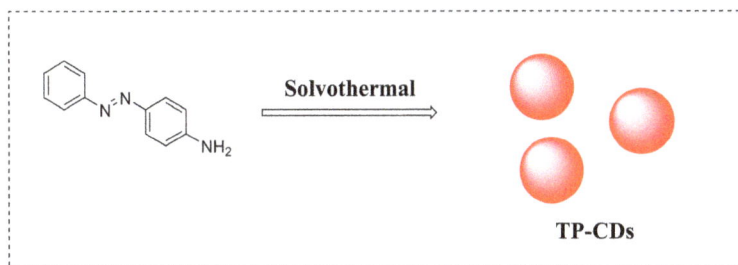

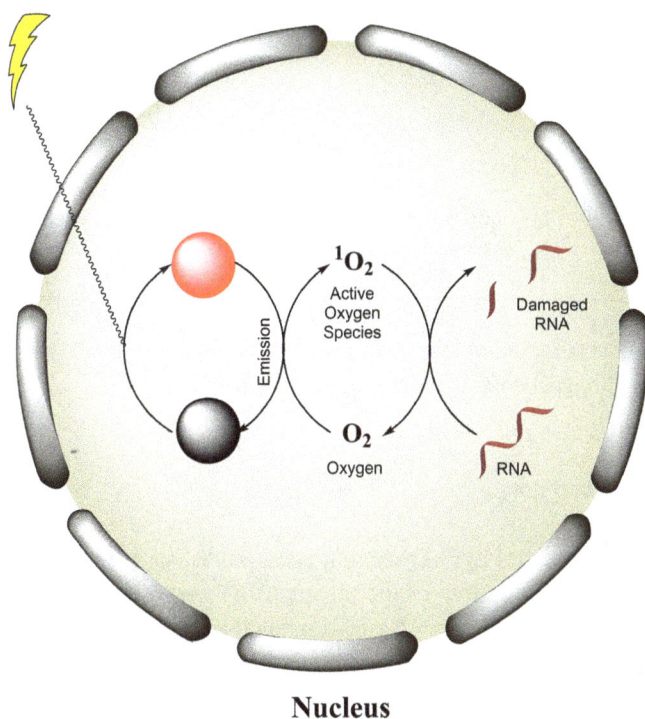

**Figure 9.6:** This schematic shows the synthesis and effect of two-photon carbon dots [43].

One of the important properties of TP-CDs is lacating near the nucleolus [43]. Fluorescence spectra confirm this subject. Three basic macromolecules were added to the TP-CDs solution, and fluorescence is measured per solution. According to the evidence, fluorescence in TP-CDs containing RNA is higher than both DNA containing TP-CDs and protein containing TP-CDs. Moreover, if RNAse is added to TP-CDs with RNA, fluorescence decreases. Three results suggest that RNA interacts with N doping or without N doping has an essential role in interaction with RNA and TP-CDs [43]. (1) If the N doping is removed, fluorescence is not changed in TP-CDs without N doping and RNA. (2). TP-CDs have positive charge, and RNA has negative. Consequently, there is an electrostatic interaction between them. (3) When cells were incubated with TP-CDs and irradiation, their RNA was extracted and RNA electrophoresis was done; there was no evident bond for the reason of effect of TP-CDs damages the RNA [43] (Figure 9.7).

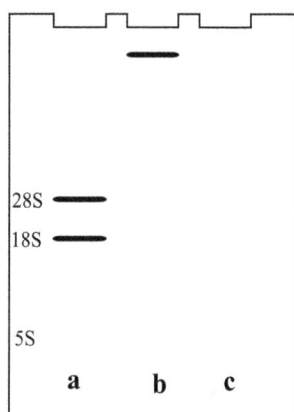

**Figure 9.7:** RNA electrophoresis of (a) cells that do not incubate with TP-CDs, (b) cells that incubate with TP-CDs and irradiation, and (c) cells that incubate with RNAse [43].

TP-CDs have effective ROS inducing in irradiation at 638 nm [43]. And MTT assay suggests that TP-CDs don't have cytotoxicity [43].

### 9.4.3.2 Se/N-doped carbon dots

We will explain practical studies by Ning Xu that were published in the carbon journal field [44]. Se/N-CDs are synthesized by the solvothermal method. Their diameter is about 3.6 nm. They have radiated emissions at 591 nm [44].

ROS made in photodynamic therapy by photosensitizer have a short lifetime; therefore, in order to increased photodynamic therapy performance, ROS should make near overriding organelle, for instance, nucleus or mitochondrial field [44, 45]. Se/N-doped CDs are located in the nucleus and intricate with RNA. RNA is expected to be a carrier for Se/N-doped CDs [44]. Se/N-doped CDs enter nucleus after they arrive near it; this subject will reveal by staining the nucleus and Se/N-doped CDs. This structure enters the nucleus after being illuminated. It is assumed that Se/N-CDs, by inducing ROS, destroyed nucleus membrane pores.

Furthermore, the use of Si nanoparticle suggests that after irradiation, fluorescent increases in the nucleus. Staining and using RNAse and DNAse with incubated cells by Se/N-CDs confirmed that this structure interacts with RNA more than DNA [44].

To study cytotoxicity, cell viability, and inducing ROS, some cell lines such as HeLa and 4T1 were employed. Standard MTT assay and Calcein-AM/PI test are efficient methods. For this purpose, cells were incubated at different concentrations of Se/N-doped CDs for a definite time both with irradiation and without irradiation field [44]. This study shows that in 7.5 µg mL$^{-1}$ of Se/N-CDs without irradiation, cell viability was about 90%. This means Se/N-CDs have low cytotoxicity. 2′,7′-dichlorofluorescein diacetate (DCFH-DA) is a structure that reacts with ROS. After irradiation, the green fluorescence that has been used is high, so that Se/N-CDs can induce ROS [44].

In vivo studies reveal proper biocompatibility and helpful effect of Se/N-doped CDs on tumour. For this purpose, BALB/c tumour-bearing mice were used. These mice were divided into four groups. The first group is injected with PBS, the second ones are injected with PBS and irradiation. The third ones are injected with Se/N-CDs, and the fourth ones are injected with Se/N-CDs and irradiation. (Time of irradiation in each group was 20 min, the wavelength was 550 nm, and the amount of each injection was 50 mW cm$^{-2}$) [44]. After that, two factors were measured; body weight and tumour size. No weight loss was observed, although tumour size was decreased in group four. This was an influential act in cancer treatment [44].

## 9.4.4 Copper-doped carbon dots

We will explain valuable studies by Jingmin Wang et al., which were published in inorganic chemistry journals. Such metal ions have an essential role in cells. Therefore, this idea is born that synthesized CDs structure with this metal ion, for instance, copper-doped CDs (CU-CDs) [46]. Electron microscopic study and spectroscopy methods reveal some properties of CU-CDs like, diameters (about 2.8 nm). Since CU-CDs have different bonds, they have emission peaks between 200 and 650 nm [46]. CU-CDs have bigger diameters than CDs [46].

Cytotoxicity measurement of photosensitizer in photodynamic therapy is critical. For measuring the cytotoxicity of CU-CDs, MTT assay was used in different concentrations of CU-CDs with or without irradiation. It demonstrates that the cytotoxicity of CU-CDs is low [46].

Another property of photosensitizer is the ability to induce ROS. ESR can be used for this purpose and revealed ROS increase in cells that incubate with CU-CDs after irradiation. This ability can be attributed to copper doping [6–7].

CU-CDs with irradiation decrease cell viability [46] (Figure 9.8). Calcein-AM/PI test and MTT assay show this matter. Two-factors play a critical role in the effect of CU-

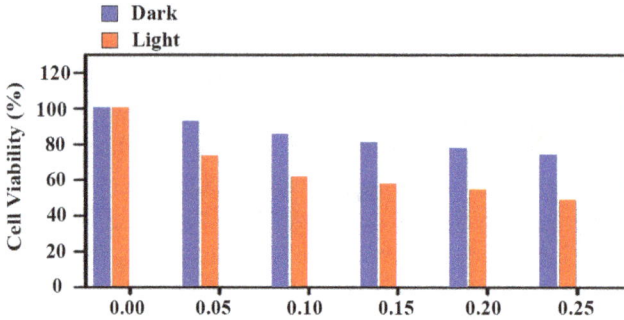

**Figure 9.8:** Illustration of effect of CU-CDs on cell viability [46].

CDs in cell viability; first one is concentration of CU-CDs and the second one is time of irradiation [46].

The decrease in cell viability of CU-CDs is higher than CDs. On the other hand, the cytotoxicity of CU-CDs is more elevated than CD, because cellular uptake of CU-CDs is more than CDs [46]. 3D microscopy studies suggest that CU-CDs influence cell growth and decrease the size of cells [46]. One advantage of CU-CDs is that they have copper. It changes $Ca^+$ concentration in cells and decreases cell viability [46]. Furthermore, copper increases stimulation and synthesizes ROS [46].

## 9.4.5 Carbon quantum dots

Carbon quantum dots have valuable properties similar to other carbon-based nano-materials, such as low cytotoxicity, good solubility, acceptable optical properties for photodynamic therapy, and high ability to induce ROS [47, 38]. They have different types, for instance, Curcumin cationic carbon dots (CCM-CDs) or carbon dots whose precursor is acetic acid [38].

Several studies reported that photodynamics by CDs can be helpful for virus diseases treatment, for instance, CDs that are synthesized from 4-aminophenyl boronic acid. It can be used to reduce virus cell entry. Another CDs can induce ROS that react with DNA, RNA, or essential viral proteins [38]. Furthermore, the CD's effects can stimulate immunological responses by increasing some cytokines or interleukins [38, 47]. Briefly, CDs can disturb different steps of the virus cycle, like attachment, and entry or replication [38].

SARS-CoV-2 is a member of Betacoronaviridae, it started a pandemic in 2019, and since it had begun, millions of people have been infected [38]. For this virus, various drugs, vaccines, and therapeutic methods have been reported and photodynamic therapy by CDs is one among them. As mentioned, CDs are an option as photosensitizer.

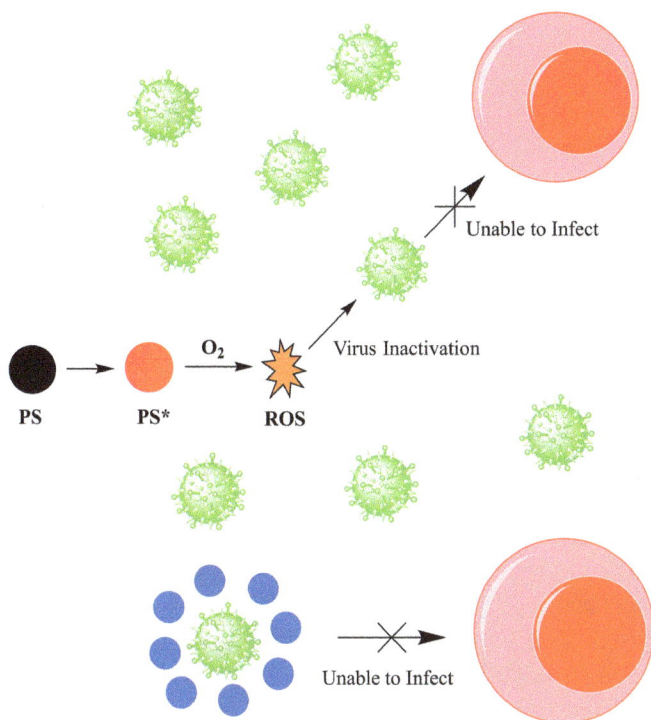

**Figure 9.9:** This schematic illustrates the effect of carbon dots in virus infection by increasing ROS [38].

CDs restrict viruses from entering cells by inducing ROS [38] (Figure 9.9). Other reports also reported CDs can help to treat herpes simplex type 1 [38, 48].

## 9.4.6 Diketopyrrolopyrrole-based carbon dots

A valuable study was done by Haozhe et al., which was about diketopyrrolopyrrole-based CDs (DPP CDs) and was published in nanoscale journal. We will explain this valuable study later.

Diketopyrrolopyrrole-based CDs are an appropriate choice for photodynamic therapy. They have unique properties such as low cytotoxicity, high cell uptake, and good tumour destruction [49]. DPP CDs are synthesized by one-pot hydrothermal method and their precursor is chitosan (Figure 9.10). DPP CDs were characterized by few methods such as TEM and UV–vis absorption and the result shows that DPP CDs' diameters are about 10 nm [49]. When the wavelength increases between 350 and 550, the absorbance decreases [49].

To measure ROS generation, 1,3-diphenyl-isobenzofuran (DPBF) UV–vis spectroscopy is used [49]. In this test, decreased absorption of DPBF means that ROS is generated.

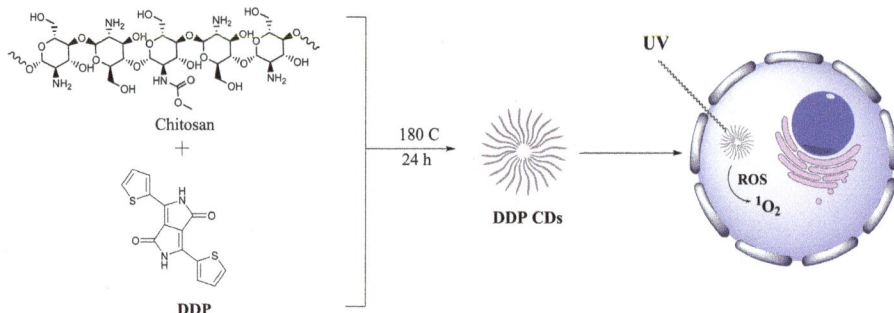

**Figure 9.10:** synthesis and mechanism of diketopyrrolopyrrole-based carbon dots [49].

Results show that DPP CDs have a good ability to generate ROS (the ability of DPP CDs in ROS generation is about 27.6%). This test was also done on the DPP; results show that DPP's ability to generate ROS is similar to DPP CDs and it was concluded that ROS generation in this structure is not affected by CDs [49].

Targeting cancer cells and cellular uptake are difficult matters in photodynamic therapy. This study used confocal laser scanning microscopy and fluorescence colocalization analyses. HepG2 cells were incubated with DPP CDs. After that, cells were fixed with formaldehyde and the nucleus was stained with Hoechst; while lysosomes were stained with Lyso-Tracker red DND-9. And result showed that there were more DPP CDs in the cytoplasm and their entry to the lysosome was by endocytosis.

For in vivo studies, female Kunming (KM) mice were suggested, and other studies were explained. Mice were divided into four groups (1) injected with PBS, (2) injected with PBS, and irradiated with laser, (3) injected with DP CDs, and (4) injected with DP CDs and irradiated with laser (wavelength was 540 nm, 20 min) [49]. The tumour size was decreased in group four (Figure 9.11). And no significant weight loss was measured. Histology studies with H&E staining were done to measure whether DPP CDs are damaged to their critical body organs [49]. And results suggest that there is no tissue damage and nor necrosis [49]. Other important matters in photodynamic therapy by CDs are tumour permeability. For this purpose, fluorescent imaging of tumour and other organs was done [49] and results revealed that the florescent signal in other tissue was low [49].

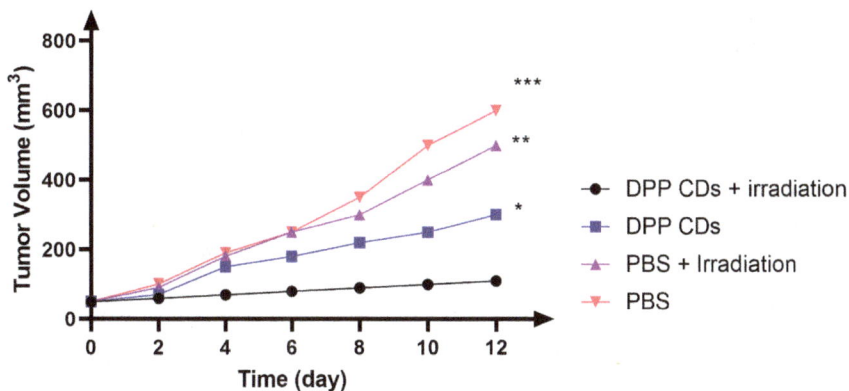

**Figure 9.11:** Tumour size decreases in mice that was injected with DPP CDs and irradiation but there was one injected with PBS with or without irradiation tumour rapidly grows [49].

## 9.5 Future suggestions

As mentioned, photodynamic therapy by CDs has some weaknesses such as low exclusivity and short time of useful ability to synthesize ROS and non-selective in most cases to select the location in cells. It is necessary to focus on these subjects in future studies, also in recent articles, molecular analysis such as PCR and proteome analysis such as western blotting and immunoblotting were used, the effect of CDs and ROS that are synthesized by them on gen expiration and protein synthesized and structures must be measured.

## References

[1]   Mintz KJ, et al. A deep investigation into the structure of carbon dots. Carbon. 2021, 173, 433–447.
[2]   Yao B, et al. Carbon dots: A small conundrum. Trends Chem. 2019, 1(2), 235–246.
[3]   El-Shafey, AM. Carbon dots: Discovery, structure, fluorescent properties, and applications. Green Process Synth. 2021, 10(1), 134–156.
[4]   Xu X, et al. Electrophoretic analysis and purification of fluorescent single-walled carbon nanotube fragments. J Am Chem Soc. 2004, 126(40), 12736–12737.
[5]   Wang W., Cheng L, Liu W. Biological applications of carbon dots. Sci China Chem. 2014, 57(4), 522–539.
[6]   Hu L, et al. Chiral evolution of carbon dots and the tuning of laccase activity. Nanoscale. 2018. 10(5), 2333–2340.
[7]   Yuan F, et al. Engineering triangular carbon quantum dots with unprecedented narrow bandwidth emission for multicolored LEDs. Nat Commun. 2018, 9(1), 1–11.
[8]   Xu D, Lin Q, Chang HT. Recent advances and sensing applications of carbon dots. Small Methods. 2020, 4(4), 1900387.

[9]    Gujar V, et al., Bioimaging applications of carbon dots (C. dots) and its cystamine functionalization for the sensitive detection of Cr (VI) in aqueous samples. J Fluoresc. 2019, 29(6), 1381–1392.

[10]   Ji C, et al. Recent developments of carbon dots in biosensing: A review. ACS Sensors. 2020, 5(9), 2724–2741.

[11]   Gaikwad A, et al. Fluorescent carbon-dots thin film for fungal detection and bio-labeling applications. ACS Applied Bio Mater. 2019, 2(12), 5829–5840.

[12]   Xu S, et al. One-step fabrication of boronic-acid-functionalized carbon dots for the detection of sialic acid. Talanta. 2019, 197, 548–552.

[13]   Khan MS, et al. Exploring the ability of water soluble carbon dots as matrix for detecting neurological disorders using MALDI-TOF MS. Int J Mass Spectrom. 2015, 393, 25–33.

[14]   Wu Y-F, et al. Multi-functionalized carbon dots as theranostic nanoagent for gene delivery in lung cancer therapy. Sci Rep. 2016, 6(1), 1–12.

[15]   Peng Z, et al. Carbon dots: Promising biomaterials for bone-specific imaging and drug delivery. Nanoscale. 2017, 9(44), 17533–17543.

[16]   Cui L, et al. Carbon dots: Synthesis, properties and applications. Nanomaterials. 2021, 11(12), 3419.

[17]   Sun X, Lei Y. Fluorescent carbon dots and their sensing applications. Trends Analyt Chem. 2017, 89, 163–180.

[18]   Pillar-Little TJ, et al. Superior photodynamic effect of carbon quantum dots through both type I and type II pathways: Detailed comparison study of top-down-synthesized and bottom-up-synthesized carbon quantum dots. Carbon. 2018. 140, 616–623.

[19]   Wu ZL, Liu ZX, Yuan YH. Carbon dots: Materials, synthesis, properties and approaches to long-wavelength and multicolor emission. J Mater Chem B. 2017, 5(21), 3794–3809.

[20]   Sharma A., Das J. Small molecules derived carbon dots: Synthesis and applications in sensing, catalysis, imaging, and biomedicine. J Nanobiotechnol. 2019, 17(1), 1–24.

[21]   Yu H, et al. Preparation of carbon dots by non-focusing pulsed laser irradiation in toluene. Chem Commun. 2015, 52(4), 819–822.

[22]   Nguyen V, et al. Femtosecond laser-induced size reduction of carbon nanodots in solution: Effect of laser fluence, spot size, and irradiation time. J Appl Phys. 2015, 117(8), 084304.

[23]   Dang H, et al. Large-scale ultrasonic fabrication of white fluorescent carbon dots. Ind Eng Chem Res. 2016, 55(18), 5335–5341.

[24]   Wang F, et al. Study on ultrasonic single-step synthesis and optical properties of nitrogen-doped carbon fluorescent quantum dots. Fuller Nanotube Carbon Nanostruct. 2015, 23(9), 769–776.

[25]   Deng Y, Qian J, Zhou Y. Solvothermal synthesis and inkjet printing of carbon quantum dots. ChemistrySelect. 2020, 5(47), 14930–14934.

[26]   Dougherty TJ, et al. Photodynamic therapy. JNCI J National Cancer Inst. 1998, 90(12), 889–905.

[27]   Kessel D. Photodynamic therapy: A brief history. J Clin Med. 2019, 8(10), 1581.

[28]   Triesscheijn M, et al. Photodynamic therapy in oncology. Oncologist. 2006, 11(9), 1034–1044.

[29]   Kwiatkowski S, et al. Photodynamic therapy–mechanisms, photosensitizers and combinations. Biomed Pharmacother. 2018, 106, 1098–1107.

[30]   Borgia F, et al. Early and late onset side effects of photodynamic therapy. Biomedicines. 2018, 6(1), 12.

[31]   Dolmans, DE, Fukumura D, Jain RK, Photodynamic therapy for cancer. Nat Rev Cancer. 2003, 3(5), 380–387.

[32]   Allison RR, Moghissi K. Photodynamic therapy (PDT): PDT mechanisms. Clinical Endoscopy. 2013, 46 (1), 24–29.

[33]   Huang Z. A review of progress in clinical photodynamic therapy. Technol Cancer Res Treat. 2005, 4 (3), 283–293.

[34]   Kübler AC. Photodynamic therapy. Biomed Laser Appl, 2005. 20(1), 37–45.

[35]   Lu D, Tao R, Wang Z. Carbon-based materials for photodynamic therapy: A mini-review. Front Chem Sci Eng. 2019, 13(2), 310–323.

[36] Szygula M, et al. Combined treatment of urinary bladder cancer with the use of photodynamic therapy (PDT) and subsequent BCG-therapy: A pilot study. Photodiagn Photodyn Ther. 2004, 1(3), 241–246.

[37] Moghissi K. Endoscopic photodynamic therapy (PDT) for oesophageal cancer. Photodiagnosis Photodyn Ther. 2006, 3(2), 93–95.

[38] Sanchez de Araujo H., Ferreira F. Quantum dots and photodynamic therapy in COVID -19 treatment. Quantum Eng, 2021, 3(4).

[39] Chen S, et al. Carbon dots based nanoscale covalent organic frameworks for photodynamic therapy. Adv Funct Mater. 2020, 30(43).

[40] Zheng M, et al. Self-targeting fluorescent carbon dots for diagnosis of brain cancer cells. ACS Nano. 2015, 9(11), 11455–11461.

[41] Ge J, et al. Red-emissive carbon dots for fluorescent, photoacoustic, and thermal theranostics in living mice. Adv Mater, 2015, 27(28), 4169–4177.

[42] Li Y, et al. Porphyrin-based carbon dots for photodynamic therapy of hepatoma. Adv Healthc Mater. 2017, 6(1).

[43] Yi S, et al. Red emissive two-photon carbon dots: Photodynamic therapy in combination with real-time dynamic monitoring for the nucleolus. Carbon. 2021, 182, 155–166.

[44] Xu N, et al. Precise photodynamic therapy: Penetrating the nuclear envelope with photosensitive carbon dots. Carbon, 2020. 159, 74–82.

[45] Noh I, et al. Enhanced photodynamic cancer treatment by mitochondria-targeting and brominated near-infrared fluorophores. Adv Sci (Weinh). 2018, 5(3), 1700481.

[46] Wang J, et al. Copper-doped carbon dots for optical bioimaging and photodynamic therapy. Inorg Chem. 2019, 58(19), 13394–13402.

[47] Vaillancourt J, Vasinajindakaw P, Lu X. A high operating temperature (Hot) middle wave infrared (Mwir) quantum-dot photodetector. Opt Photonics Lett. 2011, 04(02), 57–61.

[48] Loczechin A, et al. Functional carbon quantum dots as medical countermeasures to human coronavirus. ACS Appl Mater Interfaces. 2019, 11(46), 42964–42974.

[49] He H, et al. Diketopyrrolopyrrole-based carbon dots for photodynamic therapy. Nanoscale. 2018, 10(23), 10991–10998.

Saima Ashraf, Fahmida Jabeen, Sabeen Iqbal,
Muhammad Salman Sajid, Muhammad Naeem Ashiq
and Muhammad Najam-ul-Haq*

# Chapter 10
# Carbon dots in photothermal therapy

**Abstract:** Cancer is one of the global health threats and several developed therapies are undergoing evolution with advancements in nanotechnology and awareness of tumor microenvironments. Low-dimensional carbon-based nanomaterials like carbon dots (CDs) and their derivatives have the potential of treating cancer at the preclinical level. Zero-dimensional CDs are unique in their physical, chemical, and biomedical properties. Characteristics such as high surface area, broad absorption spectrum, high hydrophilicity, tunable fluorescence, photo-stability, and biocompatibility are significant in healthcare applications like fluorescence sensing, imaging, and drug delivery. As a photosensitizer, CDs produce reactive oxygen species under light and convert light energy to heat as a photothermal agent. Thus, CDs have applications in photodynamic therapy or photothermal therapy (PTT) of cancer. In this chapter, the functionalized CDs and their synthetic approaches are discussed to provide photothermal aspects of CDs. The mechanism of PTT and scope of therapeutic efficacy of CD-based phototheranostic is highlighted. An insight into the challenges of limiting the clinical application of CDs is provided to prompt advances in phototherapy.

**Keywords:** Carbon dots, photothermal therapy (PTT), cancer, nanotechnology

## 10.1 Carbon dots

Carbonaceous materials have a role in material engineering. From common carbonic materials like carbon black and activated carbon to high-tech carbon substances (e.g. carbon nanotubes (CNTs), graphite, carbon fibres, and graphene oxide) have gained importance because of their physical and mechanical properties, improved conductivity, and biocompatibility. These materials have high surface-to-volume ratio, lightweight, high density, strength, and hardness [1]. Carbon materials are fluorescent like fullerenes, nano-diamonds, single walled CNTs (SWCNTs), and graphene sheets [2].

*Corresponding author: Muhammad Najam-ul-Haq**, Institute of Chemical Sciences,
Bahauddin Zakariya University, Multan 60800, Pakistan, e-mail: najamulhaq@bzu.edu.pk
**Saima Ashraf, Fahmida Jabeen, Sabeen Iqbal, Muhammad Salman Sajid,**
**Muhammad Naeem Ashiq**, Institute of Chemical Sciences, Bahauddin Zakariya University, Multan
60800, Pakistan

https://doi.org/10.1515/9783110799958-010

Lack of significant band gap in macroscopic carbon materials makes it difficult to be employed as an efficient fluorescent material [3].

Carbon dots (CDs) are a current addition in the group of nanotechnologies with prime properties. CDs are quasi-spherical and fluorescent with size below 10 nm [4]. CDs of size above 50 nm are also reported. CDs have high quantum yield (QY), adjustable photoluminescence, biocompatibility, low cytotoxicity, water solubility, small size, good conductivity, inexpensive precursors, and stable at room temperature [5, 6]. CDs are made from carbon. They show amorphous and crystalline properties. CDs are $sp^2$-hybridized carbon networks, whereas, in some cases, $sp^3$ hybridization is also reported [7]. Their applications are explored in various fields, including catalysis, biomedicine, bioimaging, and optoelectronic devices [8]. CDs show sensing characteristics such as selective, specific, and complex detectability. Surface passivation of CDs is easy with groups like carbonyl, hydroxyl, amino, carboxyl, and epoxy. The surface functionalities provide benefit of binding with organic and inorganic groups [9]. The functional groups on CDs result in colloidal stability and solubility in polar organic or aqueous solvents. The fluorescence properties of CDs are due to various surface groups [10]. The classification of carbon dots, different synthesis approaches, unique optical and electrochemical properties and applications in health care are given in Figure 10.1.

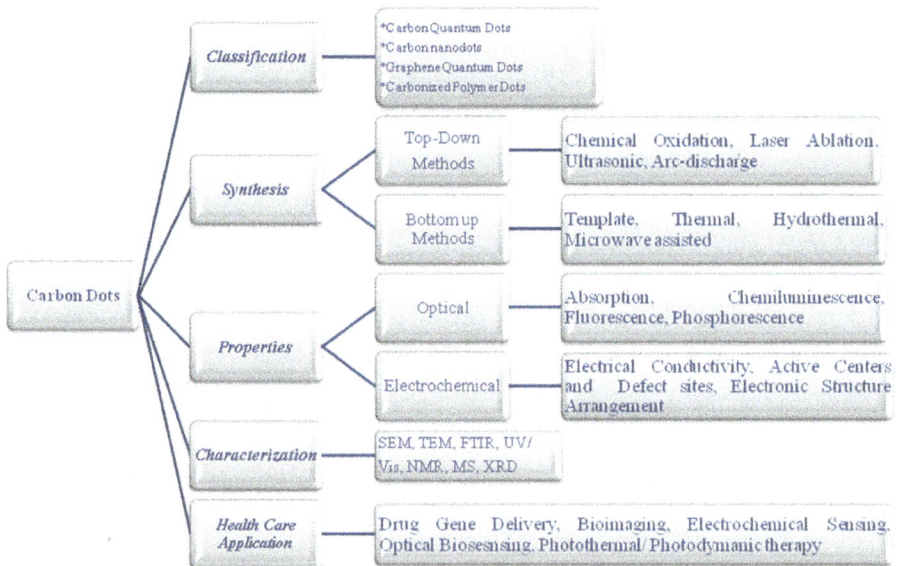

**Figure 10.1:** Outline of CD classification, synthetic approaches, properties, characterization techniques, and healthcare applications.

## 10.2 History of carbon dots

The photoluminescent CDs were discovered by Xu et al. in the purification process of SWCNTs [11]. In 2006, laser ablation method was utilized to synthesize CDs with strong photoluminescence. It was the first time when CDs were synthesized; however, some properties like low QY and complicated preparation strategies limited their applications [78]. Later, CDs having QY of 80% were made with ethylene diamine and citric acid as precursors. CDs obtained from citric acid precursor were the easiest dots to be prepared, and this helped to understand why and how synthetic methods affect CD properties. The achieved QY enhanced the chemical characteristics of CDs. These CDs can be applied to printing inks and used as precursors for synthesizing functional composites [12]. High QY, facile approach, resistance to photobleaching of CDs, and low toxicity further enhance their potential. Various strategies have been adopted for the low- and high-scale manufacturing of CDs and their applications.

## 10.3 Classification of carbon dots

Based on the fabrication strategies, surface functionalities, and properties, CDs are categorized into (a) carbonized polymer dots (CPDs), (b) carbon nanodots (CNDs), (c) carbon quantum dots (CQDs), and (d) graphene quantum dots (GQDs) [13]. This classification is attributed to their nanostructures, and characteristics can be changed by changing the graphene layer and degree of carbonization [14]. GQDs are zero-dimensional and anisotropic, which constituted single or multiple graphene sheets. GQDs exhibit properties like edge effect and quantum confinement due to the presence of chemical functionalities in their interlayer defect and on their edges. GQDs have optoelectrical properties along with excellent dispersibility, biocompatibility, and tunability. GQDs are employed in catalysis, energy devices, photodynamics, sensing, bioimaging, targeted drug delivery, and photothermal therapy (PTT) [15]. CQDs are nanocrystalline and possess functionalities that impart quantum confinement effect to CDs and intrinsic state luminescence. CNDs have high degree of carbonization with functional groups on the surface, but without showing polymeric and crystalline structure. CNDs lack the ability of quantum confinement effect [16].

CPDs are aggregated/cross-linked hybrid nanostructures of carbon and polymer. They show bright luminescence, low cytotoxicity, chemical inertness, and biocompatibility [17]. CPDs consist of cross-linked and graphitized inner core and hydrophilic polymer chain on the outer surface [18]. In CPDs, photoluminescence is due to cross-link-enhanced emission (CEE) and their surface, molecular, and subdomain states. The predominant optical property is due to molecular state and CEE effect. All other types of CDs do not show these photoluminescence features [19].

# 10.4 Synthetic methods of carbon dots

CD synthesis is influenced by factors such as surface state, molecular state, and QEE effect. These factors can be controlled by altering the synthetic strategies [20]. During CD synthesis, their surface can be modified by functional moieties such as carbonyl, epoxide, ether, carboxyl, amine, and hydroxyl [21]. The doping of CDs with heteroatoms like P, N, S, and B can be done by treating polymeric, biological, and organic substances [22]. CD modification is critical for attaining the surface properties for favourable applications [23]. The physicochemical properties such as carbonization, size, crystallinity, morphology, and photoluminescence can be modified by synthetic methods and precursor molecules. For instance, the pyrolytic method is harsher than the hydrothermal method but the former produces graphitic core structure and the latter produces incomplete carbonization, amorphous CDs, and molecular fluorophores. The surface moieties of CDs can alter the physicochemical characteristics, stability, and biocompatibility allowing modification, sensitivity, and selectivity of CDs in analytical applications [24].

CDs are synthesized by top-down and bottom-up approaches [25]. In "top-down", the carbonaceous materials such as graphite is chemically, electrochemically, or physically dissected into nano-sized fragments [26]. Bottom-up strategy involves either stepwise incorporation of small aromatic compounds or the carbonization of organic moieties. Therefore, CDs may be denoted as graphene nanodots and CNDs, respectively [27]. These synthetic approaches are classified as shown in Figure 10.2.

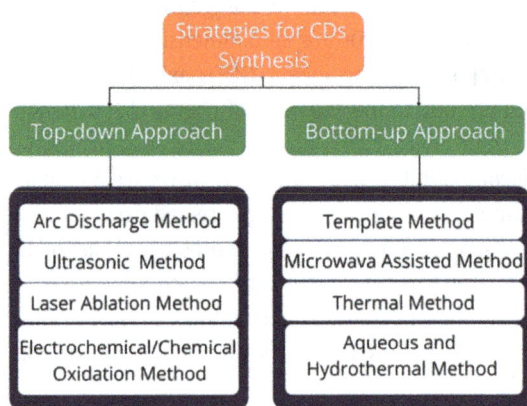

**Figure 10.2:** Synthetic approaches of carbon dots.

## 10.4.1 Top-down approach

Nowadays, carbon materials including graphite, activated carbon, and CNTs are used for the synthesis of CDs by laser ablation, ultrasonic treatment, electrochemical oxidation, and arc discharge methods [28]. These approaches are administered in conditions like high potential, energy, and acidity. Therefore, the top-down approaches are laborious as compared to bottom-up, due to harsh reaction conditions [29].

### 10.4.1.1 Laser ablation method

Laser ablation employs laser beam to irradiate carbonaceous materials in thermodynamic environment that results in high temperature and pressure. The high temperature produces plasma followed by evaporation. Subsequently, crystallization results in the conversion of generated vapour into CDs [30]. In 2006, luminescent CDs were synthesized by laser ablation in which the carbon target was ablated by argon as a carrier gas. In another study, photoluminescent CDs were prepared through laser ablation of bulk graphite as carbon source in ethanol via Nd:YAG laser irradiation [31]. Fluorescent CDs having size of 3 nm were prepared via laser ablation of carbon glassy particles immersed in polyethylene glycol (PEG) by flow jet configuration and batch process [32]. In 2021, quasi-molecular fluorophores with an average size of 5 nm were prepared by nanosecond laser ablation of bulk graphite dipped in ethylenediamine and polyethylenimine [33].

Nitrogen-doped CDs can be made by single-step laser ablation from graphite powder by using organic solvent like amino-toluene. The synthesized CDs exhibit excitation-independent emission, which is applied for monitoring the ratiometric pH because of excess surface oxygen and amine moieties [34]. CDs prepared via double-beamed laser ablation have attributes such as high surface-to-volume ratio, ultra-small size, homogeneity, and stability as compared to single-pulsed laser beam [35].

### 10.4.1.2 Arc discharge method

CDs can be prepared from crude CNT soot through the arc discharge method. The carbonaceous material is oxidized by nitric acid to introduce carboxyl groups followed by treatment with NaOH and purified by gel electrophoresis. In 2004, CDs were prepared by arc discharge using SWCNTs and multi-walled CNTs as carbon source through oxidation reaction [11]. This method is not useful for producing CDs as it produces small amount of product, and purification process is difficult.

### 10.4.1.3 Chemical oxidation method

CDs are prepared by the chemical oxidation method due to their advantages like low cost, greater yield, bulk production, purity, and controllable size [36]. CDs are chemically or electrochemically synthesized by the redox reaction under normal temperature and pressure [37]. Oxidizing agents like sulphuric acid ($H_2SO_4$), oxygen, nitric acid ($HNO_3$), and hydrogen peroxide are used. CD functionalization with hydrophilic moieties like carboxyl, amine, and hydroxyl can be done by controlling the oxidation–reduction reaction [38].

### 10.4.1.4 Ultrasonic treatment

Ultrasonic treatment is useful for producing CDs as the bulky carbonaceous materials could break down by high-energy ultrasound waves. By using single-step ultrasonic treatment, N-doped CDs were synthesized by ammonia and ascorbic acid precursor [39].

## 10.4.2 Bottom-up approach

Bottom-up methodology includes pyrolytic processes, supported synthesis, template methods, chemical oxidation, microwave-based methods, and reverse micelle processes. Bottom-up approaches are popular due to cost-effectiveness, easy instrumentation, non-toxic precursor molecules, precise control over size, convenient, facile methodology, and practical applicability.

### 10.4.2.1 Hydrothermal/solvothermal method

CDs prepared by this method are non-toxic, inexpensive, and eco-friendly. In this method, reaction occurs between carbon precursor and organic solvent in the hydrothermal reactor at high temperature and pressure. Countless raw materials can be used as precursors such as glucose, protein, chitosan, wheat bran, orange peels, cereals, grains, sugar cane brass, and citric acid. Nitrogen-doped CDs (N-CDs) are reported with carbon tetrachloride and sodium amide ($NaNH_2$) as source material via the solvothermal method [40]. Wu et al. synthesized nitrogen-doped quantum CDs (NQCDs) by using microcrystalline cellulose as carbon source and ethylenediamine as nitrogen source. The prepared NQCDs with an average size of 3 nm are applicable as a fluorescent probe for metallic detection [41]. In another work, CDs are synthesized using L-histidine and citric acid sources via the one-step hydrothermal technique [42].

### 10.4.2.2 Microwave-assisted method

Due to broad electromagnetic radiation spectrum range (1 mm to 1 m), variety of chemical bonds present in precursor molecules can be decomposed by their accelerated energies. The approach is applied for the synthesis of CDs as it has the advantage of high speed and uniform heating of precursor materials. Due to penetration of microwave radiations, uniform heating of reaction mixture is possible, which results in the formation of narrow sized crystal [43]. Using microwave-assisted synthesis, CDs produced from citric acid precursor are reported to evaluate antimicrobial photodynamic effect. By in vitro assays, effectiveness of CDs against *Staphylococcus aureus* biofilm and suspension is assessed. The results show that the use of CDs in antibacterial photodynamic therapy (PDT) is a viable treatment for *Staphylococcus aureus* infected wounds [44]. In another study, fluorescent CDs were synthesized via microwave irradiation from glucosamine@PEG@chitosan graft co-polymer. The resultant CDs show good fluorescence intensity, chemical stability, and efficacy of delivering chemotherapeutics [45].

### 10.4.2.3 Thermal method

Thermal decomposition involves the pyrolysis of bulky carbon materials at high temperatures to synthesize CDs. This method has advantages of bulk production, cost-effectiveness, ultra-fast, solvent free, choice of precursors, and easy synthesis. The fluorescence of CDs can be controlled through optimizing the pH of reaction mixture, temperature, and reflux time [46]. Photoluminescent CDs with adjustable size can be achieved by the pyrolysis of carbon microcrystal precursor in mesophase pitch. CDs exhibit high QY of ~ 87%, owing to oxygen-free character. These prepared CDs are used as a biosensor for $Fe^{3+}$ ion detection with greater sensitivity and specificity [47]. Different carbon dots fabricated by top down and bottom up approaches reported in literature are listed in Table 10.1.

**Table 10.1:** Methods for the synthesis of carbon dots, their precursors, size range, and applications in various fields.

| Carbon source | Methods | C-dot size | Applications | Ref. |
|---|---|---|---|---|
| Graphite in ethylenediamine and polyethylenimine | Laser ablation | 1–3 nm | Fluorescent CDs for different applications | [33] |
| Graphite powder | Laser ablation | Ultra-small size of 1 nm | Catalytic and sensing applications | [48] |
| $TiO_2$ nanoparticles coupled with CDs | Arc discharge | Average size of 27 nm | Sensing application | [49] |

**Table 10.1** (continued)

| Carbon source | Methods | C-dot size | Applications | Ref. |
|---|---|---|---|---|
| Boron-doped GQDs | Arc discharge | Zero-dimensional GQDs | Optical application | [50] |
| Sodium citrate + urea | Electrochemical carbonization | Approx. 2.4 nm | Sensing application | [51] |
| Citric acid, thiourea, and urea | Microwave-assisted | 10 nm | Detection of metals in water samples | [52] |
| Graphite target irradiation | Laser ablation | 2–3 nm | Potential related applications | [53] |
| Activated carbon | Ultrasonic | 5–10 nm | Biosensing and biomedical applications | [54] |
| Uric acid and ascorbic acid | Electrochemical | Nanofibres | Selective and sensitive dopamine detection | [55] |
| Wheat bran and tartaric acid | Hydrothermal | Approx. 4.85 nm | Detection of $Cu^{2+}$ ions | [56] |

# 10.5 Characteristic properties of CDs

CDs as the class of carbon materials have unique chemical, physical, and optical properties like absorption, chirality, and photoluminescence. CDs act as electron donors
and acceptors.

## 10.5.1 Optical properties

The fluorescent CDs find their ways in fields such as biosensing, molecular imaging, and
therapy. It is important to study the optical behaviour of CDs for diverse bioapplications.

### 10.5.1.1 Absorption

CDs synthesized from various sources of carbon or through artificial strategies show
excellent absorption. They show sturdy absorption in the UV region (200–400 nm).
The absorption occurs due to $\pi-\pi^*$ transition of C = C bond or n–$\pi^*$ of C = O/C = N
bonds [57]. The absorption range may change depending on the nature of CDs, surface
passivation, and functional groups [58].

### 10.5.1.2 Quantum yield

The ability of converting the absorbed light to the emitted light is known as quantum yield. Fluorophores with high QY show strong fluorescence, which decreases the quantity of fluorophores required for an application [59]. QY values for CDs, especially for bare CDs without surface passivation, are below 10%. The surface functionalization or doping can change the QY value of CDs with improved fluorescence intensity and QY. For instance, the carboxyl content of N-doped CDs is altered to improve the quality. The hydrothermal treatment of *m*-aminobenzoic acid to create CDs results in high QY of 30.7% when N is doped to CDs. A crude CD sample can be purified to separate batches of CDs with various QYs. When purified, CDs with comparable physiochemical characteristics can be separated with similar QY values. The purification of CDs functionalized with oligomeric PEG diamine (PEG1500N) does not alter the fluorescence although the QYs of each fraction vary [60]. QY values increase until they reach the levels of 55–60%. CDs with QYs up to 78% are more than double of the QY prior to fractionation. They are retrieved by purification through an aqueous gel column (Sephadex G-100) [61].

### 10.5.1.3 Fluorescence properties

The fluorescence is observed in CDs synthesized via the ultrasonic treatment. The excitation wavelength is longer than the emission wavelength due to the decrease in background autofluorescence. The luminescent process of CDs is still not fully investigated. The fluorescence of CDs is dependent on free zig-zag sites, quantum confinement effects, multi-emissive centres, special edge defects, self-trapped excitons, surface states, and their conjugated structures [62]. Fluorescence emission properties of CDs can be controlled by changing their excitation wavelength, which can be attained during CD synthesis by controlling the physicochemical parameters [63].

## 10.6 Phototherapy theory

The treatment of diseases by using light is known as phototherapy. In heliotherapy, from 1400 BC, the treatment of disease was done by sunlight. Later in the nineteenth century, scientific documentation for phototherapy could be found. The first treatment of disease (lupus vulgaris) by filtered sunlight was done by Niels Finsen in 1893 for which he received Nobel Prize in 1903 [64]. Phototherapy is classified into two types: (i) PTT and (ii) PDT. In PDT, photosensitizers (PSs) by light irradiation produce cytotoxic chemical agents. PSs can generate overheating under photoactivation [65].

## 10.6.1 Photothermal therapy (PTT)

PTT is an effective and non-invasive therapeutic treatment of cancer. It is a particular type of hyperthermia in which body tissues are exposed to higher temperatures to induce the elimination of abnormal cells. Light energy produced from laser is used to produce localized heat which causes heat ablation of tumour cell [66]. In PTT, the targeted abnormal cells are labelled with photoabsorbent nanoscale particles (also named as photothermal agent (PTA) or photosensitizing agent) [67]. Activation of PTA is done by pulsed near-infrared (NIR) laser light. The use of PTA boosts the light absorption by targeted malignant cell without damaging the surrounding healthy tissues and increases the penetration of NIR laser light. Absorption of light by the photosensitizing agent increases the temperature and causes local hyperthermia, which results in thermal ablation of the targeted tumour cell [68].

For developing an effective PTT strategy, an ideal PTA should have good biocompatibility, large absorption cross section, and high photothermal conversion efficiency (in transparent NIR window of 700–1,100 nm) along with the ease in functionalization. PTT has an advantage over conventional therapeutic treatments (like chemotherapy and radiotherapy) as biocompatible nano-sized particles (PTA) cause less side effects and selective thermal destruction of cancerous cell present deep in the tissue. PTT can be coupled with chemotherapy to improve therapeutic effectiveness [69].

There are three operating modes of PTT: (a) light only, (b) light with endogenous substances, and (c) light with exogenous substances. The third option is the most operative for heat generation due to metal's high photothermal conversion efficiency. It should be highlighted, however, that using nanoparticles (NPs) as enhancing PTT agents raises concerns regarding their targeted delivery as well as side effects of metallic nanostructures and potential cytotoxicity. Metal nanostructures can be targeted and delivered to the specific cancer tissues using mechanisms such as antibody–antigen or ligand–receptor interaction. Noble metals like silver and gold are less toxic and biocompatible, making them ideal for PTT applications [70].

## 10.6.2 Mechanism of heat generation in photothermal therapy

Laser light interaction with tissue generates heat, which is the major principle of heat generation mechanism in PTT. Microscopic and macroscopic level studies can be performed to determine the interaction between laser light and tissue. Laser contains energy in the form of high-frequency electromagnetic waves which on interaction with tissue gets converted to heat energy [71]. According to the macroscopic model, heat generation is linked by the penetration of laser light into tissue during laser–tissue interaction. When tissue is irradiated with laser, few beams are reflected and rest of them penetrate deeply into the tissue. The light that gets through is susceptible to absorption and scattering. The absorption coefficient is defined as the loss of incident radiation with depth. The scattering coefficient is characterized as the loss of laser energy due to scattering per penetration length unit. These two coefficients are tissue specific and laser wavelength dependent. Beer's law describes the beam attenuation in the case of laser light. To begin, consider an absorber rather than a scattering sample. Beer's law describes the beam attenuation according to which the heat generated in tissue is dependent on the absorbed energy. Commonly, the light falling on the sample is simultaneously absorbed and scattered. Beer's law still applies to this beam attenuation. The attenuation coefficient, also referred to as the total attenuation coefficient, is the sum of absorption and scattering coefficients [72].

In case of microscopic model, excitation of the molecule occurs from normal to excited state when energy in the form of photons hits a molecule, as shown in Figure 10.3 [73]. By the neighbouring M1 (a molecule, an atom, or an electron), the excited molecule experiences an inelastic collision, which transfers some of its energy and causes it to decay into a stable state molecule. Due to microscopic increase in temperature, thermal vibration of M1 molecule also increases. The concentrated M-type molecules absorb energy from large photon flux by laser beam, and this energy is converted to thermal vibration, and heat is responsible for temperature increase.

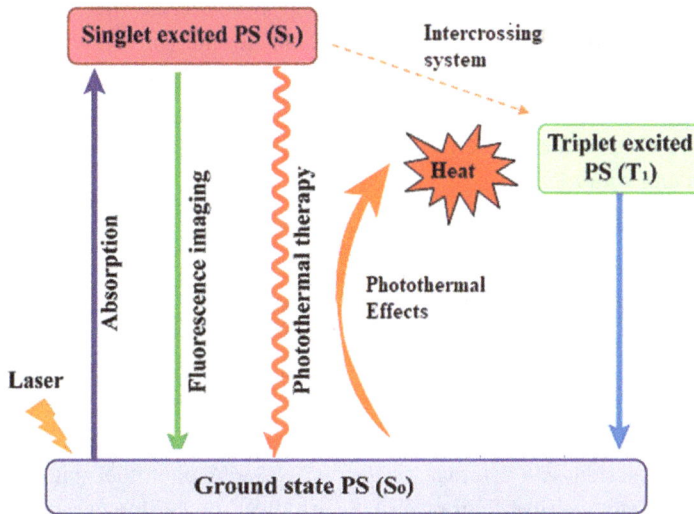

**Figure 10.3:** Representation of mechanism involved in photothermal therapy (*Lasers in Medical Science* 2008:23 217–228, originally published by and used with permission from Springer Nature).

Both necrosis and apoptosis are common causes of cell death in PTT. Necrosis is defined as the uncontrolled cell death due to interior and exterior pressures such as chemical pathogens or mechanistic injuries. This results in breakdown of plasma lemma integrity and subsequent leakage of protoplasm. Necrosis is an undesirable natural process of cell death as aberrant release can cause harmful inflammatory and immunogenic responses

[74]. Apoptosis is a programmed cell death when cells preserve their membrane integrity by phagocytosis and prevent inflammation. Apoptosis can result in secondary necrosis, causing losses to the cell membrane integrity and releases damage-associated molecular patterns but does not activate the engulfment process [75]. To date, necrosis has been described as the common in vitro immune response to phototherapy; however, few studies indicate apoptosis as the main mechanism of cell death by light exposure [76].

If no external agent is utilized, the PTT process is unable to distinguish between normal and malignant cells, thus causing damage to both. Superficial healthy tissue will be subject to more damage because of more light exposure (i.e. larger energy) as compared to diseased tissue present deep inside. It is important to mark tumour cells with NPs carrying light absorption agents to vary the response of tumour cell by increasing its energy absorption for selective photothermal treatment. The use of NPs enables the variety of unique features and specific heat responses from body cells. Due to enhanced permeability and retention effect, nanomaterials in PTT are injected either intravenously or intratumorally [77].

The conduction-band electrons in NPs create synchronized oscillation which generates heat when tumour is exposed to light at resonant energy. This heat can bring permanent cellular harm and reduction in tumour size by increasing the temperature. The rise in temperature of this tumour depends upon three factors: the amount of delivered light, NP concentration in tumour, and photothermal conversion efficiency of NPs. NPs which can absorb NIR light of wavelengths (650–1,064 nm) have greater chances to cause thermal ablation. This specific wavelength range can deeply penetrate to biological tissues than visible wavelength due to its minimum absorption by haemoglobin and water [78].

## 10.7 Carbon dots in photothermal therapy

CDs act as PTAs due to the following reasons: (a) photothermal conversion efficiency, (b) rich in π-electrons, and (c) substantial temperature variations when irradiated [79]. In recent years, CDs have been used as photosensitizing agents like red-emissive CDs (R-CDs), which show red emission, low cytotoxicity, good QY (22.9%), photothermal efficiency (43.9%), and two-photon excited fluorescence (Figure 10.4). The R-CDs can convert laser energy to heat energy very quickly upon irradiating for 10 min with laser. When the concentration of R-CDs increases by 20–200 μg/mL, the growth of MCF-7 cells is significantly reduced [80].

Geng et al. prepared N-O-CDs (nitrogen- and oxygen-doped) which are irradiated by laser of low power density. It causes 100% reduction of tumour with minimum side effects. The prepared N-O-CDs show photostability and biocompatibility, and has strong NIR absorbance [81]. S- and Se-co-doped CDs were prepared by using diphenyl diselenide and polythiophene as source of carbon. The doping enhances photothermal

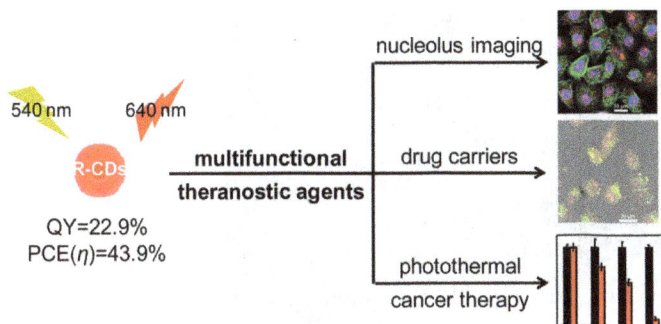

**Figure 10.4:** Red-emissive CDs and their applications (*Carbon* 2020:162,220–233, originally published by and used with permission from Elsevier Ltd.).

conversion efficiency to 58.2%, making applicable as multidimensional phototheranostic agent for cancer treatment [82].

Wang et al. prepared nitrogen- and boron-doped CDs (N-B-CDs) showing NIR absorption (1,000 nm) to enhance deep tissue penetration. The conversion of NIR to heat by N-B-CDs exhibits photothermal therapeutic effect by inhibiting tumour growth and killing cancer cells. N-B-GQDs exhibit safe profile, rapid excretion in mice, and prolonged half-life in blood, which make them compatible for biomedical applications [47]. In another study, iron-doped CDs (Fe-CDs) with size of ~3 nm are synthesized, which exhibit effective photo-enhanced enzyme-like characteristics and photothermal conversion. Fe-CDs act as PTA and nanozyme that exhibits antibacterial ratio against *E. coli* and *S. aureus*. This study shows the wound healing efficacy of Fe-CDs by increasing fibroblast proliferation and angiogenesis, and preventing infection and collagen deposition [83].

CDs with photostability, controlled size, and low biotoxicity are excellent contrast agents for optical imaging. Mostly, CDs activated by ultraviolet light exhibit visible/NIR emissions of less than 820 nm, hence, reducing bioimaging applications due to low penetration depth. It is necessary to synthesize photothermal-based CDs with NIR-II emission I in the range of 1,000–1,700 nm. Li et al. hydrothermally prepared NIR-II-emissive CDs from watermelon which possessed biocompatibility, photothermal conversion efficiency, photostability, QY, and renal clearance [84], as depicted in Figure 10.5.

In another study, sulphur- and nitrogen-co-doped NIRCDs were prepared via solvothermal method from citric acid, dimethyl sulphoxide, and urea as precursors for carbon, nitrogen, and sulphur, respectively. The prepared S,N-CDs exhibit excellence in photo-caustic imaging, photoluminescence imaging and in PTT [85]. Permatasari et al. prepared pyrrolic-N-rich CDs derived from concentrated urea by microwave-assisted hydrothermal method. These N-CDs have first NIR absorption peak at 650 nm and have negative charge on surface. They act as multifunctional carriers in cell imaging, targeted drug delivery, and PTT [86].

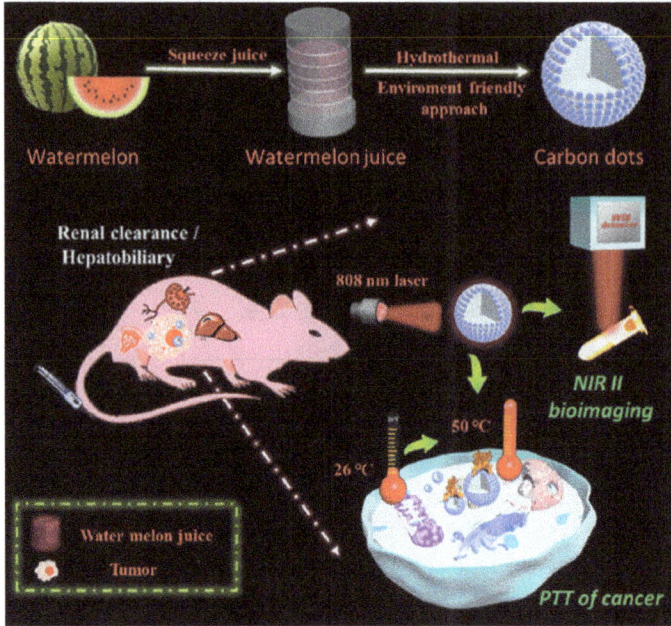

**Figure 10.5:** Synthesis mechanism of NIR-II-emissive CDs from watermelon (*Bioactive Material* 2022:12,246–256, originally published by and used with permission from Elsevier Ltd.).

Zhang et al. designed an approach by combining PDT and PTT. In their study, a nanostructure has been developed from the hybrid mixture of iron oxide CDs ($Fe_3O_4$-CDs) and phosphorene quantum dots (BPQDs) known as genipin [GP]-polyglutamic acid [PGA]-$Fe_3O_4$-CDs@BPQDs, as shown in Figure 10.6. This nanostructure has biocompatibility and photodegradability, and exhibits sturdy light absorption band [87].

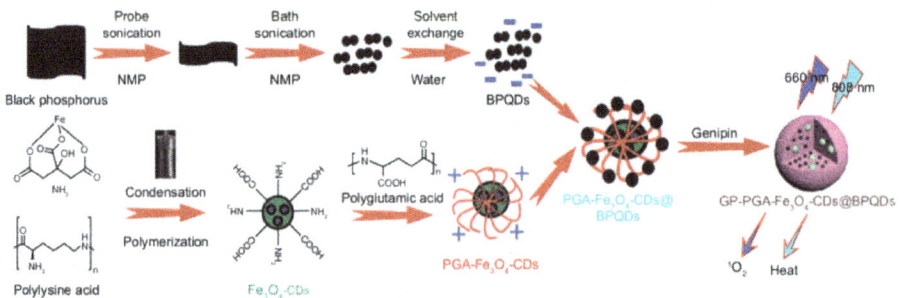

**Figure 10.6:** Preparation mechanism of GP-PGA-$Fe_3O_4$-CD@BPQD nanocomposite (*International Journal of Nanomedicine* 2018:13 2803–2819, originally published by and used with permission from Dove Medical Express Ltd.).

Huang et al. hydrothermally formulated S-, Se-CDs (sulphur, selenium co-doped CDs) with good fluorescence QY. These nanostructures have antioxidant property due to SH and Se-SH groups which enhance the reactive oxygen species scavenging [88]. Peng et al. introduced simple, ecological, and economic method to synthesize CDs decorated with Prussian blue dye (CDs@PBNP). The nanocomposite has combined characteristics of CDs such as biocompatibility, photoluminescence, and photothermal conversion ability of Prussian blue [89]. Knowing the importance of CDs as photothermal agents, a comprehensive view of different materials reported is given in Table 10.2 with photothermal efficiency, quantum yield and applications.

**Table 10.2:** CDs acting as photothermal agent, photothermal efficiency, quantum yield, and applications.

| Photothermal therapy (PTT) agents | Photothermal efficiency | Quantum yield (QL) | Applications | Ref |
|---|---|---|---|---|
| NIR–II–CDs | ~81.3% | -N/A- | Tumor treatment | [77] |
| Red-emissive CDs (R-CDs) | 43.9% | (22.9%) | Nucleolus imaging, cancer therapy and drug carrier | [78] |
| N-O-CDs | 38.3% | 16.1% | Cancer therapy | [79] |
| S-, Se-co-doped CDs | ~58.2% | ~0.2% | Photothermal therapy of cancer | [80] |
| Near-IR fluorescent CyCD | 38.7% | 5.7% | Cancer imaging and therapy | [90] |
| Fe-CDs | 35.11% | -N/A- | Wound healing and antibiotic therapy | [83] |
| NIR-II CDs | 30.6% | 0.5% | Bioimaging and photothermal therapy of cancer | [82] |
| S-, N-CDs | 59% | -N/A- | Biomedical applications | [83] |
| Pyrrolic-N-rich CDs | 54.3% | -N/A- | Bioimaging, drug delivery, and photothermal therapy | [84] |

## 10.8 Clinical challenges of CD-based photothermal therapy

CD-based materials designed for PTT have gained growth, but smart CDs of clinical value remain in infancy. CDs hold the potential but there are clinical challenges that hinder their applications. Focused efforts are required to implement cost-effective, low-toxic, and biocompatible carbon nanomaterials such as PTT solutions in biomedical applications. The challenges focused on commercial and clinical aspects of PTT based on CDs are discussed:

i.   CD nanomaterials should be developed via simple and low-cost approaches, in-
     cluding the diversity of mass production at industrial scale. Issues such as con-
     trolling size, batch-to-batch reproducibility, and purification must be considered
     for large-scale synthesis.
ii.  The development of multifunctional CDs will endure expansion of CD-related bio-
     medical significance. This can be perceived by designing new precursors as car-
     bon sources with properties like strong emission in NIR-I (700–900 nm) or NIR-II
     (1,100–1,600 nm) with high QY and multimodal bio-imaging functions in combina-
     tion with imaging technologies such as MRI and CT. Such hyphenation will extend
     the application of CDs for cardiovascular, cerebrovascular, and respiratory sys-
     tem diseases.
iii. Introduction of CD assemblies with improved hydrophobicity or hydrophilicity
     and light response in NIR region is vital. Such self-assembled CDs system can re-
     move limitations as short circulation time in blood, in vivo instability, and rela-
     tively low tumor-homing ability.
iv.  PTT with CDs must focus on increasing the efficacy, accuracy, and safety of the
     treatment. This demands systematic and detailed studies for exploration of their
     biological effects as biodistribution, metabolism, biodegradation, and prolonged
     toxicology and secretion of CDs.

These challenges require concern from researchers and clinicians to translate current
CD nanomaterials from the bench to the bedside.

## 10.9 Conclusion

CD development has offered practical approaches in PTT. It is believed that advancements in potential CDs will contribute to PTT; however, challenges will continue to hinder
the clinical applications. The research in nanotechnology is required to develop CD-based
PTAs to overcome the hurdles in cancer treatments. The preliminary works reported for
CDs with functionalized surface and various synthetic approaches are promising because
of their properties such as small size, functionalization potential, and the ability to be applied as PTA. The enhanced photoluminescence of CDs can have an advantage in treating
cancer. Combining the therapeutic functionality with improved cancer treatment can address the challenges of PTT. CD-based nanomaterials as PTA are still in the development
phase. They hold the technical capability to develop future therapeutics of targeted and
complete cancer treatment.

# References

[1]    Mansuriya BD, Altintas Z. Carbon dots: Classification, properties, synthesis, characterization, and applications in health care-an updated review (2018–2021). Nanomaterials (Basel) 2021, 11, 2525.

[2]    Molaei MJ. Carbon quantum dots and their biomedical and therapeutic applications: A review. RSC Adv. 2019, 9, 6460–6481.

[3]    Liu J, Li R, Yang B. Carbon dots: A new type of carbon-based nanomaterial with wide applications. ACS Central Sci 2020, 6, 2179–2195.

[4]    Sun YP, Zhou B, Lin Y, Wang W, Fernando KS, Pathak P, Wang H. Quantum-sized carbon dots for bright and colorful photoluminescence. J Am Chem Soc. 2006, 128, 7756–7757.

[5]    Gayen B, Palchoudhury S, Chowdhury J. Carbon dots: A mystic star in the world of nanoscience. J Nanomater 2019, 2019, 3451307.

[6]    Li J, Wang B, Zhang H, Yu J. Carbon dots-in-matrix boosting intriguing luminescence properties and applications. Small 2019, 15, 1805504.

[7]    Zheng XT, Ananthanarayanan A, Luo KQ, Chen P. Glowing graphene quantum dots and carbon dots: Properties, syntheses, and biological applications. Small 2015, 11, 1620–1636.

[8]    Yan Y, Gong J, Chen J, Zeng Z, Huang W, Pu K, Chen P. Recent advances on graphene quantum dots: From chemistry and physics to applications. Adv Mater 2019, 31, 808283.

[9]    Xia C, Zhu S, Feng T, Yang M, Yang B. Evolution and synthesis of carbon dots: From carbon dots to carbonized polymer dots. Adv Sci 2019, 6, 1901316.

[10]   Yuan F, Li S, Fan Z, Meng X, Fan L, Yang S. Shining carbon dots: Synthesis and biomedical and optoelectronic applications. Nano Today 2016, 11, 565–586.

[11]   Xu X, Ray R, Gu Y, Ploehn HJ, Gearheart L, Raker K, Scrivens WA. Electrophoretic analysis and purification of fluorescent single-walled carbon nanotube fragments. J Am Chem Soc 2004, 126, 12736–12737.

[12]   Ren J, Malfatti L, Innocenzi P. Citric acid derived carbon dots, the challenge of understanding the synthesis-structure relationship. C, 2020, 7, 2.

[13]   Namdari P, Negahdari B, Eatemadi A. Synthesis, properties and biomedical applications of carbon-based quantum dots: An updated review. Biomed Pharmacother 2017, 87, 209–222.

[14]   Jiang K, Sun S, Zhang L, Lu Y, Wu A, Cai C, Lin H. Red, green, and blue luminescence by carbon dots: Full-color emission tuning and multicolor cellular imaging. Angew Chem Int Ed 2015, 54, 5360–5363.

[15]   Yoon H, Chang, YH, Song SH, Lee ES, Jin SH, Park C, Kim YH. Intrinsic photoluminescence emission from subdomained graphene quantum dots. Adv Mater 2016, 28, 5255–5261.

[16]   Li H, Kang Z, Liu Y, Lee ST. Carbon nanodots: Synthesis, properties and applications. J Mater Chem 2012, 22, 24230–24253.

[17]   Yu Y, Tang P, Barnych B, Zhao C, Sun G, Ge M. Design and synthesis of core–shell carbon polymer dots with highly stable fluorescence in polymeric materials. ACS Appl Nano Mater 2019, 2, 6503–6512.

[18]   Tao S, Feng T, Zheng C, Zhu S, Yang B. Carbonized polymer dots: A brand new perspective to recognize luminescent carbon-based nanomaterials. J Phys Chem Let, 2019, 10, 5182–5188

[19]   Song Y, Zhu S, Zhang S, Fu Y, Wang L, Zhao X, Yang B. Investigation from chemical structure to photoluminescent mechanism: A type of carbon dots from the pyrolysis of citric acid and an amine. J Mater Chem C 2015, 3, 5976–5984.

[20]   Sun X, Lei Y. Fluorescent carbon dots and their sensing applications. TrAC Trends Anal Chem, 2017, 89, 163–180.

[21]   Tian L, Li Z, Wang P, Zhai X, Wang X, Li T. Carbon quantum dots for advanced electrocatalysis. J Energy Chem 2021, 55, 279–294.

[22]   Miao S, Liang K, Zhu J, Yang B, Zhao D, Kong B. Hetero-atom-doped carbon dots: Doping strategies, properties and applications. Nano Today 2020, 33, 100879.

[23] Ding H, Li XH, Chen XB, Wei JS, Li XB, Xiong HM. Surface states of carbon dots and their influences on luminescence. J Appl Phys 2020, 127, 231101.

[24] Frank BP, Sigmon LR, Deline AR, Lankone RS, Gallagher MJ, Zhi B, Fairbrother DH. Photochemical transformations of carbon dots in aqueous environments. Environ Sci Technol 2020, 54, 4160–4170.

[25] Tajik S, Dourandish Z, Zhang K, Beitollahi H, Van Le Q, Jang HW, Shokouhimehr M. Carbon and graphene quantum dots: A review on syntheses, characterization, biological and sensing applications for neurotransmitter determination. RSC Adv. 2020, 10, 15406–15,429.

[26] Roy P, Chen PC, Periasamy AP, Chen YN, Chang HT. Photoluminescent carbon nanodots: Synthesis, physicochemical properties and analytical applications. Mater Today 2015, 18, 447–458.

[27] Martindale BC, Hutton GA, Caputo CA, Reisner E. Solar hydrogen production using carbon quantum dots and a molecular nickel catalyst. J Am Chem Soc, 2015, 137, 6018–6025.

[28] Sagbas S, Sahiner N. Carbon Dots: Preparation, Properties, and Application. In: Nanocarbon and Its Composites, an Diego, CA, USA: Academic Press, Elsevier. 2019, 651–676.

[29] Essner JB, Baker GA. The emerging roles of carbon dots in solar photovoltaics: A critical review. Environ Sci 2017, 4, 1216–1263.

[30] Singh I, Arora R, Dhiman H, Pahwa R. Carbon quantum dots: Synthesis, characterization and biomedical applications. Turkish J Pharm Sci 2018, 15, 219.

[31] Thongpool V, Asanithi P, Limsuwan P. Synthesis of carbon particles using laser ablation in ethanol. Procedia Eng. 2012, 32, 1054–1060.

[32] Doñate-Buendia C, Torres-Mendieta R, Pyatenko A, Falomir E, Fernández-Alonso M, Mínguez-Vega G. Fabrication by laser irradiation in a continuous flow jet of carbon quantum dots for fluorescence imaging. ACS Omega 2018, 3, 2735–2742.

[33] Kaczmarek A, Hoffman J, Morgiel J, Mościcki T, Stobiński L, Szymański Z, Małolepszy A. Luminescent carbon dots synthesized by the laser ablation of graphite in polyethylenimine and ethylenediamine. Materials 2021, 14, 729.

[34] Xu H, Yan L, Nguyen V, Yu Y, Xu Y. One-step synthesis of nitrogen-doped carbon nanodots for ratiometric pH sensing by femtosecond laser ablation method. Appl Surf Sci 2017, 414, 238–243.

[35] Cui L, Ren X, Wang J, Sun M. Synthesis of homogeneous carbon quantum dots by ultrafast dual-beam pulsed laser ablation for bioimaging. Mater Today Nano, 2020, 12, 100091

[36] Sharma A, Das J. Small molecules derived carbon dots: Synthesis and applications in sensing, catalysis, imaging, and biomedicine. J Nanobiotech 2019, 17, 1–24.

[37] Pan M, Xie X, Liu K, Yang J, Hong L, Wang S. Fluorescent carbon quantum dots—synthesis, functionalization and sensing application in food analysis. J Nanomater 2020, 10, 930.

[38] Wang F, Stahl SS. Electrochemical oxidation of organic molecules at lower overpotential: Accessing broader functional group compatibility with electron– proton transfer mediators. Acc Chem Res 2020, 53, 561–574.

[39] Wang F, Wang S, Sun Z, Zhu H. Study on ultrasonic single-step synthesis and optical properties of nitrogen-doped carbon fluorescent quantum dots. Fuller Nanotub Carbon Nanostruct 2015, 23, 769–776.

[40] Dong Y, Pang H, Yang HB, Guo C, Shao J, Chi Y, Yu T. Carbon-based dots co-doped with nitrogen and sulfur for high quantum yield and excitation-independent emission. Angew Chem Int Ed 2013, 52, 7800–7804

[41] Wu P, Li W, Wu Q, Liu Y, Liu S. Hydrothermal synthesis of nitrogen-doped carbon quantum dots from microcrystalline cellulose for the detection of Fe 3+ ions in an acidic environment. RSC Adv 2017, 7, 44144–44153.

[42] Han Z, He L, Pan S, Liu H, Hu X. Hydrothermal synthesis of carbon dots and their application for detection of chlorogenic acid. Luminescence 2020, 35, 989–997.

[43] Yin H, Yamamoto T, Wada Y, Yanagida S. Large-scale and size-controlled synthesis of silver nanoparticles under microwave irradiation. Mater Chem Phys 2004 83, 66–70.

[44]  Romero MP, Alves F, Stringasci MD, Buzzá HH, Ciol H, Inada NM, Bagnato VS. One-pot microwave-assisted synthesis of carbon dots and in vivo and in vitro antimicrobial photodynamic applications. Front Microbiol 2021, 12, 1455.

[45]  Chung S, Zhang M. Microwave-assisted synthesis of carbon dot–iron oxide nanoparticles for fluorescence imaging and therapy. Front Bioeng Biotechnol 2021, 9, 576.

[46]  Dager A, Baliyan A, Kurosu S, Maekawa T, Tachibana M. Ultrafast synthesis of carbon quantum dots from fenugreek seeds using microwave plasma enhanced decomposition: Application of C-QDs to grow fluorescent protein crystals. Sci Rep 2020, 10, 1–15.

[47]  Wang H, Mu Q, Wang K, Revia RA, Yen C, Gu X, Zhang M. Nitrogen and boron dual-doped graphene quantum dots for near-infrared second window imaging and photothermal therapy. Appl Mater Today 2019, 14, 108–117.

[48]  Nguyen V, Zhao N, Yan L, Zhong P, Le PH. Double-pulse femtosecond laser ablation for synthesis of ultrasmall carbon nanodots. Mater Res Express 2020, 7, 015606.

[49]  Biazar N, Poursalehi R, Delavari H. Optical and structural properties of carbon dots/TiO2 nanostructures prepared via DC arc discharge in liquid. Paper presented at the AIP Conference Proceedings 2018.

[50]  Dey S, Govindaraj A, Biswas K, Rao C. Luminescence properties of boron and nitrogen doped graphene quantum dots prepared from arc-discharge-generated doped graphene samples. Chem Phys Lett 2014, 595, 203–208.

[51]  Hou Y, Lu Q, Deng J, Li H, Zhang Y. One-pot electrochemical synthesis of functionalized fluorescent carbon dots and their selective sensing for mercury ion. Anal Chim Acta 2015, 866, 69–74.

[52]  Tabaraki R, Sadeghinejad N. Microwave assisted synthesis of doped carbon dots and their application as green and simple turn off–on fluorescent sensor for mercury (II) and iodide in environmental samples. Ecotoxicol Environ Saf 2018, 153, 101–106.

[53]  Małolepszy A, Błonski S, Chrzanowska-Giżyńska J, Wojasiński M, Płocinski T, Stobinski L, Szymanski Z. Fluorescent carbon and graphene oxide nanoparticles synthesized by the laser ablation in liquid. Appl Phys A 2018, 124, 1–7.

[54]  Li H, He X, Liu Y, Huang H, Lian S, Lee ST, Kang Z. One-step ultrasonic synthesis of water-soluble carbon nanoparticles with excellent photoluminescent properties. Carbon 2011, 49, 605–609.

[55]  Fang J, Xie Z, Wallace G, Wang X. Co-deposition of carbon dots and reduced graphene oxide nanosheets on carbon-fiber microelectrode surface for selective detection of dopamine. Appl Surf Sci 2017, 412, 131–137.

[56]  Xu J, Wang C, Li H, Zhao W. Synthesis of green-emitting carbon quantum dots with double carbon sources and their application as a fluorescent probe for selective detection of Cu2+ ions. RSC Adv 2020, 10, 2536–2544.

[57]  Demchenko AP. Excitons in carbonic nanostructures. C 2019, 5, 71.

[58]  Bhartiya P, Singh A, Kumar H, Jain T, Singh BK, Dutta P. Carbon dots: Chemistry, properties and applications. J Ind Chem Soc 2016, 93, 759–766.

[59]  Wang X, Qu K, Xu B, Ren J, Qu X. Microwave assisted one-step green synthesis of cell-permeable multicolor photoluminescent carbon dots without surface passivation reagents. J Mater Chem 2011, 21, 2445–2450.

[60]  Wang Y, Kim SH, Feng L. Highly luminescent N, S-Co-doped carbon dots and their direct use as mercury (II) sensor. Anal Chim Acta 2015, 890, 134–142.

[61]  Anilkumar P, Wang X, Cao L, Sahu S, Liu JH, Wang P, Sun YP. Toward quantitatively fluorescent carbon-based "quantum" dots. Nanoscale 2011, 3, 2023–2027.

[62]  El-Shabasy RM, Farouk Elsadek M, Mohamed Ahmed B, Fawzy Farahat M, Mosleh KN, Taher MM. Recent developments in carbon quantum dots: Properties, fabrication techniques, and bio-applications. Processes 2021, 9, 388.

[63]  Liu M. Optical properties of carbon dots: A review. Nanoarchitectonics 2020, 1–12.

[64] Chung E, Vitkin A. Photon mayhem: New directions in diagnostic and therapeutic photomedicine. Biomed Eng Lett, 2019, 9, 275–277.

[65] Yoo SW, Oh G, Ahn JC, Chung E. Non-oncologic applications of nanomedicine-based phototherapy. Biomedicines 2021, 9, 113.

[66] Wagalgave SM, Birajdar SS, Malegaonkar JN, Bhosale SV. Patented AIE materials for biomedical applications. Prog Mol Biol Transl Sci 2021, 185, 199–223.

[67] Nomura S, Morimoto Y, Tsujimoto H, Arake M, Harada M, Saitoh D, Takayama E. Highly reliable, targeted photothermal cancer therapy combined with thermal dosimetry using a near-infrared absorbent. Sci Rep 2020, 10, 1–7.

[68] Eskiizmir G, Baskın Y, Yapıcı K. Graphene-based Nanomaterials in Cancer Treatment and Diagnosis. In: Fullerens, Graphenes and Nanotubes, Elsevier, 2018, 331–374.

[69] Fong JF, Ng YH, Ng SM. Carbon Dots as a New Class of Light Emitters for Biomedical Diagnostics and Therapeutic Applications. In: Fullerens, Graphenes and Nanotubes, ed, Press, Elsevier, 2018, 227–295.

[70] Li Z, Chen Y, Yang Y, Yu Y, Zhang Y, Zhu D, Zhao Y. Recent advances in nanomaterials-based chemo-photothermal combination therapy for improving cancer treatment. Front Bioeng Biotechnol 2019, 7, 293.

[71] Fong JFY, Ng YH, Ng SM. Carbon Dots as a New Class of Light Emitters for Biomedical Diagnostics and Therapeutic Applications. In: AM. Grumezescu (Ed.), Fullerens, Graphenes and Nanotubes, William Andrew Publishing, 2018, 227–295.

[72] Gellci K, Mehrmohammadi M. Photothermal Therapy. In: M. Schwab (Ed.), Encyclopedia of Cancer Berlin, Heidelberg: Springer Berlin Heidelberg, 2014, 1–5.

[73] Huang X, Jain PK, El-Sayed IH, El-Sayed MA. Plasmonic photothermal therapy (PPTT) using gold nanoparticles. Lasers Med Sci 2008, 23, 217–228.

[74] Cullen JM. 9.07 - Histologic Patterns of Hepatotoxic Injury*. In: CA. McQueen (Ed.), Comprehensive Toxicology, Second Edition, Oxford: Elsevier, 2010, 141–173.

[75] Pfeffer CM, Singh ATK. Apoptosis: A target for anticancer therapy. Int J Mol Sci 2018, 19, 448.

[76] Melamed JR, Edelstein RS, Day ES. Elucidating the fundamental mechanisms of cell death triggered by photothermal therapy. ACS Nano 2015, 9, 6–11.

[77] Wang, C., Li, Y., Shi, X., Zhou, J., Zhou, L., & Wei, S. (2018). Use of an NIR-light-responsive CO nanodonor to improve the EPR effect in photothermal cancer treatment. Chem Comm, 54 (95),13403–13,406. doi:10.1039/C8CC07873D

[78] Vines JB, Yoon JH, Ryu NE, Lim DJ, Park H. Gold nanoparticles for photothermal cancer therapy. Front Chem 2019, 7, 167.

[79] Geng B, Shen W, Fang F, Qin H, Li P, Wang X, Shen L. Enriched graphitic N dopants of carbon dots as F cores mediate photothermal conversion in the NIR-II window with high efficiency. Carbon 2020 162, 220–233.

[80] Sun S, Zhang L, Jiang K, Wu A, Lin H. Toward high-efficient red emissive carbon dots: Facile preparation, unique properties, and applications as multifunctional theranostic agents. Chem Mater 2016, 28, 8659–8668.

[81] Geng B, Yang D, Pan D, Wang L, Zheng F, Shen W, Li X. NIR-responsive carbon dots for efficient photothermal cancer therapy at low power densities. Carbon 2018, 134, 153–162.

[82] Lan M, Zhao S, Zhang Z, Yan L, Guo L, Niu G, Zhang W. Two-photon-excited near-infrared emissive carbon dots as multifunctional agents for fluorescence imaging and photothermal therapy. Nano Res 2017, 10, 3113–3123.

[83] Liu Y, Xu B, Lu M, Li S, Guo J, Chen F, Zhou D. Ultrasmall Fe-doped carbon dots nanozymes for photoenhanced antibacterial therapy and wound healing. Bioact Mater 2022, 12, 246–256.

[84] Li Y, Bai G, Zeng S, Hao J. Theranostic carbon dots with innovative NIR-II emission for in Vivo renal-excreted optical imaging and photothermal therapy. ACS Appl Mat Interf 2019, 11, 4737–4744.

[85] Bao X, Yuan Y, Chen J, Zhang B, Li D, Zhou D, Qu S. In vivo theranostics with near-infrared-emitting carbon dots—highly efficient photothermal therapy based on passive targeting after intravenous administration. Light Sci Appl, 2018, 7, 91.

[86] Permatasari FA, Fukazawa H, Ogi T, Iskandar F, Okuyama K. Design of pyrrolic-N-rich carbon dots with absorption in the first near-infrared window for photothermal therapy. ACS Appl Nano Mater 2018, 1, 2368–2375.

[87] Zhang M, Wang W, Cui Y, Zhou N, Shen J. Near-infrared light-mediated photodynamic/photothermal therapy nanoplatform by the assembly of Fe3O4 carbon dots with graphitic black phosphorus quantum dots. Int J Nanomedicine 2018, 13, 2803–2819.

[88] Huang G, Lin Y, Zhang L, Yan Z, Wang Y, Liu Y. Synthesis of sulfur-selenium doped carbon quantum dots for biological imaging and scavenging reactive oxygen species. Sci Rep 2019, 9, 19651.

[89] Peng X, Wang R, Wang T, Yang W, Wang H, Gu W, Ye L. Carbon dots/Prussian blue satellite/core nanocomposites for optical imaging and photothermal therapy. ACS Appl Mater Interfaces 2018, 10, 1084–1092.

[90] Zheng M, Li Y, Liu S, Wang W, Xie Z, Jing X. One-pot to synthesize multifunctional carbon dots for near infrared fluorescence imaging and photothermal cancer therapy. ACS Appl Mat Interf 2016, 8, 23533–23541.

Shokoh Parham, Seyedeh-Shirin Parham and Hadi Nur

# Chapter 11
# Carbon dots in antibacterial, antiviral, antifungal, and antiparasitic agents

**Abstract:** One of the main causes of death across the world is infections by microorganisms, including bacterial, fungal, viral, or parasitic agents. Further development of pathogens into multidrug-resistant (MDR) agents leads to serious challenges in the treatment of the above-mentioned diseases, leading to an increase in mortality rates, as well as medical costs. Several studies have recently focused on examining safe antimicrobials with strong antimicrobial effects. Carbon dots (C-dots) are highly biocompatible and less toxic, making them promising candidates due to their favourable antimicrobial characteristics. Another promising strategy to deal with the problem of MDR pathogens is photodynamic inactivation of bacteria using photosensitizers, with C-dots as one of their effective members in detecting and inactivating various bacteria species. Hence, C-dots can be applied as efficient alternatives. This chapter aims to investigate the characteristics of C-dots as antibacterial, antiviral, antifungal, and antiparasitic agents.

**Keywords:** Antibacterial agents, antifungal agents, antiparasitic agents, antiviral agents, carbon dots

## 11.1 Introduction

Many researchers have been interested in the possible application of different carbon allotropes at a range of electronic devices to biosensing and bioimaging agents. Accordingly, considerable research has been conducted on graphitic structural materials

**Acknowledgement:** The authors would like to express their sincere thanks to the University of Technology Malaysia (UTM) for providing the platform and research facilities, along with the necessary support.

**Shokoh Parhama,** Department of Biomaterials, Nanotechnology and Tissue Engineering, Faculty of Advanced Technologies in Medicine, Isfahan University of Medical Sciences, Isfahan, Iran; Centre for Sustainable Nanomaterials, Ibnu Sina Institute for Scientific and Industrial Research, Universiti Teknologi Malaysia, 81310 UTM Skudai, Johor, Malaysia
**Seyedeh-Shirin Parham,** Department of Veterinary, Shahrekord Branch, Islamic Azad University, Shahrekord, Iran
**Hadi Nur,** Department of Chemistry, Faculty of Mathematics and Natural Sciences, Universitas Negeri Malang, Malang, Indonesia; Central Laboratory of Minerals and Advanced Materials, Faculty of Mathematics and Natural Science, Universitas Negeri Malang, Malang, Indonesia

https://doi.org/10.1515/9783110799958-011

such as zero-dimensional spherical fullerene, diamond nanocrystals and carbon dots (C-dots), 1D cylindrical carbon nanotubes (CNTs), and 2D graphene quantum dots, and graphene in the past few years [1–7].

C-dots belong to the fluorescent carbon materials possessing a diameter of <10 nm and providing good alternatives to quantum dots based on metals as they have a unique composition and are biocompatible. The distinctive fluorescence characteristics, along with being biocompatible and less toxic, have led to different investigations on these agents in biosensing, gene transmission, drug delivery, and bioimaging probes. Research has focused on the use of C-dots in analytical chemistry, particularly environmental as well as biological sensing and imaging due to their excellent fluorescence characteristics [8–13].

Hence, the fast outspread of infections is constantly threatening the health of humans, complicated by microbial resistance to antibiotics. These conditions make it necessary to develop new microbial probes and antimicrobial agents to deal with the problem of antibiotic resistance. Fortunately, antimicrobial nanomaterials are currently showing promising results in the treatment of infection. One of these agents is a carbon nanoparticle with a size range of $\leq 10$ nm, known as C-dots, and contributes to microbial imaging, detecting, and inactivating because of its great optical characteristics, modifiability of surface, and highly biocompatible nature [14–18]. Totally, C-dots can be described as carbon quantum dots at a size of $\leq 10$ nm, and luminescent carbon nanoparticles represent surface passivation and contribute as good candidates in the fields of medicine, bioimaging, sensing, electronic devices, and catalysis [14]. To the best knowledge of authors, the existing studies on the application of these materials in imaging and eliminating microorganisms have been primarily conducted on bacteria with scant research on fungal and viral agents [14, 19–29]. This chapter has focused on the latest breakthroughs in the application of C-dots to eliminate bacterial, fungal, viral, and parasitic microorganisms.

## 11.2 Carbon dots (C-dots)

One of the famous materials worldwide is carbon, whose nanosized structures show significantly unique characteristics after preparation. C-dots can be characterized as nanostructures, possessing fundamental oxygen and hydrogen fractions. Researchers have become more interested in these materials because of their optical characteristics and particularly their fluorescence emissions and usages. C-dots have strong fluorescent and non-blinking properties, along with convenient and cost-effective synthesis. It is possible to tune their emission colour using a variety of excitation wavelengths [30]. C-dots have interesting physicochemical characteristics such as extremely small size (<10 nm), high functional groups, and fluorescence while being chemically stable, biocompatible, and non-toxic, making them another epoch-making carbon-based nanomaterial after fullerene, nanotubes, and graphene [31]. C-dots are a new member of the carbon family, attracting a lot of interest in chemistry because of their rich characteristics [32].

From the historical perspective, many great scientific achievements have come true by accident. Xu et al. [33] initially introduced and characterized C-dots as "fluorescent nanoparticles" when purifying single-walled CNTs by the preparative electrophoresis approach. Yet, researchers did not pay much attention to this topic until Sun et al. [34] introduced the term "carbon quantum dots", used for their precise distinction from the broad group of carbon nanoparticles, including carbon black [31, 35].

The accurate definition and classification of C-dots are difficult because of their diverse nanostructures. Some studies have provided very loose definitions of C-dots as "fluorescent nanoparticles" and "carbogenic dots" [33, 36]. According to different studies in the relevant literature, C-dots have been primarily classified into five groups, taking into account their diverse carbon cores. This classification has introduced graphene quantum, graphitic carbon nitride quantum, carbon quantum, carbon nanodots, and carbonized polymer dots [31, 37–40].

C-dots belong to novel nanomaterials with promising applications in research conducted on next-generation biomedical imaging techniques and mechanisms to deliver drugs. Thus, familiarity with the precise interaction mechanisms of these materials with biomolecules can lead to better design of biosensors [41]. Research has also focused on the antimicrobial effects of C-dots, indicating their significantly great activities in this regard [41–43]. Different classes of C-dots are presented in Figure 11.1 [31].

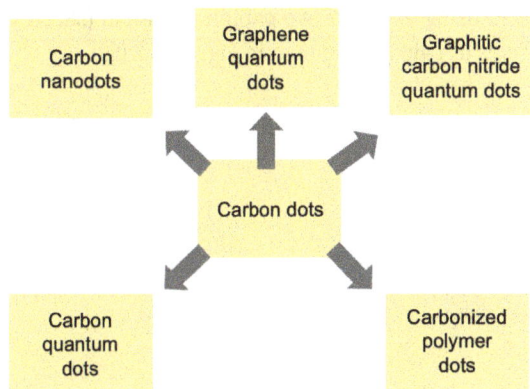

**Figure 11.1:** Different classes of C-dots [31].

## 11.3 Medicinal characteristics of carbon dots

According to a typical definition, C-dots are nanomaterials based on carbon having, different physicochemical properties, and desirable characteristics. They are biocompatible with distinctive optical characteristics, cost-effective, eco-friendly, and highly stable, with numerous functional groups (including amino, hydroxyl, and carboxyl). C-dots have wide applications in biomedicine, including biomedical imaging (in vivo and in vitro), phototherapy (photodynamic and photothermal therapies), delivery of drugs/genes, and nanomedicine. The chemical process of charring is complicated and covers the fields of chemistry, pharmacology, biochemistry, and so on. Scientists believe that performing this process at high temperatures would make it possible to derive a material basis for the impacts of charcoal medications [44–46].

Accordingly, the incorporation of nanomaterial science and herbal products can possibly lead to extensive procedures for the diagnosis and treatment of different health problems, including cancers and neurological complications. C-dots derived from natural products (NPdCDs) possess significant characteristics in the diagnosis and treatment of cancers, microbial imaging, and sensing, as well as delivery of drugs. These plant-derived materials reveal excellent medicinal characteristics while also being biocompatible, photostable, and easily functionalized due to the variety of bioactive phytomolecules, leading to their extensive range of applications [47]. Besides, the medicinal applications of C-dots have been reported in several studies [48–56], and one of which provided a technique to prepare C-dots from medicinal plants and examined the products' possible antibacterial impacts on four ordinary infectious pathogens [57]. Yet, research has also reported considerable fluorescent characteristics of C-dots for in vitro cellular imaging applications. The photo trigger abilities of quinoline were used to efficiently release anticancer medications utilizing one- and two-photon excitation [58]. The medicinal characteristics and applications of C-dots are shown in Figure 11.2 [59].

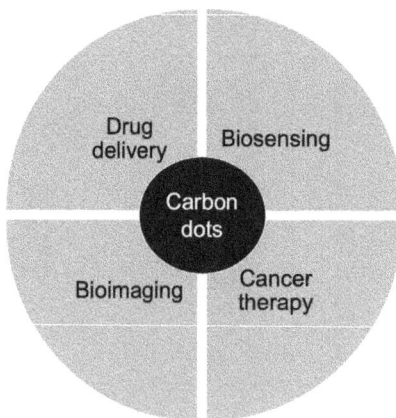

**Figure 11.2:** Medicinal characteristics and applications of C-dots [59].

## 11.3.1 The antibacterial properties of C-dots

One of the serious threats to the safety and health of humans is related to pathogenic diseases caused by bacteria. Research on the use of nanoparticles for imaging and diagnosis has currently shown the significant role of these materials in the management of infections. In the meantime, C-dots have been the centre of attention because of their distinctive optical and physicochemical characteristics, along with their eco-friendly nature, making them good alternatives to perform imaging and detect pathogenic bacterial diseases [60]. Several studies have focused on the antibacterial effects of C-dots [61–65]. A study focused on the preparation of the amphiphilic cow milk-derived C-dot (ACMCD)-supported silver nanoparticles (ACMCD-Ag nanocomposites) and their application against gram-positive (*Staphylococcus aureus*) and gram-negative (*E. coli*) bacteria [66]. Other studies have introduced facile active structure preservation methods for the direct development of novel self-functional graphitic C-dots, exhibiting efficient antibacterial effects on the above-mentioned bacteria [67]. Yet, another study focused on the henna plant for the synthesis of C-dots, suggesting that these C-dots can eliminate gram-positive and gram-negative bacteria through mechanisms similar to antibiotics. It was also found that C-dots exert their antibacterial effects in considerably lower concentrations [68]. Another study highlighted different dimensions of C-dots in terms of their visible/natural light-induced microbicidal characteristics while reviewing several research works on the antimicrobial effects of these materials and providing more insights concerning the challenges and chances for developing and practical use of these antimicrobial agents [69]. Research has also focused on the separation of chlorhexidine gluconate-derived C-dots in three groups of small-, middle-, and large-scale particle sizes, utilizing various molecular weight cut-off membranes, exhibiting considerable antibacterial effects on *E. coli* and *S. aureus*, increasing the permeability of bacterial cells, leading to synergistic destabilization, and breaking the plasma membrane integrity. It is noteworthy that according to the study results, the antibacterial effects increased with a decrease in the size of C-dots [70]. Hence, these nanomaterials were introduced as antibacterial alternatives.

## 11.3.2 The antifungal properties of C-dots

Over the past few years, yeast infections have significantly contributed to nosocomial infectious conditions because of changes made in the patient's immune systems. As the use of antifungal agents increases, there is also an increase in the cases of resistance to drugs [71]. C-dots, as novel nanomaterials possessing antimicrobial and antifungal effects, have been recently the centre of attention by the public [72]. Many studies have referred to the antifungal effects of these materials [62–75]. This research has focused on the toxicological impacts of a new hydrophilic nanoconjugate gold@carbon dot (Au@CD) and C-dots on *Candida albicans* [76]. Other studies have shown evidence for the preparation of C-dots with the use of glucose as a carbon source (GCD) and doped by heteroatoms, including boron, sulphur, and nitrogen for their performance enhancement. GCDs doped with nitrogen show strong antifungal impacts on *Ammophilus fumigatus*, *Penicillium citrinum*, and *C. albicans*. On the other hand, GCDs doped with sulphur show greater inhibition impacts on the *Fusarium solani* development [77]. The synthesis of multifunctional C-dots has also been performed using hydrothermal processes from metabolites (postbiotics) secreted from baker's yeast [78]. There have been also some reports on the synthesis of C-dots doped with nitrogen and iodine as peroxidase mimics to disinfect against *C. albicans* [79]. Yet, another study conducted the synthesis of multifunctional C-dots, showing anticancer and antifungal impacts. Lower cytotoxic features, fluorescent stability, and significant effectiveness in photothermal conversion have enabled C-dots to play a role in imaging-guided photothermal therapies. Besides, C-dots exhibit internal antifungal impacts, which can be evident even at lower concentrations [80]. Recent reports by Li et al. have shown that vitamin C-derived C-dots with less toxicity (at particle sizes of 1.03–1.11 nm) are considerably biodegradable. The complete degradation of these biodegradable C-dots into $CO_2$, CO, and $H_2O$ has been exposed to visible light in the atmosphere or at 37 °C following a 20 days period while maintaining significant antibacterial and antifungal effects at lower concentrations (50–100 µg mL$^{-1}$). Easy degradation of C-dots into $CO_2$ and CO using various growth media suggests the possibility of their in vivo degradation [81, 82], leading to their application as antifungal agents.

### 11.3.3 The antiviral and antiparasitic properties of C-dots

Infectious diseases have always posed considerable health challenges worldwide, particularly because of the growing virus resistance and unfavourable side-effects of long-term medication consumption, which adversely affect the efficacy of antiviral treatments. Hence, it is necessary to develop safe and strong therapeutic agents to replace ordinary antiviral medications [83]. As C-dots are gradually used in antiviral studies, it is hoped to introduce novel antiviral C-dots, which are highly biocompatible and effective [84].

The antiviral effects of C-dots on a variety of viral agents such as coronavirus have been shown in several studies [85–89]. The activities of C-dots against viral replication were examined in a study, in which pseudorabies and porcine reproductive and respiratory syndrome viruses were used as the models of DNA and RNA viruses, correspondingly [90]. One type of benzoxazine monomer-derived C-dots was examined in a different study, demonstrating that they could block the infections and inhibit the fatal Flaviviridae (Japanese encephalitis, Zika, and dengue viral agent) and non-enveloped viral agents (porcine parvovirus and adeno-associated viruses) [91]. In another study, it was sought to briefly describe the possible applications of C-dots in antiviral therapies, especially emphasizing their potential contribution to combating the coronavirus outspread, along with their role in biosensing [92]. Yet, theranostic C-dots were synthesized from *Allium sativum* in another study to introduce new therapeutic armamentariums to manage the COVID-19 pandemic while simultaneously diagnosing infections with this virus [93]. As shown, the antiviral effects of C-dots on human norovirus-like particles, including GI.1 and GII.4, were assessed, by primarily inhibiting the virus binding to HBGA receptors and moderating the inhibition of the virus binding to their antibodies. However, there were no effects on the viral capsid protein and particle integrity [94]. Yet, despite all these studies, the activities of C-dots against viral agents need to become clear in future studies [90]. Although the field of chemotherapy for and prophylaxis of parasite-induced infections has currently faced significant achievements [95], little research has been conducted on the antiparasitic effects of C-dots [96–100]. Another study focused on developing new highly efficient and nontoxic therapies with lower drug resistance against leishmaniasis using C-dots and gallium-doped C-dots (Ga@CDs). Accordingly, preparation of the nanoscale materials at a range of 4–7 nm took place using ultrasonication with no catalysts, followed by their immersion in a commercial ointment. More significant effects of the prepared ointments with C-dots and Ga@CDs were reported with regard to the commercial samples [101], confirming their antiviral and antiparasitic effects. Various compounds of C-dots contributing as antimicrobial agents are shown in Table 11.1 [102–117].

**Table 11.1:** Various carbon dot compounds utilized as antimicrobial agents.

| Materials | Application | Results | Research condition | Reference |
|---|---|---|---|---|
| C-dot-modified $TiO_2$ nanorod (C-$TiO_2$ NR) | Antibacterial effects | Antibacterial effects on *Staphylococcus aureus* | In vitro and in vivo | [102] |
| Glycosylated C-dot-epigallocatechin-3-gallate (gCDs-E) | Antifungal effects | Effective inhibition of the effects of *Candida albicans* | In vitro | [103] |
| Nitrogen-doped C-dots functionalized with copper centres by Cu–N coordination (Cu/NCD) | Antibacterial effects | Antibacterial effects on gram-negative *Escherichia coli* (*E. coli*) | In vitro | [104] |
| Pd/CDs/$Ti^{3+}$-$TiO_2$ | Antibacterial and antifungal effects | Considerably effective antibacterial and antifungal function | In vitro | [105] |
| C-dot nanoparticles | Antiviral effects | Prevention of HIV-1 infections | In vitro | [106] |
| Carbon nanodots (CNDs) from metronidazole | Antibacterial effects | Inhibition of the development of obligate anaerobes, including *Porphyromonas gingivalis* (*P. gingivalis*) directly | In vitro | [107] |

**Table 11.1** (continued)

| Materials | Application | Results | Research condition | Reference |
|---|---|---|---|---|
| Glycyrrhizic acid-based C-dots | Antiviral activity | Highly biocompatible and considerable antiviral effects | In vitro | [108] |
| C-dot-kanamycin sulphate (CD-Kan) | Antibacterial effects | Acceptable inhibition of gram-negative *E. coli* and gram-positive *S. aureus* | In vitro | [109] |
| C-dot-releasing hydrogels | Antibacterial effects | Long-term strong broad-spectrum antibacterial effects (even on bacteria that show drug resistance) | In vitro and in vivo | [110] |
| Copper oxide/carbon (CuO/C) with *Adhatoda vasica* leaf extracts | Antibacterial and antifungal effects | Antibacterial effects on the pathogenic bacterial strains *E. coli*, *Pseudomonas aeruginosa*, *Klebsiella pneumoniae*, and *S. aureus* and antifungal effects on the fungi *Aspergillus niger* and *C. albicans* | In vitro | [111] |
| PVA/C-dot hydrogel | Antibacterial effects | Significant enhancement of the antibacterial effects on *S. aureus* and *E. coli* | In vitro | [112] |
| Silver-graphene quantum dots (Ag-GQDs) | Antibacterial effects | The antibacterial effects on both gram-negative and gram-positive bacteria, utilizing *P. aeruginosa* and *S. aureus* as model bacteria, respectively | In vitro | [113] |
| Amine-terminated C-dots (CDs-NH$_2$) functionalized with ampicillin (AMP) | Antibacterial effects | A significant contribution to inactivating *E. coli* | In vitro | [114] |
| *p*-Phenylenediamine serves as both the carbon source and the origin for the functional group anchored on the derived C-dots | Antibacterial effects | Antibacterial activity against *S. aureus* and *E. coli* growth | In vitro | [115] |
| Amino phenylboronic acid-modified C-dots (APBA-CDs) | Antiviral effects | Inhibition of HIV-1 | In vitro | [116] |
| CDs from levofloxacin hydrochloride (named F-CDs) | Antibacterial effects | Antibacterial effects on *E. coli*, *P. aeruginosa*, *S. aureus*, and *Bacillus subtilis* | In vitro | [117] |

## 11.4 Antimicrobial functional mechanism

C-dots lead to an inhibition of the bacterial growth or eliminate them by using complicated mechanisms of action such as the production of ROS, disintegrating the structure of
the cells, and fragmenting and condensing genomic DNA, resulting in the cytoplasm leakage [82]. The excellent antibacterial effects (on both gram-positive and gram-negative bacteria) were reported by Dou et al. for multifunctional C-dots derived from glucose and
poly(ethylenimine) and quaternized with benzyl bromide, although the synthesis took
place over a 12-h period. As the quaternized C-dots have positive surface charges, they attach to the cell membrane of bacteria and disrupt it [118]. These effects are attributed to
the surface charges of C-dots and ROS production. As Bing et al. indicated in their study,
spermine-functionalized C-dots with positive charges exerted antibacterial effects on
*E. coli* primarily by ROS production as they mainly disrupted the electrostatic interactions
[119]. Accordingly, different potential pathways have been proposed in various studies for
the antimicrobial effects of these materials, including physical/mechanical damage, ROS-
induced, oxidative stress, photocatalytic impacts, and bacterial metabolism inhibition
[120–122]. The first mechanism, which includes physical/mechanical destruction of the bacterial cell wall or external membrane, results in cytoplasm leakage, subsequently leading
to its dysfunction and bacteriostatic as well as bactericidal impacts as the most prevalent
antibacterial mechanisms. As an example, C-dots with less toxicity and higher biodegrad-
ability were provided by Li et al., using a one-step electrochemical therapeutic method of
vitamin C [24, 120]. Antimicrobial mechanisms of C-dots are presented in Figure 11.3 [82].

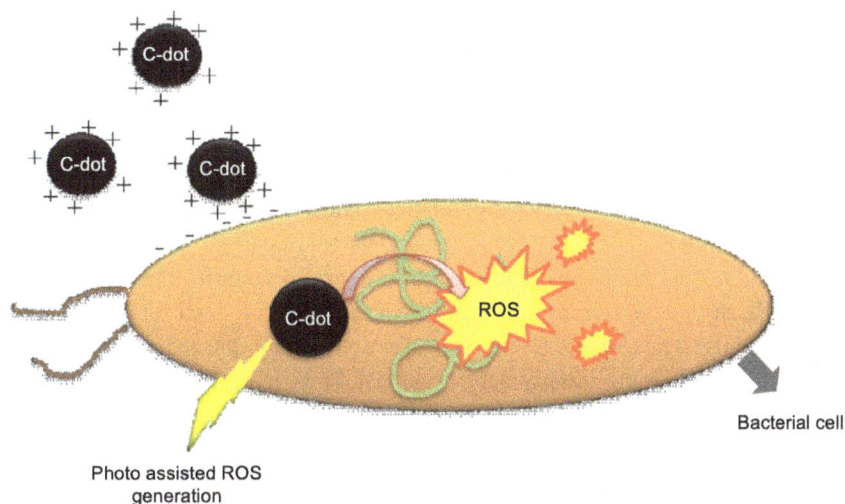

**Figure 11.3:** Antimicrobial mechanisms of C-dots [82].

## 11.5 Conclusion

Researchers on C-dots emphasize the improvement of antimicrobial characteristics of these materials. Although C-dots are carbon nanomaterials showing superior antimicrobial effects, there are several main technological problems that limit the efficient design and applicability of these materials. A variety of antimicrobial effects on various bacteria has been reported for C-dots depending on their size and surface functionalization [82]. The production of ROS is among the antimicrobial mechanisms shown by C-dots [119]. ROS possesses significant toxicity and can lead to severe diseases, including cancers [123–128]. Thus, it can be an excellent antimicrobial with antifungal and antibacterial effects. It should be noted that this material has also antiviral effects for a variety of viruses. Yet, little has been known about the activity of C-dots against viral agents so far [90]. C-dots have been also shown to possess antiparasitic effects, but little research has been conducted in this area. However, C-dots have recently been the focus of attention as innovative nanoparticles because of their distinguished optical, as well as physicochemical characteristics and high biosafety. Thus, it seems that C-dots can contribute as promising antimicrobial agents.

## 11.6 Research directions for future

Given the growing resistance to drugs in the present age, it is vital to develop novel antimicrobial agents which possess unique characteristics and high specificity. In many cases, the available antibiotics are structurally or functionally modified to produce new antibiotics [124]. C-dots belong to a class of quasi-spherical carbon-based fluorescent nanomaterials typically at a size of $\leq 10$ nm. C-dots are significantly stable and highly soluble in water in terms of their chemical features, while they also show superior photoelectric characteristics. Besides, these materials are less toxic and considerably biocompatible [129]. The mentioned features, along with their high antimicrobial effects, have made C-dots as good candidates. However, little research has been conducted on the antiparasitic effects of these useful materials, necessitating a focus on this area of research in the future. Besides, there is a need for studies that use animal models to investigate C-dots. It is also noteworthy that ROS production is among the antimicrobial mechanisms of C-dots [119]. Thus, more research is required on C-dots using natural materials that show antioxidant effects, including medicinal herbs as strong antioxidants [124]. This study suggests the use of various herbs with considerable antioxidant effects to modify C-dots.

# References

[1]    Wang J, Qiu J. A review of carbon dots in biological applications. J Mater Sci. 2016, 51(10), 4728–4738.
[2]    Sall M, Monnet I, Moisy F, Grygiel C, Jublot-Leclerc S, Della–Negra S, Toulemonde M, Balanzat E. Track formation in III-N semiconductors irradiated by swift heavy ions and fullerene and re-evaluation of the inelastic thermal spike model. J Mater Sci. 2015, 50(15), 5214–5227.
[3]    Yu M, George C, Cao Y, Wootton D, Zhou J. Microstructure, corrosion, and mechanical properties of compression-molded zinc-nanodiamond composites. J Mater Sci. 2014, 49(10), 3629–3641.
[4]    Hola K, Bourlinos AB, Kozak O, Berka K, Siskova KM, Havrdova M, Tucek J, Safarova K, Otyepka M, Giannelis EP, Zboril R. Photoluminescence effects of graphitic core size and surface functional groups in carbon dots: COO⁻ induced red-shift emission. Carbon. 2014, 70, 279–286.
[5]    De Volder MF, Tawfick SH, Baughman RH, Hart AJ. Carbon nanotubes: Present and future commercial applications. science. 2013, 339(6119), 535–539.
[6]    Zubair M, Mustafa M, Ali A, Doh YH, Choi KH. Improvement of solution based conjugate polymer organic light emitting diode by ZnO–graphene quantum dots. J Mater Sci Mater Electron. 2015, 26(5), 3344–3351.
[7]    Liu M, He L, Liu X, Liu C, Luo S. Reduced graphene oxide and CdTe nanoparticles co-decorated TiO₂ nanotube array as a visible light photocatalyst. J Mater Sci. 2014, 49(5), 2263–2269.
[8]    Tuerhong M, Yang XU, Xue-Bo YI. Review on carbon dots and their applications. Chin J Anal Chem. 2017, 45(1), 139–150.
[9]    Baker SN, Baker GA. Luminescent carbon nanodots: Emergent nanolights. Angew Chem Int Edition. 2010, 49(38), 6726–6744.
[10]   Lin L, Rong M, Luo F, Chen D, Wang Y, Chen X. Luminescent graphene quantum dots as new fluorescent materials for environmental and biological applications. TrAC Trends Anal Chem. 2014, 54, 83–102.
[11]   Liu C, Zhang P, Zhai X, Tian F, Li W, Yang J, Liu Y, Wang H, Wang W, Liu W. Nano-carrier for gene delivery and bioimaging based on carbon dots with PEI-passivation enhanced fluorescence. Biomaterials. 2012, 33(13), 3604–3613.
[12]   Lai CW, Hsiao YH, Peng YK, Chou PT. Facile synthesis of highly emissive carbon dots from pyrolysis of glycerol; gram scale production of carbon dots/mSiO₂ for cell imaging and drug release. J Mater Chem. 2012, 22(29), 14403–14409.
[13]   Choi Y, Kim S, Choi MH, Ryoo SR, Park J, Min DH, Kim BS. Highly biocompatible carbon nanodots for simultaneous bioimaging and targeted photodynamic therapy in vitro and in Vivo. Adv Funct Mater. 2014, 24(37), 5781–5789.
[14]   Lin F, Bao YW, Wu FG. Carbon dots for sensing and killing microorganisms. C. 2019, 5(2), 33.
[15]   D'Costa VM, King CE, Kalan L, Morar M, Sung WW, Schwarz C, Froese D, Zazula G, Calmels F, Debruyne R, Golding GB. Antibiotic resistance is ancient. Nature. 2011, 477(7365), 457–461.
[16]   Gao G, Jiang YW, Sun W, Wu FG. Fluorescent quantum dots for microbial imaging. Chin Chem Lett. 2018, 29(10), 1475–1485.
[17]   Sun W, Wu FG. Two-dimensional materials for antimicrobial applications: Graphene materials and beyond. Chem–Asian J. 2018, 13(22), 3378–3410.
[18]   Feng H, Qian Z. Functional carbon quantum dots: A versatile platform for chemosensing and biosensing. Chem Rec. 2018, 18(5), 491–505.
[19]   Kasibabu BSB, D'souza SL, Jha S, Singhal RK, Basu H, Kailasa SK. One-step synthesis of fluorescent carbon dots for imaging bacterial and fungal cells. Anal Methods. 2015, 7, 2373–2378.
[20]   Mehta VN, Jha S, Kailasa SK. One-pot green synthesis of carbon dots by using Saccharum officinarum juice for fluorescent imaging of bacteria (*Escherichia coli*) and yeast (*Saccharomyces cerevisiae*) cells. Mater Sci Eng C. 2014, 38, 20–27.

[21]  Kasibabu BSB, D'souza SL, Jha S, Kailasa SK. Imaging of bacterial and fungal cells using fluorescent carbon dots prepared from Carica papaya juice. J Fluoresc. 2015, 25, 803–810.

[22]  Hua X-W, Bao Y-W, Wang H-Y, Chen Z, Wu F-G. Bacteria-derived fluorescent carbon dots for microbial live/dead differentiation. Nanoscale. 2017, 9, 2150–2161.

[23]  Wu LN, Yang YJ, Huang LX, Zhong Y, Chen Y, Gao YR, Lin LQ, Lei Y, Liu AL. Levofloxacin-based carbon dots to enhance antibacterial activities and combat antibiotic resistance. Carbon. 2022,186, 452–64.

[24]  Li H, Huang J, Song Y, Zhang M, Wang H, Lu F, Huang H, Liu Y, Dai X, Gu Z et al. Degradable carbon dots with broad-spectrum antibacterial activity. ACS Appl Mat Interf. 2018, 10, 26936–26946.

[25]  Barras A, Pagneux Q, Sane F, Wang Q, Boukherroub R, Hober D, Szunerits S. High efficiency of functional carbon nanodots as entry inhibitors of herpes simplex virus type 1. ACS Appl Mat Interf. 2016, 8, 9004–9013.

[26]  Du T, Liang J, Dong N, Liu L, Fang L, Xiao S, Han H. Carbon dots as inhibitors of virus by activation of type I interferon response. Carbon. 2016, 110, 278–285.

[27]  Dong X, Moyer MM, Yang F, Sun YP, Yang L. Carbon dots' antiviral functions against noroviruses. Sci Rep. 2017, 7, 519.

[28]  Ting D, Dong N, Fang L, Lu J, Bi J, Xiao S, Han H. Multisite inhibitors for enteric coronavirus: Antiviral cationic carbon dots based on curcumin. ACS Appl Nano Mater. 2018, 1, 5451–5459.

[29]  Huang S, Gu J, Ye J, Fang B, Wan S, Wang C, Ashraf U, Li Q, Shao L, Song Y et al. Benzoxazine monomer derived carbon dots as a broad-spectrum agent to block viral infectivity. J Colloid Interface Sci. 2019, 542, 198–206.

[30]  Zuo P, Lu X, Sun Z, Guo Y, He H. A review on syntheses, properties, characterization and bioanalytical applications of fluorescent carbon dots. Microchim Acta. 2016, 183(2), 519–542.

[31]  Li S, Li L, Tu H, Zhang H, Silvester DS, Banks CE, Zou G, Hou H, Ji X. The development of carbon dots: From the perspective of materials chemistry. Mater Today. 2021, 21.

[32]  Baker SN, Baker GA. Luminescent carbon nanodots: Emergent nanolights. Angew Chem Int Edition. 2010, 49(38), 6726–6744.

[33]  Xu X, Ray R, Gu Y, Ploehn HJ, Gearheart L, Raker K, Scrivens WA. Electrophoretic analysis and purification of fluorescent single-walled carbon nanotube fragments. J Am Chem Soc. 2004, 126(40), 12736–12737.

[34]  Sun YP, Zhou B, Lin Y, Wang W, Fernando KS, Pathak P, Meziani MJ, Harruff BA, Wang X, Wang H, Luo PG. Quantum-sized carbon dots for bright and colorful photoluminescence. J Am Chem Soc. 2006, 128(24), 7756–7757.

[35]  Nekoueian K, Amiri M, Sillanpää M, Marken F, Boukherroub R, Szunerits S. Carbon-based quantum particles: An electroanalytical and biomedical perspective. Chem Soc Rev. 2019, 48(15), 4281–4316.

[36]  Bourlinos AB, Stassinopoulos A, Anglos D, Zboril R, Karakassides M, Giannelis EP. Surface functionalized carbogenic quantum dots. small. 2008, 4(4), 455–458.

[37]  Hu C, Li M, Qiu J, Sun YP. Design and fabrication of carbon dots for energy conversion and storage. Chem Soc Rev. 2019, 48(8), 2315–2337.

[38]  Zhu S, Song Y, Zhao X, Shao J, Zhang J, Yang B. The photoluminescence mechanism in carbon dots (graphene quantum dots, carbon nanodots, and polymer dots): Current state and future perspective. Nano Res. 2015, 8(2), 355–381.

[39]  Xia C, Zhu S, Feng T, Yang M, Yang B. Evolution and synthesis of carbon dots: From carbon dots to carbonized polymer dots. Adv Sci. 2019, 6(23), 1901316.

[40]  Shen J, Xu B, Wang Z, Zhang J, Zhang W, Gao Z, Wang X, Zhu C, Meng X. Aggregation-induced room temperature phosphorescent carbonized polymer dots with wide-range tunable lifetimes for optical multiplexing. J Mater Chem C. 2021, 9(21), 6781–6788.

[41]  Jhonsi MA, Ananth DA, Nambirajan G, Sivasudha T, Yamini R, Bera S, Kathiravan A. Antimicrobial activity, cytotoxicity and DNA binding studies of carbon dots. Spectrochim Acta A. 2018, 196, 295–302.

[42] Al Awak MM, Wang P, Wang S, Tang Y, Sun YP, Yang L. Correlation of carbon dots' light-activated antimicrobial activities and fluorescence quantum yield. RSC Adv. 2017, 7(48), 30177–30184.

[43] Qing W, Chen K, Yang Y, Wang Y. Liu X. $Cu^{2+}$-doped carbon dots as fluorescence probe for specific recognition of Cr (VI) and its antimicrobial activity. Microchem J. 2020, 152, 104262.

[44] Li D, Xu KY, Zhao WP, Liu MF, Feng R, Li DQ, Bai J, Du WL. Chinese medicinal herb-derived carbon dots for common diseases: efficacies and potential mechanisms. Front Pharmacol. 2022, 22(13), 815479.

[45] Liu J, Li R, Yang B. Carbon dots: A new type of carbon-based nanomaterial with wide applications. ACS Central Sci. 2020, 6(12), 2179–2195.

[46] Chen Z, Ye SY, Yang Y, Li ZY. A review on charred traditional Chinese herbs: Carbonization to yield a haemostatic effect. Pharm Biol. 2019, 57(1), 498–506.

[47] Naik GG, Shah J, Balasubramaniam AK, Sahu AN. Applications of natural product-derived carbon dots in cancer biology. Nanomedicine. 2021, 16(7), 587–608.

[48] Shukla D, Pandey FP, Kumari P, Basu N, Tiwari MK, Lahiri J, Kharwar RN, Parmar AS. Label-free fluorometric detection of adulterant malachite green using carbon dots derived from the medicinal plant source Ocimum tenuiflorum. Chem Select. 2019, 4(17), 4839–4847.

[49] Mohammed LJ, Omer KM. Carbon dots as new generation materials for nanothermometer. Nanoscale Res Lett. 2020, 15(1), 1–21.

[50] Gaddam RR, Mukherjee S, Punugupati N, Vasudevan D, Patra CR, Narayan R, Kothapalli RV. Facile synthesis of carbon dot and residual carbon nanobeads: Implications for ion sensing, medicinal and biological applications. Mater Sci Eng C. 2017, 73, 643–652.

[51] Naik GG, Alam M, Pandey V, Mohapatra D, Dubey PK, Parmar AS, Sahu AN. Multi-Functional carbon dots from an ayurvedic medicinal plant for cancer cell bioimaging applications. J Fluoresc. 2020, 30(2), 407–418.

[52] Konwar A, Gogoi N, Majumdar G, Chowdhury D. Green chitosan–carbon dots nanocomposite hydrogel film with superior properties. Carbohydr Polym. 2015, 115, 238–245.

[53] Vasimalai N, Vilas-Boas V, Gallo J, de Fátimacerqueira M, Menéndez-Miranda M, Costa-Fernández JM, Diéguez L, Espiña B, Fernández-Argüelles MT. Green synthesis of fluorescent carbon dots from spices for in vitro imaging and tumour cell growth inhibition. Beilstein J Nanotechnol. 2018, 9(1), 530–544.

[54] Wang H, Bi J, Zhu BW, Tan M. Multicolorful carbon dots for tumor theranostics. Curr Med Chem. 2018, 25(25), 2894–2909.

[55] Meena R, Singh R, Marappan G, Kushwaha G, Gupta N, Meena R, Gupta JP, Agarwal RR, Fahmi N, Kushwaha OS. Fluorescent carbon dots driven from ayurvedic medicinal plants for cancer cell imaging and phototherapy. Heliyon. 2019, 5(9), e02483.

[56] Zheng M, Liu S, Li J, Qu D, Zhao H, Guan X, Hu X, Xie Z, Jing X, Sun Z. Integrating oxaliplatin with highly luminescent carbon dots: An unprecedented theranostic agent for personalized medicine. Adv Mater. 2014, 26(21), 3554–3560.

[57] Saravanan A, Maruthapandi M, Das P, Luong JH, Gedanken A. Green synthesis of multifunctional carbon dots with antibacterial activities. Nanomaterials. 2021, 11(2), 369.

[58] Karthik S, Saha B, Ghosh SK, Singh NP. Photoresponsive quinoline tethered fluorescent carbon dots for regulated anticancer drug delivery. Chem Comm. 2013, 49(89), 10471–10473.

[59] Ge G, Li L, Wang D, Chen M, Zeng Z, Xiong W, Wu X, Guo C. Carbon dots: Synthesis, properties and biomedical applications. J Mat Chem B. 2021, 9, 6553–6575

[60] Cui F, Ye Y, Ping J, Sun X. Carbon dots: Current advances in pathogenic bacteria monitoring and prospect applications. Biosens Bioelectron. 2020, 156, 112085.

[61] Suner SS, Sahiner M, Ayyala RS, Bhethanabotla VR, Sahiner N. Nitrogen-doped arginine carbon dots and its metal nanoparticle composites as antibacterial agent. C. 2020, 6(3), 58.

[62] Li H, Huang J, Song Y, Zhang M, Wang H, Lu F, Huang H, Liu Y, Dai X, Gu Z, Yang Z. Degradable carbon dots with broad-spectrum antibacterial activity. ACS Appl Mater Interfaces. 2018, 10(32), 26936–26946.

[63] Rabe DI, Al Awak MM, Yang F, Okonjo PA, Dong X, Teisl LR, Wang P, Tang Y, Pan N, Sun YP, Yang L. The dominant role of surface functionalization in carbon dots' photo-activated antibacterial activity. Int J Nanomed. 2019, 14, 2655.

[64] Wang H, Lu F, Ma C, Ma Y, Zhang M, Wang B, Zhang Y, Liu Y, Huang H, Kang Z. Carbon dots with positive surface charge from tartaric acid and m-aminophenol for selective killing of Gram-positive bacteria. J Mat Chem B. 2021, 9(1), 125–130.

[65] Dong X, Awak MA, Tomlinson N, Tang Y, Sun YP, Yang L. Antibacterial effects of carbon dots in combination with other antimicrobial reagents. PloS One. 2017, 12(9), e0185324.

[66] Han S, Zhang H, Xie Y, Liu L, Shan C, Li X, Liu W, Tang Y. Application of cow milk-derived carbon dots/Ag NPs composite as the antibacterial agent. Appl Surf Sci. 2015, 328, 368–373.

[67] Hou P, Yang T, Liu H, Li YF, Huang CZ. An active structure preservation method for developing functional graphitic carbon dots as an effective antibacterial agent and a sensitive pH and Al (III) nanosensor. Nanoscale. 2017, 9(44), 17334–17341.

[68] Shahshahanipour M, Rezaei B, Ensafi AA, Etemadifar Z. An ancient plant for the synthesis of a novel carbon dot and its applications as an antibacterial agent and probe for sensing of an anti-cancer drug. Mater Sci Eng C. 2019, 98, 826–833.

[69] Dong X, Liang W, Meziani MJ, Sun YP, Yang L. Carbon dots as potent antimicrobial agents. Theranostics. 2020, 10(2), 671.

[70] Sun B, Wu F, Zhang Q, Chu X, Wang Z, Huang X, Li J, Yao C, Zhou N, Shen J. Insight into the effect of particle size distribution differences on the antibacterial activity of carbon dots. J Colloid Interface Sci. 2021, 584, 505–519.

[71] Yang YL, Lo HJ. Mechanisms of antifungal agent resistance. J Microbiol Immunol Infect = Wei Mianyugan Ran Zazhi. 2001, 34(2), 79–86.

[72] Gao Z, Li X, Shi L, Yang Y. Deep eutectic solvents-derived carbon dots for detection of mercury (II), photocatalytic antifungal activity and fluorescent labeling for *C. albicans*. Spectrochim Acta A. 2019, 220, 117080.

[73] Tejwan N, Saini AK, Sharma A, Singh TA, Kumar N, Das J. Metal-doped and hybrid carbon dots: A comprehensive review on their synthesis and biomedical applications. J Control Release. 2021, 330, 132–150.

[74] Wang ZX, Wang Z, Wu FG. Carbon dots as drug delivery vehicles for antimicrobial applications: A mini review. Chem Med Chem. 2022, 17.

[75] Borse V, Thakur M, Sengupta S, Srivastava R. N-doped multi-fluorescent carbon dots for 'turn off-on' silver-biothiol dual sensing and mammalian cell imaging application. Sens Actuators B Chem. 2017, 248, 481–492.

[76] Priyadarshini E, Rawat K, Prasad T, Bohidar HB. Antifungal efficacy of Au@ carbon dots nanoconjugates against opportunistic fungal pathogen, Candida albicans. Coll Surf Biointerf. 2018, 163, 355–361.

[77] Ezati P, Rhim JW, Molaei R, Priyadarshi R, Roy S, Min S, Kim YH, Lee SG, Han S. Preparation and characterization of B, S, and N-doped glucose carbon dots: Antibacterial, antifungal, and antioxidant activity. Sustainable Mater Technol. 2022, 32, e00397.

[78] Ghorbani M, Tajik H, Moradi M, Molaei R, Alizadeh A. One-pot microbial approach to synthesize carbon dots from baker's yeast-derived compounds for the preparation of antimicrobial membrane. J Environ Chem Eng. 2022, 10(3), 107525.

[79] Li X, Wu X, Yuan T, Zhu J, Yang Y. Influence of the iodine content of nitrogen-and iodine-doped carbon dots as a peroxidase mimetic nanozyme exhibiting antifungal activity against *C. Albicans* Biochem Eng J. 2021, 175, 108139.

[80]  Zhao S, Huang L, Xie Y, Wang B, Wang F, Lan M. Green synthesis of multifunctional carbon dots for anti-cancer and anti-fungal applications. Chin J Chem Eng. 2021, 37, 97–104.
[81]  Li H, Huang J, Song Y, Zhang M, Wang H, Lu F, Huang H, Liu Y, Dai X, Gu Z, Yang Z. Degradable carbon dots with broad-spectrum antibacterial activity. ACS Appl Mater Interfaces. 2018, 10(32), 26936–26946.
[82]  Anand A, Unnikrishnan B, Wei SC, Chou CP, Zhang LZ, Huang CC. Graphene oxide and carbon dots as broad-spectrum antimicrobial agents –a minireview. Nanoscale Horiz. 2019, 4(1), 117–137.
[83]  Galdiero S, Falanga A, Vitiello M, Cantisani M, Marra V, Galdiero M. Silver nanoparticles as potential antiviral agents. Molecules. 2011, 16(10), 8894–8918.
[84]  Tong T, Hu H, Zhou J, Deng S, Zhang X, Tang W, Fang L, Xiao S, Liang J. Glycyrrhizic-acid-based carbon dots with high antiviral activity by multisite inhibition mechanisms. Small. 2020, 16(13), 1906206.
[85]  Serrano-Aroca Á, Takayama K, Tuñón-Molina A, Seyran M, Hassan SS, Pal Choudhury P, Uversky VN, Lundstrom K, Adadi P, Palù G, Aljabali AA. Carbon-based nanomaterials: Promising antiviral agents to combat COVID-19 in the microbial-resistant era. ACS Nano. 2021, 15(5), 8069–8086.
[86]  da Silva Júnior AH, Macuvele DL, Riella HG, Soares C, Padoin N. Are carbon dots effective for ion sensing and antiviral applications? A state-of-the-art description from synthesis methods to cost evaluation. J Mater Res Technol. 2021, 12, 688–716.
[87]  Chen L, Liang J. An overview of functional nanoparticles as novel emerging antiviral therapeutic agents. Mater Sci Eng C. 2020, 112, 110924.
[88]  Ting D, Dong N, Fang L, Lu J, Bi J, Xiao S, Han H. Multisite inhibitors for enteric coronavirus: Antiviral cationic carbon dots based on curcumin. ACS Appl Nano Mater. 2018, 1(10), 5451–5459.
[89]  Manivannan S, Ponnuchamy K. Quantum dots as a promising agent to combat COVID-19. Appl Organomet Chem. 2020, 34(10), e5887.
[90]  Du T, Liang J, Dong N, Liu L, Fang L, Xiao S, Han H. Carbon dots as inhibitors of virus by activation of type I interferon response. Carbon. 2016, 110, 278–285.
[91]  Huang S, Gu J, Ye J, Fang B, Wan S, Wang C, Ashraf U, Li Q, Wang X, Shao L, Song Y. Benzoxazine monomer derived carbon dots as a broad-spectrum agent to block viral infectivity. J Colloid Interface Sci. 2019, 15(542), 198–206.
[92]  Kotta S, Aldawsari HM, Badr-Eldin SM, Alhakamy NA, Md S, Nair AB, Deb PK. Exploring the potential of carbon dots to combat COVID-19. Front Mol Biosci. 2020, 428.
[93]  Kalkal A, Allawadhi P, Pradhan R, Khurana A, Bharani KK, Packirisamy G. *Allium sativum* derived carbon dots as a potential theranostic agent to combat the COVID-19 crisis. Sens Int. 2021, 2, 100102.
[94]  Dong X, Moyer MM, Yang F, Sun YP, Yang L. Carbon dots' antiviral functions against noroviruses. Sci Rep. 2017, 7(1), 1–0.
[95]  Rosenblatt JE. Antiparasitic Agents. In: Mayo Clinic Proceedings, vol. 74, No. 11, Elsevier, 1999, 1161–1174.
[96]  Bartolomei B, Dosso J, Prato M. New trends in nonconventional carbon dot synthesis. Trends Chem. 2021, 3(11), 943–953.
[97]  Singh NK, Chakma B, Jain P, Goswami P. Protein-induced fluorescence enhancement based detection of Plasmodium falciparum glutamate dehydrogenase using carbon dot coupled specific aptamer. ACS Comb Sci. 2018, 20(6), 350–357.
[98]  Fu X, Lv R, Su J, Li H, Yang B, Gu W, Liu X. A dual-emission nano-rod MOF equipped with carbon dots for visual detection of doxycycline and sensitive sensing of $MnO_4^-$. RSC Adv. 2018, 8(9), 4766–4772.
[99]  Larki A. A novel application of carbon dots for colorimetric determination of fenitrothion insecticide based on the microextraction method. Spectrochim Acta A. 2017, 15(173), 1–5.
[100] Murugan K, Nataraj D, Jaganathan A, Dinesh D, Jayashanthini S, Samidoss CM, Paulpandi M, Panneerselvam C, Subramaniam J, Aziz AT, Nicoletti M. Nanofabrication of graphene quantum dots

with high toxicity against malaria mosquitoes, Plasmodium falciparum and MCF-7 cancer cells: Impact on predation of non-target tadpoles, odonate nymphs and mosquito fishes. J Cluster Sci. 2017, 28(1), 393–411.

[101]  Kumar VB, Dolitzky A, Michaeli S, Gedanken A. Antiparasitic ointment based on a biocompatible carbon dot nanocomposite. ACS Appl Nano Mater. 2018, 1(4), 1784–1791.

[102]  He D, Zhang X, Yao X, Yang Y. In vitro and in vivo highly effective antibacterial activity of carbon dots-modified $TiO_2$ nanorod arrays on titanium. Coll Surf Biointerf. 2022, 6, 112318.

[103]  Yan C, Wang C, Shao X, Shu Q, Hu X, Guan P, Teng Y, Cheng Y. Dual-targeted carbon-dot-drugs nanoassemblies for modulating Alzheimer's related amyloid-β aggregation and inhibiting fungal infection. Mater Today Bio. 2021, 1(12), 100167.

[104]  Jijie R, Barras A, Bouckaert J, Dumitrascu N, Szunerits S, Boukherroub R. Enhanced antibacterial activity of carbon dots functionalized with ampicillin combined with visible light triggered photodynamic effects. Coll Surf Biointerf. 2018, 170, 347–354.

[105]  Zhang J, Liu S, Wang X, Yao J, Zhai M, Liu B, Liang C, Shi H. Highly efficient $Ti^{3+}$ self-doped $TiO_2$ co-modified with carbon dots and palladium nanocomposites for disinfection of bacterial and fungi. J Hazard Mater. 2021, 413, 125318.

[106]  Fahmi MZ, Sukmayani W, Khairunisa SQ, Witaningrum AM, Indriati DW, Matondang MQ, Chang JY, Kotaki T, Kameoka M. Design of boronic acid-attributed carbon dots on inhibits HIV-1 entry. RSC Adv. 2016, 6(95), 92996–93002.

[107]  Liu J, Lu S, Tang Q, Zhang K, Yu W, Sun H, Yang B. One-step hydrothermal synthesis of photoluminescent carbon nanodots with selective antibacterial activity against Porphyromonas gingivalis. Nanoscale. 2017, 9(21), 7135–7142.

[108]  Tong T, Hu H, Zhou J, Deng S, Zhang X, Tang W, Fang L, Xiao S, Liang J. Antiviral carbon dots: glycyrrhizic-acid-based carbon dots with high antiviral activity by multisite inhibition mechanisms (Small 13/2020). Small. 2020, 16(13), 2070068.

[109]  Luo Q, Qin K, Liu F, Zheng X, Ding Y, Zhang C, Xu M, Liu X, Wei Y. Carbon dots derived from kanamycin sulfate with antibacterial activity and selectivity for $Cr^{6+}$ detection. Analyst. 2021, 146(6), 1965–1972.

[110]  Cui F, Sun J, Ji J, Yang X, Wei K, Xu H, Gu Q, Zhang Y, Sun X. Carbon dots-releasing hydrogels with antibacterial activity, high biocompatibility, and fluorescence performance as candidate materials for wound healing. J Hazard Mater. 2021, 15(406), 124330.

[111]  Bhavyasree PG, Xavier TS. Green synthesis of copper oxide/carbon nanocomposites using the leaf extract of Adhatoda vasica Nees, their characterization and antimicrobial activity. Heliyon. 2020, 6(2), e03323.

[112]  Hu M, Gu X, Hu Y, Deng Y, Wang C. PVA/carbon dot nanocomposite hydrogels for simple introduction of Ag nanoparticles with enhanced antibacterial activity. Macromol Mater Eng. 2016, 301(11), 1352–1362.

[113]  Habiba K, Bracho-Rincon DP, Gonzalez-Feliciano JA, Villalobos-Santos JC, Makarov VI, Ortiz D, Avalos JA, Gonzalez CI, Weiner BR, Morell G. Synergistic antibacterial activity of PEGylated silver–graphene quantum dots nanocomposites. Appl Mater Today. 2015, 1(2), 80–87.

[114]  Jijie R, Barras A, Bouckaert J, Dumitrascu N, Szunerits S, Boukherroub R. Enhanced antibacterial activity of carbon dots functionalized with ampicillin combined with visible light triggered photodynamic effects. Coll Surf Biointerf. 2018, 170, 347–354.

[115]  Ye Z, Li G, Lei J, Liu M, Jin Y, Li B. One-step and one-precursor hydrothermal synthesis of carbon dots with superior antibacterial activity. ACS Appl Bio Mater. 2020, 3(10), 7095–7102.

[116]  Aung YY, Kristanti AN, Khairunisa SQ, Nasronudin N, Fahmi MZ. Inactivation of HIV-1 infection through integrative blocking with amino phenylboronic acid attributed carbon dots. ACS Biomater Sci Eng. 2020, 6(8), 4490–4501.

[117] Liang J, Li W, Chen J, Huang X, Liu Y, Zhang X, Shu W, Lei B, Zhang H. Antibacterial activity and synergetic mechanism of carbon dots against gram-positive and-negative bacteria. ACS Appl Bio Mater. 2021, 4(9), 6937–6945.
[118] Dou Q, Fang X, Jiang S, Chee PL, Lee TC, Loh XJ. Multi-functional fluorescent carbon dots with antibacterial and gene delivery properties. RSC Adv. 5, 46817–46822, 2015.
[119] Bing W, Sun H, Yan Z, Ren J, Qu X. Programmed bacteria death induced by carbon dots with different surface charge. Small. 2016, 12(34), 4713–4718.
[120] Long C, Jiang Z, Shangguan J, Qing T, Zhang P, Feng B. Applications of carbon dots in environmental pollution control: A review. Chem Eng J. 2021, 15(406), 126848.
[121] Song Y, Lu F, Li H, Wang H, Zhang M, Liu Y, Kang Z. Degradable carbon dots from cigarette smoking with broad-spectrum antimicrobial activities against drug-resistant bacteria. ACS Appl Bio Mater. 2018, 1, 1871–1879.
[122] Jian HJ, Wu RS, Lin TY, Li YJ, Lin HJ, Harroun SG, Lai JY, Huang CC. Super-cationic carbon quantum dots synthesized from spermidine as an eye drop formulation for topical treatment of bacterial keratitis. ACS Nano. 2017, 11(7), 6703–6716.
[123] Parham S, Kharazi AZ, Bakhsheshi-Rad HR, Nur H, Ismail AF, Sharif S, RamaKrishna S, Berto F. Antioxidant, antimicrobial and antiviral properties of herbal materials. Antioxidants. 2020, 9(12), 1309.
[124] Cheung EC, Vousden KH. The role of ROS in tumour development and progression. Nat Rev Cancer. 2022, 22(5), 280–97.
[125] Parham S, Wicaksono DH, Bagherbaigi S, Lee SL, Nur H. Antimicrobial treatment of different metal oxide nanoparticles: A critical review. J Chin Chem Soc. 2016, 63(4), 385–393.
[126] Parham S, Nemati M, Sadir S, Bagherbaigi S, Wicaksono DH, Nur H. In situ synthesis of silver nanoparticles for Ag-NP/cotton nanocomposite and its bactericidal effect. J Chin Chem Soc. 2017, 64(11), 1286–1293.
[127] Yu W, Tu Y, Long Z, Liu J, Kong D, Peng J, Wu H, Zheng G, Zhao J, Chen Y, Liu R. Reactive oxygen species bridge the gap between chronic inflammation and tumor development. Oxid Med Cell Longev. 2022, 2022, 1–22.
[128] Parham S, Wicaksono DH. Nur H.A proposed mechanism of action of textile/Al$_2$O$_3$TiO$_2$ bimetaloxide nanocomposite as an antimicrobial agent. J Tex Inst. 2019, 110(5), 791798.
[129] Ghirardello M, Ramos-Soriano J, Galan MC. Carbon dots as an emergent class of antimicrobial agents. Nanomaterials. 2021, 11(8), 1877.

Golnar Bayatani, Mahdie Matin, Alireza Alikhanian,
Mahtab Mirhoseinian, Mohammad Nazari Montazer, Burak Tüzün*,
Mohammad Mahdavi, Parham Taslimi

# Chapter 12
# Carbon dots in antibiosis: disinfection and sterilization

**Abstract:** Antibiosis is the antagonism caused by the toxicity of secondary metabolites produced by the microorganisms. Our understanding of microbiology has improved as a result of research on antibiosis and its role in antibiotics. Nowadays Major bacterial diseases are becoming more resistant to standard antibiotic treatments, and multidrug-resistant bacteria are emerging at an alarming rate. High rates of morbidity and mortality are a challenge caused by antibiotic resistance in bacterial infections hence it is essential to synthesize and introduce new antimicrobial agents. Many studies have reported that Carbon Dots (CDs) show useful functions in antibiotic, antitumor, and antiviral activities. In this chapter several CD-based structures and their effects on antibacterial resistant bacteria's have been investigated.

## 12.1 Antibiosis

Antibiosis is an association between two microorganisms that are harmful to at least one of them, and in fact, the antagonism is caused by the toxicity of secondary metabolites produced by the microorganisms [1, 2]. For example, antibiosis includes the association between antibiotics and bacteria or animals and disease-causing pathogens [3].

Many bacteria mention antibiosis as a survival and dominance tactic. The emergence of new medications and antibiotics has finally been linked to this process of bacteria [4].

*Corresponding author: Burak Tüzün**, Department of Chemistry, Faculty of Science, Cumhuriyet University, 58140, Sivas, Turkey; Plant and Animal Production Department, Technical Sciences Vocational School of Sivas, Sivas Cumhuriyet University, e-mail: theburaktuzun@yahoo.com, http://orcid.org/0000-0002-0420-2043
**Golnar Bayatani, Mahdie Matin, Alireza Alikhanian, Mahtab Mirhoseinian,
Mohammad Nazari Montazer, Mohammad Mahdavi,** Endocrinology and Metabolism Research Center, Endocrinology and Metabolism Clinical Sciences Institute, Tehran University of Medical Sciences, Tehran, Iran
**Parham Taslimi,** Department of Biotechnology, Faculty of Science, Bartin University, 74100 Bartin, Turkey

https://doi.org/10.1515/9783110799958-012

### 12.1.1 History

At first, everyone was acquainted with symbiosis as the partnership of two living organisms for mutual benefit. For example, in the nodules on the roots of peas, clover, and other plants, there are bacteria known as common root tubercles. By working together, bacteria identify a suitable home, work to accumulate nitrogen molecules, and eventually improve their hosts' ecosystems. In 1899, an article titled "Symbiosis" published in *Annals of Botany* noted that "if one of two associated organisms injures the other, this condition is called antibiosis" [5]. Furthermore, the experiments by Corneil and Babes, then after them, the experiment of Garre, showed that microbial antagonism is caused by a substance that can be released from a living organism. At the end of the nineteenth century, this issue was utterly accepted as "antibiosis" [3].

In the end, influential researchers determined the great possibilities involved in antagonistic interactions between microorganisms and drew attention to their possible applications, and obtained significant results in this regard, including Much in 1924, Pasteur and Joubert in 1877, then Cantani in 1885, Bouchard, Vaudremer in 1913, Emmerich and Low in 1889, Fleming in 1929, and finally Dubus in 1939 (Figure 12.1) [5].

### 12.1.2 Antibiosis in plants and insects

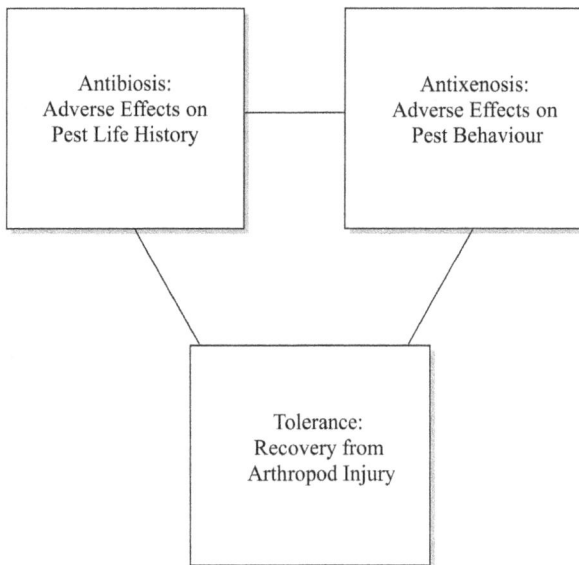

**Figure 12.1:** Categorizing styles of plant resistance [6].

A vital global problem is safeguarding agricultural crops from pest attack. Biological traits and the attraction of plants to herbivores are influenced by chemical and morphological defensive mechanisms found in plants. Herbivores may be impacted by these systems either directly or indirectly, for instance, by drawing in natural enemies. In addition, a lot of insect predators consume plant life. In vegetarian species the impact of diets containing both plant and animal food sources on the survival and longevity of insect predators have been researched in particular. Predators that cannot be found at a single trophic level are given critical nutrients or amino acids by this combination of meals. As a result, predators with a diet rich in plants and animals exhibit improved reproductive and biological traits. However, due to antibiosis, plant secondary chemicals can weaken the effectiveness of natural enemies [7, 8].

We can use the rice leaffolder as an illustration; under the right circumstances, *Cnaphalocrocis medinalis* (Guenee) can do significant harm. An affordable and environmentally friendly method of addressing this pest is host plant resistance. Six antibiosis indices and five biochemical components were measured in seven rice genotypes in order to pinpoint the causes of resistance to leaffolder. These genotypes had a total larval life span of 15.09–19.57 days. Antibiosis in W1263, IET22155, and TKM6 resulted in longer larval life (18.46–19.57 days), despite the fact that these genotypes had much lower percentage pupation (54–74%). There was no evidence of a genetic influence on adult longevity, adult emergence, or pupal length. The expression of biochemical elements like free amino acids, reducing sugars and total soluble, and total soluble proteins was lower in W1263, IET22155, and TKM6, both at 70 and 50 days old, but the total phenol content was higher. Free amino acids ($r = -0.76$), total soluble sugars ($r = -0.69$), and reducing sugars ($r = -0.80$) all showed negative associations with larval length. Similar to this, total phenol content showed a positive correlation with larval time ($r = 0.68$) and a negative correlation with pupal weight ($r = -0.79$) and larval survival ($r = -0.62$). It is feasible to understand the underlying processes of antibiosis resistance and create robust leaffolder-resistant variations by taking advantage of the difference in the expression of antibiosis parameters and biochemical components among genotypes against leaffolder [9].

For example, in order to assess the insecticidal ability of TSP14 to defend crops against insect damage, chimeric gene constructs of TSP14 were generated in transgenic plants. The modified PCISV full-length transcript (FLt) promoter with duplicated enhancer domains and the rbcSE9 terminator sequence were located in the middle of the TSP14 gene's coding region. According to molecular research, these chimeric genes were expressed in transgenic tobacco and were stably passed down to succeeding plant generations (R0, R1, and R2 offspring). Plant extracts underwent a Western blot analysis to reveal the existence of a polypeptide of the anticipated size that interacted with TSP14-specific antibodies. Several separate homozygous transgenic plant lines (R2 progeny) fed to tobacco budworm and tobacco hornworm larvae resulted in lower growth rates and higher mortality than plants transformed by a vector control. The study's findings demonstrate the possibility of introducing the TSP14 gene into plants as a lepidopteran pest defence. Additionally, the TSP14 gene's expression in transgenic plants results in antibiosis-type resistance to insect feeding and may offer a cutting-edge and reliable technique for managing insect populations that harm crops. The study's findings demonstrate the possibility of introducing the TSP14 gene into plants as a lepidopteran pest defence [7].

### 12.1.3 Antibiosis in fungi

The use of chemical compounds produced by endophytic fungi to prevent the spread of pathogens as biological control agents is expressed as a type of antibiosis [10]. For example, a panel of test microorganisms were used to assess the antibacterial activity of fermentation extracts made from fungi isolated from vineyard habitats. Evaluation of the antibacterial activity of fermentation extracts from 182 different fungi grown in eight different mediums. Extracts exhibiting antibacterial activity against at least one harmful bacterium were produced by 71 different fungi. The examination of internal transcribed spacer rRNA sequences for the identification of active fungi showed that Pleosporales, Hypocreales, and Xylariales species predominated. Utilizing liquid chromatography–mass spectrometry, several of the compounds detected in the active extracts were in some cases tentatively identified. While having little to no impact on probiotic bacteria, antimicrobial compounds produced by fungi in a vineyard ecosystem may prevent food-borne pathogens from colonizing and spoiling food products [11].

Foods frequently include pathogenic bacteria and yeasts that can degrade food and infect people with food-borne illnesses. The use of chemically produced preservatives is being discouraged in favour of natural alternatives, which guarantee a sufficiently long shelf life for foods and assure food safety in terms of viruses that can be transmitted through food. A reliable source of natural ingredients that might be utilized as preservatives to ensure the preservation and safety of food has arisen from microorganisms [11].

### 12.1.4 Antibiosis in bacteria

The bioactivity of numerous BCAs, including fluorescent Trichoderma spp., Pseudomonas spp., Streptomyces spp., and Bacillus spp., is attributable to antibiosis, a common and well-known phenomenon [1].

Different secondary metabolites with diverse roles and antifungal efficacy against different kinds of fungal infections may be produced by different BCA strains. For example, the *Pseudomonas fluorescens* strain CHAO produces siderophores, 2,4-diacetylphloroglucinol; and phenazines, cyanide; diverse combinations are the cause of the host-parasite antagonistic behaviour shown by this strain towards *gaeumannomyces graminis* var. tritici and *Chalara elegans*. Trichoderma spp. strains produce a wide range of secondary metabolites, including antibiotics and CWDEs, which have been implicated in biocontrol activity. It is important to understand that not all of BCA's antagonistic activity may be attributed to a single antifungal metabolite. A certain strain of BCA may produce several secondary metabolites that have antagonistic effects on various pathogen types. Therefore, it is impossible to generalize from one patho-system to another or from one BCA to another. The finest illustration comes from the work of Woo and Lorito, who discovered that a strain of *T. harzianum* produced different secondary metabolites based on the host plant it was administered to as well as the target pathogen infecting that specific host plant [1].

Moreover, *Staphylococcus aureus* (*S. aureus*) significantly reduced the growth of biofilms created by *A. fumigatus* regardless of biofilm formation stage or bacterial inoculum. The bacterium had an antibiosis effect on the fungus, which resulted in reduced *A. fumigatus* biofilm generation, disorganized fungal structures, abortive hyphae, and restricted hyphal growth. Conversely, conidia were also reduced, showed surface changes, and contained lyses [2].

## 12.2 Antibiotic

Our understanding of microbiology has improved as a result of research on antibiosis and its role in antibiotics. To better understand molecular processes including cell wall production and recycling, research into how antibiotics affect beta-lactam creation via the antibiosis relationship and interaction of the particular drugs with the bacteria exposed to the molecule [3]. Antibiosis between microbes was documented prior to the discovery of penicillin, including by Louis Pasteur, who postulated that germs might emit chemicals that kill other bacteria. It was noted around the start of the twentieth century that bacteria created heat- and diffusion-stable chemicals, and that their potential for treating infectious disorders had been researched. In the 1890s, Emmerich and Low treated hundreds of patients with an extract of *Pseudomonas aeruginosa* (then known as *Bacillus pyocyaneus*); this extract, known as pyocyanase, was used until the 1910s [12]. One of the most crucial turning moments in human history may have been the discovery of antibiotics. The use of antibiotics has produced incalculable benefits for biology, physiology, and medicine. Our understanding of microbial physiology and pathogenicity has evolved over time as a result of our knowledge of antibiotics and resistance to them. Additionally, this has given us the tools to make it easier for us to pursue new knowledge. The ampicillin (AMP) resistance gene, for instance, is frequently utilized as a selectable marker in recombinant DNA technology [3].

More than two thousand years ago, traditional mouldy bread poultices were applied to open wounds in Serbia, China, Greece, and Egypt as part of the usage of antibiotic-producing bacteria to prevent disease. The Eber papyrus, which was created about 1550 B.C. and contains a list of treatments that includes mouldy bread and therapeutic soil, making it the oldest extant medical record. Salvarsan, the first antibiotic, was introduced in 1910. In just over a century, antibiotics have changed contemporary medicine and extended the average human lifetime by 23 years. With the 1928 discovery of penicillin, the golden age of natural product antibiotic research began in the middle of the 1950s. Since then, a progressive reduction in the discovery and development of antibiotics as well as the emergence of drug resistance in several human infections have contributed to the current antimicrobial resistance dilemma [12].

## 12.2.1 Antibiotic resistance

Major bacterial diseases are becoming more resistant to standard antibiotic treatments, and multidrug-resistant bacteria are emerging at an alarming rate. High rates of morbidity and mortality are a challenge caused by antibiotic resistance in bacterial infections. Gram-positive and Gram-negative bacteria with multidrug resistance patterns can be difficult to treat and sometimes resistant to standard treatments. Associated with high rates of morbidity and mortality, antimicrobial resistance in bacterial pathogens is a problem that affects the entire world. Gram-positive and Gram-negative bacterial multidrug resistance patterns have resulted in infections that are challenging to cure. Important bacterial diseases are becoming increasingly resistant to basic antimicrobial treatments, and multidrug-resistant bacteria are emerging at an alarming rate, leading to the appearance of infections that cannot be treated with traditional antimicrobials [13]. The ultimate example of bacterial adaptability and evolution is the bacterial reaction to the "attack" of antibiotics. The concept of "survival of the fittest" stems from the enormous genetic plasticity of bacterial pathogens, which causes specific responses that result in mutational adaptations, acquisition of genetic material, or modification of gene expression, and thereby produces resistance to nearly all antibiotics currently used in clinical settings [14].

The life sciences and public health face a significant global challenge in the development of strategies to combat the emergence of antibiotic resistance. In the past few decades, human-pathogenic bacteria that are resistant to one or multiple antibiotics have increased dramatically on a global scale. Biofilms contribute to multidrug resistance and can complicate infection control [13]. After only eight decades of antibiotic use, once-treatable bacterial infections are now incurable. Antimicrobials have facilitated the development of numerous medical disciplines. Antibiotic prophylaxis and the ability to address infection problems are prerequisites for the success of numerous surgical operations and immunosuppressive therapies. Antibiotic resistance thus poses a serious danger to a sizable section of healthcare as we currently practice it. Gram-negative microbes that produce many forms of carbapenemase, gonorrhea, and multidrug-resistant tuberculosis are some of the areas of greatest concern. Improved antimicrobial stewardship, such as better infection prevention and diagnosis, can help preserve the antimicrobial medications currently on the market because antibiotic resistance is correlated with antibiotic use. If new anti-infectives are to be developed at a rate that keeps up with the rise in antimicrobial resistance, significant worldwide action and investment, from both the public and private sectors, are necessary [15].

Antibiotic resistance genes (ARGs) discovered in clinical infections are not the only ones that matter, it is becoming more widely accepted. Instead, all types of bacteria – pathogenic, commensal, and environmental – along with bacteriophages and mobile genetic elements – form the resistome, a source of ARGs from which pathogenic bacteria can horizontally transfer resistance (HGT). As shown for a number of clinically relevant ARGs, HGT has caused the transmission of antibiotic resistance from commensal and ambient organisms to pathogenic ones. Understanding the size of the resistome and how pathogenic bacteria deploy it is essential for efforts to restrict the dissemination of these genes.

In addition, because of its inherent antibiotic resistance, pathogenic microbial biofilm is seen as a global issue. Along with treatment restrictions, biofilms that develop on medical equipment can be an infection source. For the purpose of creating an efficient dosing schedule and preventing the emergence of antimicrobial tolerance and resistance in biofilm infections, historical pharmacokinetic (PK) and pharmacodynamic (PD) profiles of an antimicrobial drug are essential. It is important to note that none of the biofilm regimens have been approved for clinical use or shown to be helpful in guiding antibiotic therapy [12].

## 12.2.2 Carbon dots derived from kanamycin sulphate

A valuable study was done by Qian Luo et al. and published in *Royal Society of Chemistry* journal. That synthesized new carbon dots (CDs) base structure called carbon dots kanamycin sulphate (CDs-Kan) [16]. Antibiotics are one of the important ways to fight pathogens. However, many pathogens become resistant to some traditional antibiotics; hence, it is essential to synthesize and introduce new antimicrobial agents. Many studies have reported that CDs show useful functions in antitumor [17] and antiviral activities; plus, some studies reveal that they are antioxidants. Furthermore, CDs-Kan shows exclusive antibacterial properties. The Fourier transform infrared (FTIR) spectrum of CDs-Kan suggests seven peaks that show in Table 12.1. CDs-Kan was characterized by fluorescence emission spectroscopy and UV–vis [16], and they have an absorption peak at 280 nm and emission peaks at 315 and 425 nm. Moreover, transmission electron microscopy (TEM) imaging shows that the average diameter of CDs-Kan is about 3.87 nm (Figure 12.2).

*Escherichia coli* (*E. coli*) and *S. aureus* were used to determine the antibacterial activity of CDs-Kan; for this purpose, $1 \times 10^9$ (bacteria mL$^{-1}$) of *S. aureus* with OD600 = 1.2 and ($5 \times 108$ bacteria mL$^{-1}$) of *E. coli* incubate on LB solid agar plates at 37 °C for 16 h,

**Table 12.1:** The FTIR spectrum of CDs-Kan (peak unit is cm$^{-1}$) [16].

| peaks | 3433 | 1630 | 1559 | 1457 | 1399, 1053, 624 |
|---|---|---|---|---|---|
| bond | N–H or O–H | C = O | C–N | C–O | C–H |

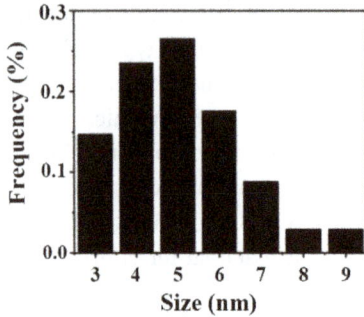

**Figure 12.2:** variety of particle size in CDs-Kan [16].

after added difference concentration of CDs-Kan and Kan. Briefly IDZ test reveal at 100 µg mL$^{-1}$ of CDs-Kan IDZ diameters was 12 mm for *S. aureus* and 10 nm for *E. coli* (Figure 12.3). The control group that didn't have CDs-Kan and Kan did not spectacle any antibacterial effect [16]. Furthermore, the colony-forming unit (CFU) counting method also showed bacterial growth density decreases with increasing concentration of CDs-Kan. These two tests show the antibacterial efficiency of CDs-Kan similar to Kan.

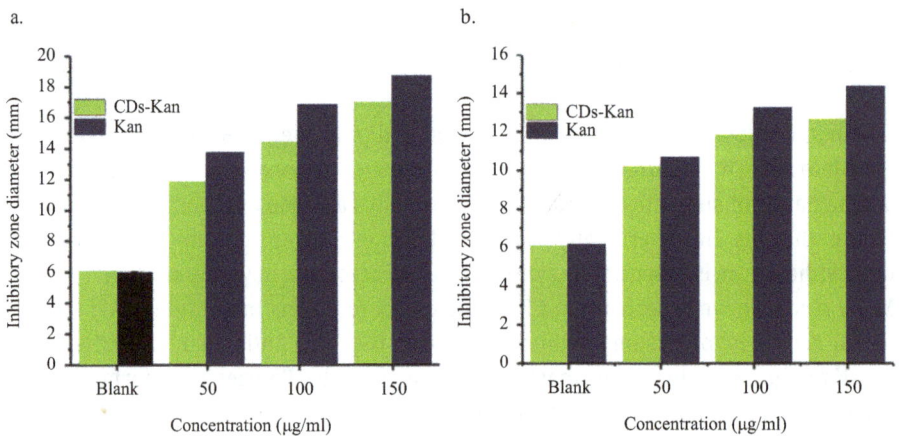

**Figure 12.3:** (a) Illustrates IDZ diameter of *E. coli* histogram. (b) Demonstrate IDZ diameters of *S. aureus* histogram [16].

Another experiment that helps to understand antibacterial effects is morphological analyses. In this study, scanning electron microscopy was used, and the morphology of *E. coli* and *S. aureus* *before* and *after* incubation with CDs-Kan and Kan (Figure 12.4).

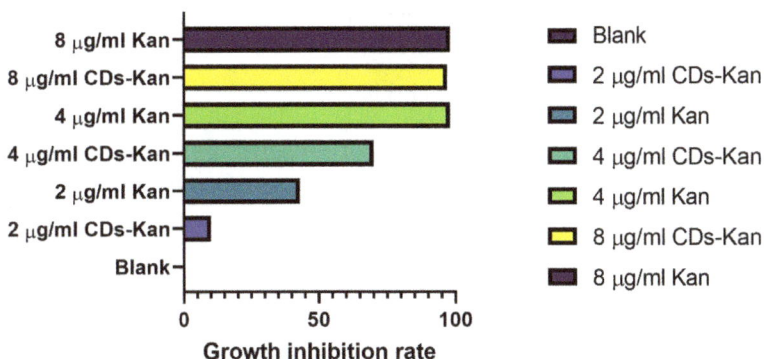

**Figure 12.4:** This schematic illustrates the growth inhabitation rate of *E. coli* and *S. aureus* in different concentrations of CDs-Kan and Kan.

After treatment by CDs-Kan, in *E. coli*, the strength of the cell wall's thickness decreased. *S. aureus* is Gram positive. Consequently, its peptidoglycan layer is thicker than Gram-negative bacteria. Therefore, no morphology change was seen. Nonetheless, cell shrank was reduced.

CDs derived from kanamycin sulphate (CDs-Kan) manufactured by a one-step hydrothermal method. More than antibacterial effects have good biocompatibility and fluorescence stability; thus, they can be used for specific staining of plants and bacterial cells.

## 12.2.3 Fluorescent carbon dots with a high nitric oxide

An excellent study was done by Shixin Liu et al. and published in *Royal Society of Chemistry* journal. And introduced new structure which synthesized N-diazeniumdiolate (NONOate) and chitosan-graft- poly(amidoamine) dendrimer (CPA). This structure was made by the facial hydrothermal method. High-resolution TEM studies suggest that the size of CPA-CDs was about 10 nm and CPA-CDs/NONOate was about 5 nm [4]. This size reduces perhaps for high pressure of NO. Correspondingly, FTIR spectra were done for CPA, CPA-CDs, and CPA-CDs/NONOate (Figure 12.5) [18].

Many NO-based antibiotics have been developing because NO induces reactive by-products such as dinitrogen trioxide ($N_2O_3$) and peroxy- nitrite (ONOO −). These reactive by-products react with bacterial membranes and DNA, but NO-based antibiotics have a non-existent diagnostic function; nonetheless, CPA-CDs/NONOate have valuable properties [18]. In this study *P. aeruginosa* was used. It's a Gram-negative bacterium. The antibacterial effect of the CPA-CDs/NONO was measured by CFU counting assay, and results suggest that the antibacterial effect of this structure has concentration-dependent; for instance, in 0.0625 mg mL$^{-1}$ of CPA-CDs/NONOate, more than 1 log of bacterial viability was reduced.

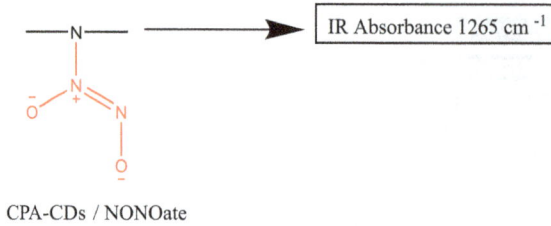

CPA-CDs / NONOate

**Figure 12.5:** FTIR spectra study of CPA, CPA-CDs, and CPA-CDs/NONOate [18].

Making a wound is one of the valuable experiments done in this study. Male SD rats (average body weight was 200 g) were divided into three groups, and two wounds in each rat's back were created two days before the experiment started. The size of injuries was about 10 nm. 100 μL of *P. aeruginosa* was injected into wounds, and group one treated with CPA-CDs, group two was treated with CPA-CDs/NONOate, and group three was not treated. Imaging of the wounds was done for ten days. And result suggests that CPA-CDs/NONOate play an influential role in fighting bacteria and wound healing (Figure 12.6).

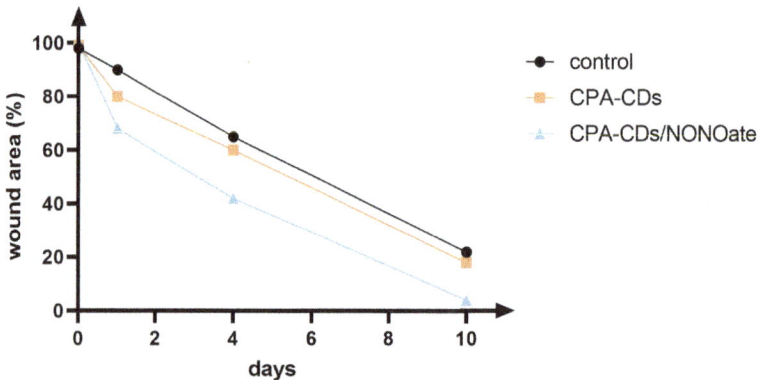

**Figure 12.6:** Effect of CPA-CDs/NONOate in wound healing [18].

Good biocompatibility is one of the essential properties of each therapeutic agent; therefore, biocompatibility of CPA-CDs/NONOate must be measured. For this purpose, rat's body weight was measured, and no significant body weight change was reported. Furthermore, histological studies were done on critical organs like the heart, liver, kidney, lung, and spleen. The result revealed that treatment with CPA-CDs/NONOate and CPA-CDs does not have toxicity and side effects on other organs [18].

## 12.2.4 CDs-C12

Jing Yang et al. studied about CDs-$C_{12}$, and it was published in ACS's *Applied Materials & Interfaces* journal. CDs-$C_{12}$ [19] they synthesized CDs-C12 and experimented with their antibacterial effect with their precure that were CDs(synthesized by a combination of AEEA and glycerol) and BS-12 (which were made with 1-(3-dimethylaminopropyl)-3-ethylcarbodiimide.

N-hydroxysulfosuccinimide (Sulfo-NHS) in 2-(N-morpholino) ethanesulfonic acid (MES) buffer solution). The result suggest that CDs-$C_{12}$ is practical in antibacterial and antibiosis. We explain this excellent article below briefly.

For characterized many nanostructures, some analyses have been done. One of them is TEM image. This method was used in this study, and their consequences reveal that the middling size s CDs is about 2.7 nm, and the average size of CDs-$C_{12}$ is bigger, approximately 4 nm. Moreover, the hydrodynamic diameter of CDs-$C_{12}$ is larger than CDs[19].

Furthermore, to achieve more detail about the structure of CDs-$C_{12}$, CDs and BS-12 FTIR spectra were used. This method suggests that signals in the range of ~ 1010–1160 cm$^{-1}$ attributed to Si – O and C – O stretching vibration were joint between CDs-$C_{12}$ and CDs. Moreover, three signals (1467 cm$^{-1}$, 2922 cm$^{-1}$, 2852 cm$^{-1}$) were similar between CDs-$C_{12}$ and BS-12. UV–vis show that BS-12 does not have an absorption pick; CDs have two wide picks at 280 and 340 nm, and CDs-$C_{12}$ has a new pick at 510 nm. This new pick may be attributed to changes in the structure's functional group surface. One important future of many nanostructures used for antibacterial is their stability in different aqueous solutions [19]. For this purpose, CDs-$C_{12}$ was scattered in PBS solution, normal saline, deionized water, and LB medium. Results suggest CDs-C12 has a stable structure and is valid for biomedical application, and this theory arose that excellent water-dispersibility for abundant hydroxyl group on their surface [19].

Some tests had to be done to confirm that CDs-C12 is a practical antimicrobial agent. one of them was measuring optical density (OD), and for this purpose, *S. aureus* was used. The result revealed that at 30 ug/mL of CDs-C12 (at a defined time), the growth of bacteria was wholly inhibited, whereas the same concentration of CDs and BS-12 did not inhibit bacterial growth. Furthermore, a cell counting kit-8 (CCK-8) assay was used and suggested that CDs-C12 play a role in decreasing *S. aureus* viability while neither BS12 nor CDs cannot decrease *S. aureus* viability [19].

Four plates were selected for the CFU counting method, and *S. aureus* was grown on them, and one-to-one was treated with BS-12(,10 µg/mL) CDs, CDs-C12, and one for the control group not treated. And results suggest that the number of colonies treated with CDs-C12 significantly decreased (Figure 12.7) [19].

In another experiment in vitro, mice were used as animal models and made a wound on their back in this study. After that, they were divided into to two groups [19], group one was treated with white PBS solution, and group two was treated with CDs-CD12 result after quantification revealed that in group one, bacteria in the

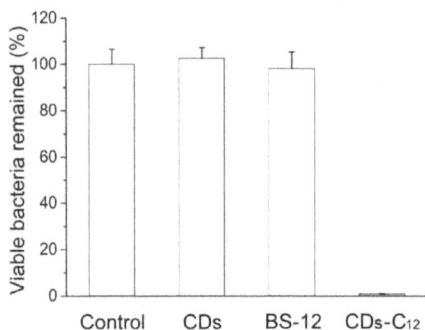

**Figure 12.7:** CFU counting method after treated with CDs, CD-C12, BS12 [19].

infectious tissue showed $2.2 \times 108$ CFU/g, and in group two, the bacterial burden was $2.8 \times 106$ CFU/g.(Figure 12.8)

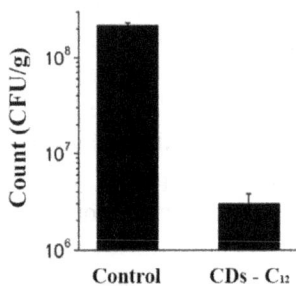

**Figure 12.8:** The bacterial burden in the wound of mice [19].

For the experiment on gram-negative bacteria, *E. coli* was used, and some tests, like the CCK-8 assay, were done. The result revealed that BS-12 and CDs do not have an antibacterial effect on Gram-negative bacteria. Additionally, CDs-C12 at a concentration of 200 µg/mL does not show significantly antibacterial properties. Moreover, optical density studies confirm that CDs-C12 can inhibit Gram-positive bacteria more than Gram-negative bacteria [19].

## 12.2.5 Nitrogen-doped carbon quantum dots

Antibiotic resistance has been important and perilous subject in recent years; the possibility exists that if antibiotic resistance expands and does not invent other fighting methods against bacteria, perhaps come back to the time before antibiotics. Accordingly, Chengfei Zhao et al. studied nitrogen-doped carbon quantum dots and broadcasted it in *Biointerfaces* journal. In this study, for synthesized nitrogen-doped carbon quantum dots (NCQDs), glucose and diethylenetriamine were used as a precursor and made by the heat fusion method. TEM microscopy suggests that the average diameter of NCQDs was approximately 5 nm. FTIR spectrum analysis of NCQDs is summarized in the Table 12.2 [20].

**Table 12.2:** FTIR spectrum of nitrogen-doped carbon quantum dots (NCQDs) (peak unit is cm$^{-1}$) [20].

| Bond | −OH/NH | CH | C = C | −CH2 | C − N | N − H |
|------|--------|----|----|----|----|-----|
| Peak | 3425 | 2926, 2854 | 1637 | 1491, 1439 | 1344, 1237, 1140, 1078 | 755, 567 |

To measure the antibacterial activity of NCQDs, some experiments were done. The plate-diffusion method is one of them, with eight agar plates, including:
1. *S. aureus (ATCC6538)*
2. *S. aureus (ATCC43300)*
3. *S. epidermidis*
4. *MRSA*
5. *E. coli*
6. *Salmonella paratyphi-β (S. paratyphi-β)*
7. *Pseudomonas aeruginosa (P. aeruginosa)*
8. *Enterococcus faecalis (E. faecalis)*

after that, three disks were created on each plate containing NCQDs, DETA, and glucose. DETA and glucose were used as control and inhabitation zone measured in each plate, *S. aureus (ATCC6538)*, *S. aureus (ATCC43300)*, *S. epidermidis*, and *MRSA*. The inhibition zones on the agar plates incubating *S. aureus (ATCC43300)* were approximately 15.5 mm, and the result for *S. aureus (ATCC6538)*, *S. epidermidis*, and *MRSA* was about 14.5 mm. The result of the plate-diffusion method suggest that NCQDs have more antibacterial effect on Gram-positive bacteria, especially *Staphylococcus* more than others [20].

Another experiment was done to confirm the antibacterial activity of NCQDs; in this experiment, the negative control groups *MRSA* and *E. coli* were treated with normal saline. In another group, *MRSA* and *E. coli* were treated with NCQDs. In the third group, *S. aureus (ATCC6538)* was treated with NCQDs and TEM imaging was done on bacteria in each group [20]. Results show that no cell death or damage was evident in group one. In group two, rupture and loss of integrity were observed in *MRSA;* on the other hand, the morphology of *E. coli* in groups one and two did not differ. It can be considered that NCQD does not have an antibacterial effect against *E. coli*, and in the third group, the cell structure of *S. aureus* collapsed [20].

A wound formed in the back of the SD rats was excised and infected with *MRSA* in this article, which also measured the therapeutic effect NCQDs. Wounds were treated with NCQDs after four days. As a control group, other mice received standard saline treatment, and in a third group, treatment with white lasted for seven days. In the group that received standard saline treatment, the wound surface still had pus and exudates. In the group that received treatment with NCQD and vancomycin, the size of the infected wounds shrank dramatically. The wound areas were in the normal saline group NCQD and vancomycin groups after receiving treatment for two weeks. In other words, both therapy groups had essentially healed their wounds [20].

Furthermore, the cytotoxicity of NCQDs on mammalian cells measured for this purpose HeLa and PANC-1 were used, cell viability was measured, and results were revealed when the concentration of NCQDs was 0.128 mg/mL. The viabilities of HeLa and PANC-1 cells were more than 80% [20].

## 12.2.6 Levofloxacin-based carbon dots

The grand structure was introduced by Li-Na Wu and published in an article in *Carbon* journal. Levofloxacin-based carbon dots (LCDs) synthesized to fight with antibiotic resistance. Levofloxacin is soluble in water. LCDs were made by wet-heating method mainly through bottom-up synthesis methods. TEM study suggests the size of LCDs was about $7.00 \pm 0.25$ nm. At pH 7.4, the zeta potential of LCDs was $+ 28.00 \pm 0.50$ mV. Spectroscopy studies reveal absorption peaks at 240 nm, 293 nm, and 340 nm for levofloxacin. The absorption peak in LCDs was lifted from 240 nm to 260 nm and 293 nm to 276 nm. FTIR analyses reveal more information about the structure and charm of LCDs. For instance, peaks of $C = O$ stretching vibration were at 1698.57 $cm^{-1}$. 3500–2500 $cm^{-1}$ were for COOH. 3431.78 $cm^{-1}$ was for $-OH$ (Table 12.3) [21].

The disk diffusion test was done to measure in vitro antibacterial activity of LCDs. Bacterial used were *S. aureus*, *S. epidermidis*, *E. faecalis*, *L. monocytogenes*, MRSA, *E. coli*, *P. aeruginosa*, and *S. marcescens*. four disks in each plate were created that contained: 10 mg of insoluble levofloxacin, soluble levofloxacin hydrochloride solution, LCDs solution from levofloxacin hydrochloride, and LCDs solution from levofloxacin. Results suggest that insoluble levofloxacin in the insoluble state had no antibacterial effect. And the antibacterial ability of LCDs was higher than levofloxacin. At 0.125 mg/mL of LCDs, the MIC of *S. aureus*, *E. coli*, and *S. marcescens* was more minor than others; at 0.5 mg/mL of LCDs, the MIC of *S. epidermidis*, *E. faecalis*, and *L. monocytogenes* was larger.

Another experiment to measure the antibacterial effect of LCDs was TEM microscopy studies. *S. aureus*, *E. coli*, and *MRSA* was incubated with LCDs and Levo-HCL and control group. According to the findings, bacteria that were incubated with levofloxacin hydrochloride still had fully developed cell membranes. However, the bacterial walls and membranes were torn, allowing for the leakage of intracellular materials [21].

### Cow milk-derived carbon dots

This study is done by Yu Tang et al. and published in applied *Surface Science* journal. Hydrothermal treatment of cow milk is used to arrange cow milk-derived carbon dots (CMCDs) then this prepared CMCDs get extracted by ethyl acetate for obtaining amphiphilic CMCDs (ACMCDs). By using ACMCDs as reducing agent and template, we can make ACMCDs-supported silver nanoparticles (ACMCD-Ag nanocomposites) which have a great biocidal impact on Gram-positive (*Staphylococcus aureus*) and Gram-negative (*E. coli*) bacteria. After that, we can prepare new ACMCD Ag/polymethylmethacrylate (PMMA) nanocomposite antibacterial film by solvent casting method. Due to its excellent antibacterial properties, light absorption and flexibility, nanocomposite antibacterial film has a high potential in applications[22]. The FTIR spectrum of CMCDs is shown in Table 12.1 [23]. The UV–vis spectrum of the this CMCDs shows a broad absorption band which are centred at 274 nm [24]. The size distribution of this CMCDs are represented in Figure 12.9, which showed that the CMCDs have good dispersion and narrow size distribution (average diameter = 1–5 nm) [22].

**Table 12.3:** The FTIR spectrum of CMCDs [22].

| Peaks (cm$^{-1}$) | 3200–3700 | 2923, 2850 | 1635 | 1570 | 1420 |
|---|---|---|---|---|---|
| Bond | O–H and N–H | C–H | C = O | N–H | C = C |

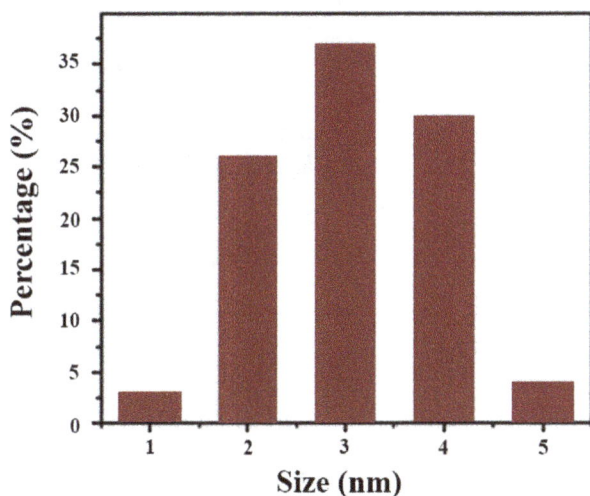

**Figure 12.9:** distribution of the CMCD particle size [22].

The minimum inhibitory concentration (MIC) test and Kirby-Bauer disk diffusion method were used to confirm the antibacterial activity of ACMCD-Ag against Gram-negative (*E. coli*) and Gram-positive (*S. aureus*) bacteria. AgNO$_3$ served as a model substance for activity comparison. For both *E. coli* and *S. aureus* strains, the disks with ACMCD-Ag were surrounded by an inhibition zone with a wider width than those with AgNO$_3$. After 24-hour incubation period, AgNO$_3$'s MIC was higher than that of

ACMCD-Ag (Table 12.4). So, at the same Ag concentration, ACMCD-Ag possesses better antibacterial properties than AgNO3 [22].

**Table 12.4:** Lists of the minimum inhibitory concentrations of AgNO3 and ACMCD-Ag nanocomposite for two bacteria ($\mu$g mL$^{-1}$, calculated by Ag) [22].

| Culture | Strain no | MIC$_a$ | MIC$_b$ |
|---------|-----------|---------|---------|
| S. aureus | ATTC 25,922 | 7 | 5 |
| E. coli | ATCC 25,923 | 13 | 10 |

Furthermore, compared to E. coli, the nanocomposite showed higher biocidal activity against the Gram-positive bacteria S. aureus. The cell shape may be the reason for the ACMCD-enhanced Ag's antibacterial effectiveness against Gram-positive bacteria. The loose cell wall of Gram-positive S. aureus makes it vulnerable to assault by nanoparticles. As a result, DNA denaturation in S. aureus occurred more quickly than in E. coli as a result of the liberated silver nanoparticles' ability to enter the cell wall and bind to DNA [25].

The thin film UV–vis absorption spectra as a function of nanocomposite doping levels are shown in Figure 12.10. The thin film attachment method was employed to assess the antibacterial property after one day of incubation, and the results revealed that the percent decrease of bacteria was near to 100% against both S. aureus and E. coli. Consequently, the ACMCD-Ag/PMMA thin film has a good antimicrobial efficiency regardless of Gram classes [22].

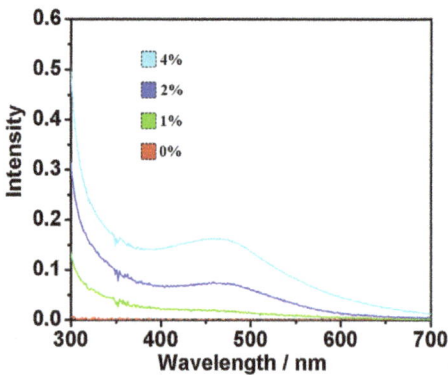

**Figure 12.10:** The UV–vis absorption spectra of PMMA composite films with various ACMCD-Ag nanocomposite doping levels [22].

**Photodynamic effects were produced by carbon dots functionalized with ampicillin when exposed to visible light**

Rabah Boukherroub et al. conducted this work, which was published in the journal of *Colloids and Surfaces B: Biointerfaces*. In this study, we demonstrate a novel strategy for antibacterial therapy using amine-terminated carbon dots (CDs-NH$_2$) functionalized with AMP. The amine-functionalized CDs are utilized to transport and immobilize AMP as well as operate as a visible light-activated antibacterial agent (CDs-AMP). Furthermore, AMP immobilized on the CDs-NH$_2$ surface is more stable in solution than free AMP. The AMP conjugated CDs platform maintains the Theranostic properties of CDs-NH$_2$ while combining them with the antibacterial properties of AMP. As a result, *E. coli* growth is effectively inhibited by the immobilized AMP on CDs-NH2 surface and the generation of reactive oxygen species under visible light irradiation [26]. While AMP absorbs light below 260 nm, CDs-NH$_2$ has three bands in its UV–vis absorption spectra at 237, 342, and 450 nm [27]. Table 12.1 displays the FTIR spectrum of the CDs-AMP conjugate (Table 12.5) [26].

**Table 12.5:** The FTIR spectrum of CDs-AMP conjugate[26].

| Peaks (cm$^{-1}$) | 3650–3200 | 3080–3030 | 3000–2840 | 1800–1650 | 1297 | 1126 | 700 |
|---|---|---|---|---|---|---|---|
| Bond | O–H | Aromatic C–H | Aliphatic C–H | C = O | C-C | C-O | C-S |

Results of cytotoxicity and cellular absorption of mammalian (HeLa) cells unequivocally demonstrate the suitability of CDs-NH2 and CDs-AMP for biomedical applications [26].

The *E. coli* K12-MG 1655 strain was used to test the bactericidal effects of the CDs-NH$_2$ and CDs-AMP conjugates with and without illumination by visible light. To count the number of viable cells, we used plating. To measure cell proliferation, we used a fluorescence-based cell dead/live assay and optical density at 600 nm. Red fluorescence indicates dead cells, while green fluorescence indicates living cells. Contrary to CDs-NH$_2$ and CDs-AMP, which demonstrated a strong bactericidal effect on *E. coli*, the amine-functionalized CDs were not dangerous [26].

Additionally, we have looked at the capacity of CDs-NH2 and CDs-AMP to produce singlet oxygen (1O$_2$) when illuminated with visible light for photodynamic therapy using antimicrobial particles (PPDT). According to the findings, 1O$_2$ levels rose with exposure time and visible light lamp intensity. It was feasible to assess how successfully these chemicals damage bacterial cells by exposing *E. coli* K12-MG 1655 to CDs-NH2 and CDs-AMP at various concentrations (for 10 or 20 min at 0.3 W). As shown in Figure 12.11, at a concentration of 400 g mL-1, CDs-NH$_2$ was able to reduce *E. coli* survival, indicating that the concentration of CDs-NH2 impacts the photodynamic bacterial killing impact. E. coli cells after 20 min of radiation by 4 log10. This is in line with how much singlet oxygen CDs-NH$_2$ produces [26].

Additionally, exposing CDs-AMP to visible light increased the conjugate's bactericidal activity (Figure 12.11B). The significant improvement in the bactericidal activity of CDs-AMP may be due to the production of ROS species by the CDs-NH2, including singlet oxygen ($^1O_2$). The CDs-AMP conjugate still has some of the CDs-NH2's natural Theranostic abilities. It follows that the CDs-AMP combination can be used as a multifunctional nanoplatform based on the combined therapeutic killing of antibiotics with visible light-generated photodynamic effects [26].

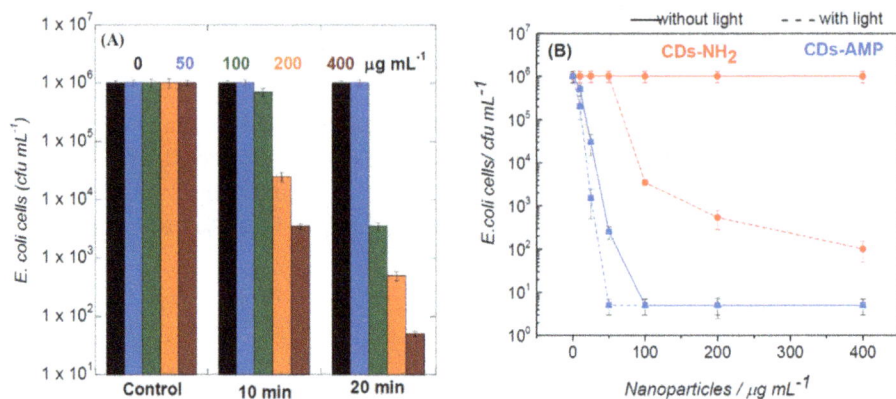

**Figure 12.11:** (A) Shows the photodynamic efficacy of CDs-NH2 for *E. coli* K12-MG 1655 inactivation after illumination at 0.3 W for 10 and 20 min. (B) Role of CDs-NH2 and CDs-AMP concentrate on effectiveness of *E. coli* treatment without (solid lines) and with (dash lines) illumination by visible light (20 min, 0.3 W). The standard deviation of three separate experiments is shown by the error bars [26].

## Titanium-based carbon quantum dots@hematite nanostructures

Omran Moradlou and colleagues conducted this investigation, which was then published in *Journal of Photochemistry and Photobiology A: Chemistry.* Under both dark and light circumstances, thin films of nanostructured hematite (-Fe2O3) and hematite combined with carbon quantum dots (CQDs@-Fe2O3) were tested for their ability to inhibit the growth of Gram-positive (*S. aureus*) and Gram-negative (*E. coli*) bacteria [28]. Samples of Ti/-Fe2O3 and Ti/CQD@-Fe2O3 were created using the hydrothermal technique[29]. With the use of FE-SEM and HRTEM, the surface morphology of the samples was examined[28].

In order to test the antibacterial properties of the Ti/-Fe2O3 and Ti/CQD@-Fe2O3 samples against Gram-positive *S. aureus* (ATCC 6538, PTCC 1112) and Gram-negative *E. coli* (ATCC 25,922, PTCC 1399) bacteria in both dark and light environments, an antibacterial drop test was used [28].

## Dark conditions

The bactericidal activity of Ti/-Fe$_2$O$_3$ against *E. coli* was minimal (10%), as shown in Table 12.1. By comparison, the Ti/-Fe$_2$O$_3$ mixture inactivate 65% of the *S. aureus* bacteria. About 20% of *E. coli* and 70% of *S. aureus* bacteria were inactivated by the Ti/CQD@-Fe$_2$O$_3$ when left in the dark condition. In fact, both the Ti/-Fe$_2$O$_3$ and Ti/CQD@-Fe$_2$O$_3$ samples have a higher propensity to inactivate *S. aureus* in comparison to *E. coli*. It suggests that bacterial strains of S. aureus are more susceptible to antibacterial metal oxide compounds than are bacterial strains of *E. coli*. The existence of an additional outer membrane layer in Gram-negative bacteria is mostly to blame for the enhanced resistance of Gram-negative strains to hematite-based compounds and the samples' improved antibacterial effectiveness against Gram-positive bacterial strains [30].

## The illumination conditions

It is anticipated that under light irradiation, more ROS would be produced and, as a result, a lower percentage of the bacteria would survive. According to the findings (Table 12.6), Ti/-Fe$_2$O$_3$ inactivates *E. coli* 50% more quickly under light illumination than it does in the dark. Additionally, photoinactivation of the *E. coli* population with Ti/CQD@-Fe$_2$O$_3$ sample is greater in the presence of light than it is in the absence of light (Table 12.6). Under conditions of light irradiation, ROS can develop and enter bacterial cell membranes and harm them. The FE-SEM pictures suggest that *S. aureus* bacteria are easily distorted when by antibacterial chemicals, and after their membrane is damaged, the contents of the cell leak outside. However, it may be said that the *E. coli* bacteria strain's cell walls are more resistant to oxidative species. Actuality, Gram-negative bacteria have an additional outer membrane that frequently provides chemical resistance [28].

**Table 12.6:** Antibacterial activity data of Ti/-Fe2O3 and Ti/CQD@-Fe2O3 samples under circumstances of darkness and visible light [28].

| Condition | Microorganism | Initial concentration (CFU/mL) | Sample | Antibacterial activity (%) |
|---|---|---|---|---|
| Dark | *E. coli* | $1.3 \times 10^6$ | Ti/α-Fe$_2$O$_3$ | 10 |
| | | $1.2 \times 10^6$ | Ti/CQD@α-Fe$_2$O$_3$ | 20 |
| | *S. aureus* | $4.1 \times 10^5$ | Ti/α-Fe$_2$O$_3$ | 65 |
| | | $7.1 \times 10^5$ | Ti/CQD@α-Fe$_2$O$_3$ | 70 |
| Visible light illumination | *E. coli* | $1.3 \times 10^6$ | Ti/α-Fe$_2$O$_3$ | 15 |
| | | $1.0 \times 10^6$ | Ti/CQD@α-Fe$_2$O$_3$ | 35 |
| | *S. aureus* | $5.3 \times 10^5$ | Ti/α-Fe$_2$O$_3$ | 70 |
| | | $3.3 \times 10^5$ | Ti/CQD@α-Fe$_2$O$_3$ | 80 |

In contrast to Gram-negative bacteria, which lack an additional outer membrane layer, the antimicrobial analyses show that -$Fe_2O_3$ and CQDs@-$Fe_2O_3$ are hazardous to the chosen microorganisms. The samples also showed long-lasting antibacterial action against Gram-positive bacterial strains. It was determined that the preferred mechanism of bactericidal activity of hematite-based thin films is both the penetration of iron cations into the bacteria cell via their membrane and the generation of reactive oxygen species based on the results of the antibacterial activity of -Fe2O3 and CQDs@-Fe2O3 under dark and light irradiation conditions. These results imply that CQDs@-Fe2O3 nanoparticles can aid in the development of visible-light antimicrobial materials for their potential bactericidal uses [28].

### Green synthesis of multifunctional carbon dots

This study is done by Aharon Gedanken et al. and published in *Nanomaterials* journal. By using a simple one-step hydrothermal process, CDs were extracted from medicinal turmeric leaves (Curcuma longa) and tested for their bactericidal effects on two Gram-negative (*E. coli, Klebsiella pneumoniae*) and two Gram-positive (*S. aureus, S. epidermidis*) bacteria. The average size of the CDs was 2.6 nm, and they had spherical forms (Figure 12.12). Spectra of UV absorption reveal a distinctive peak at 288 nm [29–31].

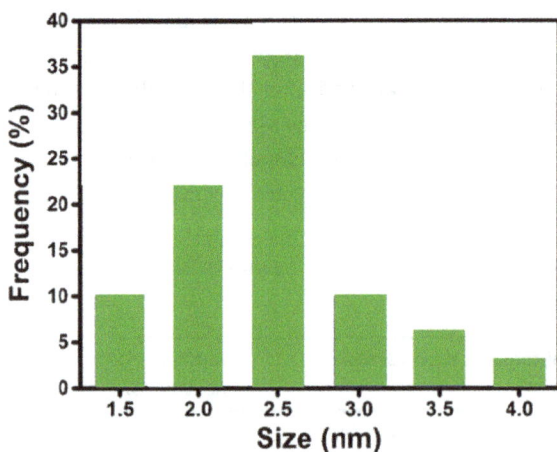

**Figure 12.12:** Mean particle size of 2.6 nm on average (b) [31].

The CDs were shown to be superior at killing Gram-positive *S. aureus* and *S. epidermidis* bacteria as well as Gram-negative *E. coli* and *K. pneumoniae* bacteria. For *E. coli* and *S. aureus*, the MIC is 0.25 mg/mL; for *K. pneumoniae* and *S. epidermidis*, it is 0.5 mg/mL. Figure 12.13a,b shows that CDs effectivity at inhibiting the growth of *E. coli* and *S. aureus* within 8 h at 0.25 mg/mL and at 0.5 mg/mL for *K. pneumoniae* and *S.*

*epidermidis.* However, following a 24-hour incubation with 1 mg/mL of the CDs, the entire eradication of bacterial cells was seen (Figure 13c, d) [31].

The findings showed that the CDs responded quickly to growth inhibition of *E. coli* and *S. aureus* with low concentration and short incubation times. But *K. pneumoniae* and *S. epidermidis* showed a substantially longer incubation period before being completely eliminated. Our manufactured CDs displayed improved antibacterial activities against all four pathogens without any atom passivation. Demethoxycurcumin and bisdemethoxycurcumin are two important chemicals that are partially retained inside or on the surface of CDs, enhancing their ability to kill bacteria [31].

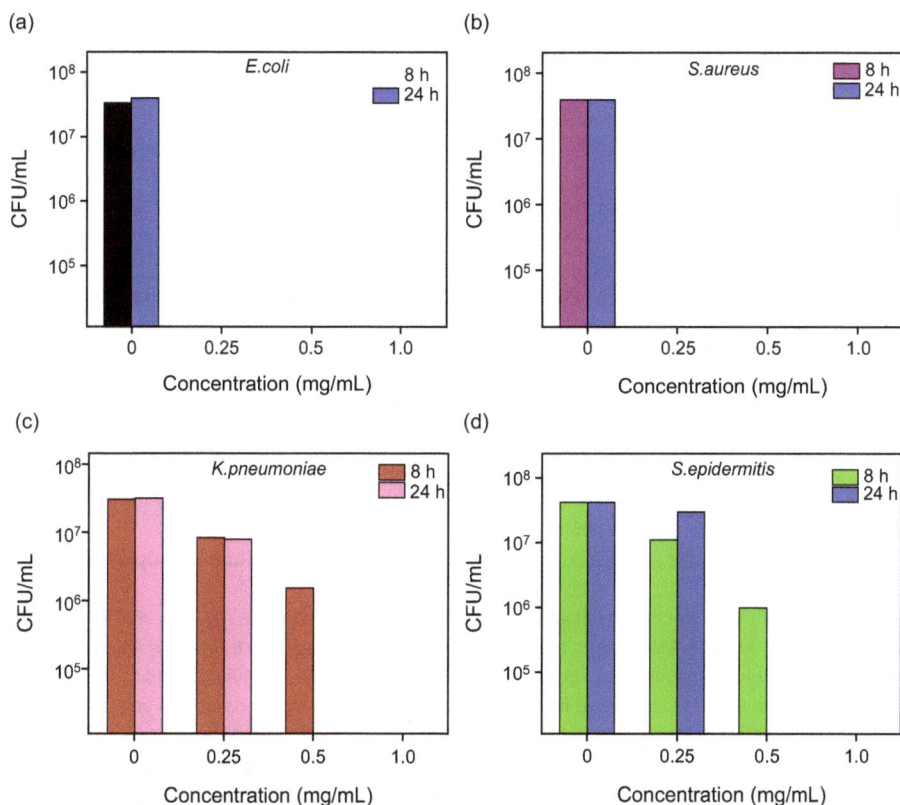

**Figure 12.13:** The illustrates the bactericidal effects of CDs on (a) *E. coli*, (b) *S. aureus*, (c) *K. pneumoniae*, and (d) *S. epidermitis* [31].

The standard MTT colorimetric test was used to assess the cytotoxicity of CDs in water. In order to determine if CDs are naturally cytotoxic, The PC-3 cell line was also used for the cell viability experiment. After a 24-hour incubation period, cell survival was still more than 95% even at a CDs concentration of 200 g/mL. However, after a 24-hour incubation, CDs at 500 g/mL decreased cell viability by 50% [31].

Our discovery might open the way for the production of naturally produced CDs from turmeric leaves as a potential novel antibacterial agent.

# References

[1]     Nehra S, et al. Bio-management of Fusarium spp. associated with fruit crops. Fungi Bio-Prospects Sustainable Agric Environ Nano-Technol. 2021, 1, 475–505.

[2]     Ramírez Granillo A, et al. Antibiosis interaction of staphylococcccus aureus on aspergillus fumigatus assessed in vitro by mixed biofilm formation. BMC Microbiol. 2015, 15(1), 1–15.

[3]     KONG KF, Schneper L, Mathee K. Beta-lactam antibiotics: From antibiosis to resistance and bacteriology. Apmis. 2010, 118(1), 1–36.

[4]     Dwivedi GR, Sisodia BS. Secondary Metabolites: Metabolomics for Secondary Metabolites. In: New and Future Developments in Microbial Biotechnology and Bioengineering, Elsevier, 2019, 333–344.

[5]     Bowman HHM. Antibiosis, University of Toledo, Toledo, Ohio, 1947.

[6]     Stout MJ. Host-plant Resistance in Pest Management. In: Integrated Pest Management, Elsevier, 2014, 1–21.

[7]     Maiti IB, et al. Antibiosis-type insect resistance in transgenic plants expressing a teratocyte secretory protein (TSP14) gene from a hymenopteran endoparasite (Microplitis croceipes). Plant Biotechnol J. 2003, 1(3), 209–219.

[8]     Holtz AM, et al. Antibiosis of Eucalyptus plants on Podisus nigrispinus. Phytoparasitica. 2010, 38(2), 133–139.

[9]     Singh H, Sarao PS, Sharma N. Quantification of antibiosis and biochemical factors in rice genotypes and their role in plant defense against Cnaphalocrocis medinalis (Gùenee)(Lepidoptera: Pyralidae). Int J Trop Insect Sci. 2022, 42(2), 1605–1617.

[10]   Aamir M, et al. Fungal Endophytes: Classification, Diversity, Ecological Role, and Their Relevance in Sustainable Agriculture. In: Microbial Endophytes, Elsevier; 2020, 291–323.

[11]   Cueva C, et al. Antibiosis of vineyard ecosystem fungi against food-borne microorganisms. Res Microbiol. 2011, 162(10), 1043–1051.

[12]   Hutchings MI, Truman AW, Wilkinson B. Antibiotics: Past, present and future. Curr Opin Microbiol. 2019, 51, 72–80.

[13]   Frieri M, Kumar K, Boutin A. Antibiotic resistance. J Infect Public Health. 2017, 10(4), 369–378.

[14]   Munita JM, Arias CA. Mechanisms of antibiotic resistance. Microbiol Spectrum. 2016, 4(2), 4.2 15.

[15]   MacGowan A, Macnaughton E. Antibiotic resistance. Medicine. 2017, 45(10), 622–628.

[16]   Luo Q, et al. Carbon dots derived from kanamycin sulfate with antibacterial activity and selectivity for Cr 6+ detection. Analyst. 2021, 146(6), 1965–1972.

[17]   Xue X, et al. Multistage delivery of CDs-DOX/ICG-loaded liposome for highly penetration and effective chemo-photothermal combination therapy. Drug Delivery. 2018, 25(1), 1826–1839.

[18]   Liu S, et al. Fluorescent carbon dots with a high nitric oxide payload for effective antibacterial activity and bacterial imaging. Biomater Sci. 2021, 9(19), 6486–6500.

[19]   Yang J, et al. Carbon dot-based platform for simultaneous bacterial distinguishment and antibacterial applications. ACS Appl Mater Interfaces. 2016, 8(47), 32170–32181.

[20]   Zhao C, et al. Nitrogen-doped carbon quantum dots as an antimicrobial agent against Staphylococcus for the treatment of infected wounds. Colloids Surf B: Biointerfaces. 2019, 179, 17–27.

[21]   Wu L-N, et al. Levofloxacin-based carbon dots to enhance antibacterial activities and combat antibiotic resistance. Carbon. 2022, 186, 452–464.

[22]   Han S, et al. Application of cow milk-derived carbon dots/Ag NPs composite as the antibacterial agent. Appl Surface Sci. 2015, 328, 368–373.

[23]  Zhou L, He B, Huang J. Amphibious fluorescent carbon dots: One-step green synthesis and application for light-emitting polymer nanocomposites. Chem Commun. 2013, 49(73), 8078–8080.

[24]  Deng Y, et al. Long lifetime pure organic phosphorescence based on water soluble carbon dots. Chem Commun. 2013, 49(51), 5751–5753.

[25]  Kong H, Jang J. Antibacterial properties of novel poly (methyl methacrylate) nanofiber containing silver nanoparticles. Langmuir. 2008, 24(5), 2051–2056.

[26]  Jijie R, et al. Enhanced antibacterial activity of carbon dots functionalized with ampicillin combined with visible light triggered photodynamic effects. Colloids Surf B: Biointerfaces. 2018, 170, 347–354.

[27]  Jia X, Li J, Wang E. One-pot green synthesis of optically pH-sensitive carbon dots with upconversion luminescence. Nanoscale. 2012, 4(18), 5572–5575.

[28]  Moradlou O, Rabiei Z, Delavari N. Antibacterial effects of carbon quantum dots@ hematite nanostructures deposited on titanium against Gram-positive and Gram-negative bacteria. J Photochem Photobiol A: Chem. 2019, 379, 144–149.

[29]  Moradlou O, et al. Carbon quantum dots as nano-scaffolds for $\alpha$-Fe2O3 growth: Preparation of Ti/CQD@ $\alpha$-Fe2O3 photoanode for water splitting under visible light irradiation. Appl Catal B: Environ. 2018, 227, 178–189.

[30]  Nor YA, et al. Engineering iron oxide hollow nanospheres to enhance antimicrobial property: Understanding the cytotoxic origin in organic rich environment. Adv Funct Mater. 2016, 26(30), 5408–5418.

[31]  Saravanan A, et al. Green synthesis of multifunctional carbon dots with antibacterial activities. Nanomaterials. 2021, 11(2), 369.

Nicole Remaliah Samantha Sibuyi*, Anelisiwe Mbengashe,
Zimkhitha Bianca Nqakala, Antoinette Alliya Ajmal, Tswellang Mgijima,
Cate Malope Mashilo, Aluwani Matshaya, Samantha Meyer,
Mervin Meyer, Martin Opiyo Onani, Abram Madimabe Madiehe
and Adewale Oluwaseun Fadaka*

# Chapter 13
# Carbon dots in drug delivery

**Abstract:** Drug delivery is an important aspect of any successful disease therapy; it ensures that the drugs reach the target site in its intact form and selectively infer its activity with reduced adverse effects on the surrounding tissues. However, selectivity and drug solubility has always been a major limitation for most therapeutic drugs. To overcome these limitations, drug delivery strategies have been devised, which include chemical modification of the drugs or the use of drug delivery systems (DDS). Although DDS are able to increase drug solubility and bioavailability based on their drug target and route of administration, they can be limited by several factors such as bystander toxicity, early drug release and clearance. In recent years, the focus has shifted to nanocarriers due to their unique physicochemical properties. Carbon dots (CDs), in particular, stand out as they are made from carbon sources and are perceived to be biocompatible. Furthermore, their smaller sizes (2–5 nm) afford them a tunable photoluminescence and fluorescent properties that can help monitor CD–drug conjugates in real time.

**Keywords:** Carbon dots, drug delivery, drug loading, drug monitoring, nanocarriers, tracking agents

---

*Corresponding authors: Nicole Remaliah Samantha Sibuyi, Department of Science and Innovation (DSI), Mintek Nanotechnology Innovation Centre (NIC), Advanced Materials Division, Health Platform, Mintek, Randburg, South Africa; DSI/Mintek NIC Biolabels Node, Department of Biotechnology, University of the Western Cape, Bellville, South Africa
*Corresponding authors: Adewale Oluwaseun Fadaka, DSI/Mintek NIC Biolabels Node, Department of Biotechnology, University of the Western Cape, Bellville, South Africa; Department of Anesthesia, Division of Pain Management, Cincinnati Children's Hospital Medical Center, Cincinnati, OH 45229, USA
Anelisiwe Mbengashe, Antoinette Alliya Ajmal, Cate Malope Mashilo, Aluwani Matshaya, Mervin Meyer, Abram Madimabe Madiehe, DSI/Mintek NIC Biolabels Node, Department of Biotechnology, University of the Western Cape, Bellville, South Africa
Zimkhitha Bianca Nqakala, Tswellang Mgijima, Martin Opiyo Onani, Organometallics and Nanomaterials, Department of Chemical Sciences, University of the Western Cape, Bellville, South Africa
Samantha Meyer, Department of Biomedical Sciences, Faculty of Health and Wellness Sciences, Cape Peninsula University of Technology, Bellville, South Africa

The original version of this chapter was revised. Unfortunately, the affiliation of the author Adewale Oluwaseun Fadaka was incorrect in the original publication. This has been corrected. We apologize for the mistake.
https://doi.org/10.1515/9783110799958-013

## 13.1 Introduction

Most drugs fail to reach their full therapeutic potential mainly due to their adverse by-stander effects and low bioavailability [1]. These limitations, to some extent, were re-solved through the use of conventional drug delivery systems (DDS), which had shown a significant improvement in the efficacy of the drugs through various modes of adminis-tration [2]. However, the DDS can be limited by poor bioavailability, fast drug metabo-lism, early drug release, early drug clearance, and poor permeability [1, 3]. Nanocarriers are among the advanced DDS that were explored in order to overcome the shortcomings of the conventional DDS. The nanomaterials present unique physicochemical properties that can be easily manipulated to produce stimuli-responsive DDS [4] for targeted and controlled drug release [5]. Nanocarriers are ideal DDS, as they are biocompatible, are non-cytotoxic, and have a prolonged residence time [6]. Moreover, their small size allows them to move freely within the diseased tissues without a need for a targeting moiety [7]. Various organic and inorganic nanoparticles (NPs) showed potential; however, the chapter focuses on the carbon dots (CDs) as an innovative tool for drug delivery.

The role of CDs as DDS for various diseases is receiving an enormous attention in recent years. Their ability to target drugs and monitor their circulation and localization brings a fresh perspective to drug delivery [3] and disease therapy [8]. Their outstanding optical, non-toxic, and size-based properties are a promising platform that can allow the CDs to be built into multifunctional systems for biomedical applications. The carbon-based CDs are more especially appealing due to their photostable fluorescent properties which can be used to track their mobility and activity when used in vivo [3, 9]. More-over, the CDs are versatile and present an opportunity to load and/or encapsulate bioac-tive materials onto them; that way they can be developed into multiplex systems [3, 10]. The chapter discusses the feasibility of the CDs as drug delivery agents, starting with the conventional DDS and their limitations. The attributes that make CDs as potential drug delivery agents are also highlighted, together with the strategies that are used to load and encapsulate the drugs. In addition to their application as drug delivery agents, the tunable photoluminescence (PL) and fluorescent properties of the CDs can be employed to monitor drug response in real time. The possible biosafety of CDs can be improved by developing smart and stimuli-responsive CD DDS.

## 13.2 Conventional DDS and their limitations

Non-specificity, biodegradation, and solubility are the major limitations of the current drugs, resulting in insufficient dose reaching the pathological tissues, and thus re-duced drug efficacy. Using higher doses to increase the amount of drugs that reach the target tissues is usually associated with adverse bystander effects. Strategies have been devised to reduce the side effects of the drugs, by delivering the drugs directly at

Intra-derma

Intra-venous

Sub-cutaneous

**Ocular**
Drops (solutions, emulsions,
suspensions), ointments, contact
lens, implants, inserts and
intravitreal

Intra-muscular

**Otic**
- Topical
- Intracochlear
- intratympanic

**Parenteral**

Intramuscular

Epidermis
Dermis
Subcutaneous fat

Muscle

**Rectal/Vaginal**
- Suppositories
- Pessaries
- Tablets
- Enema
- Creams and gels
- Foams and sponges

**Nasal**
-Drops
-Sprays

**Topical**
- Semi-solid dosage
  forms (Ointments,
  Creams, Lotions, Gels
  and Liniments)
- Sprays
- Transdermal patches

**Inhalation**
- Dry powders
- Liquid sprays
- Aerosol

**Oral**
**Solid dosage forms**
- Tablets
- Capsules
- Granules
- Sub-lingual tablets
- Buccal tablets
- Effervescent
  tablets
- Thin films
- Medical gums
- Lozenges
- Liquid dosage
  forms
- Solution
- Suspensions
- Emulsions

**Oral (Non-sterile)**
**Liquid dosage forms**
- Syrup
- Solution
- Tincture
- Suspension
- Emulsion
- Lotion
- Elixir
- Draughts
- Enemas
- Gargles

**Oral (Sterile)**
**Liquid dosage forms**
- Injectables
- IV Bolus
- Eye drops

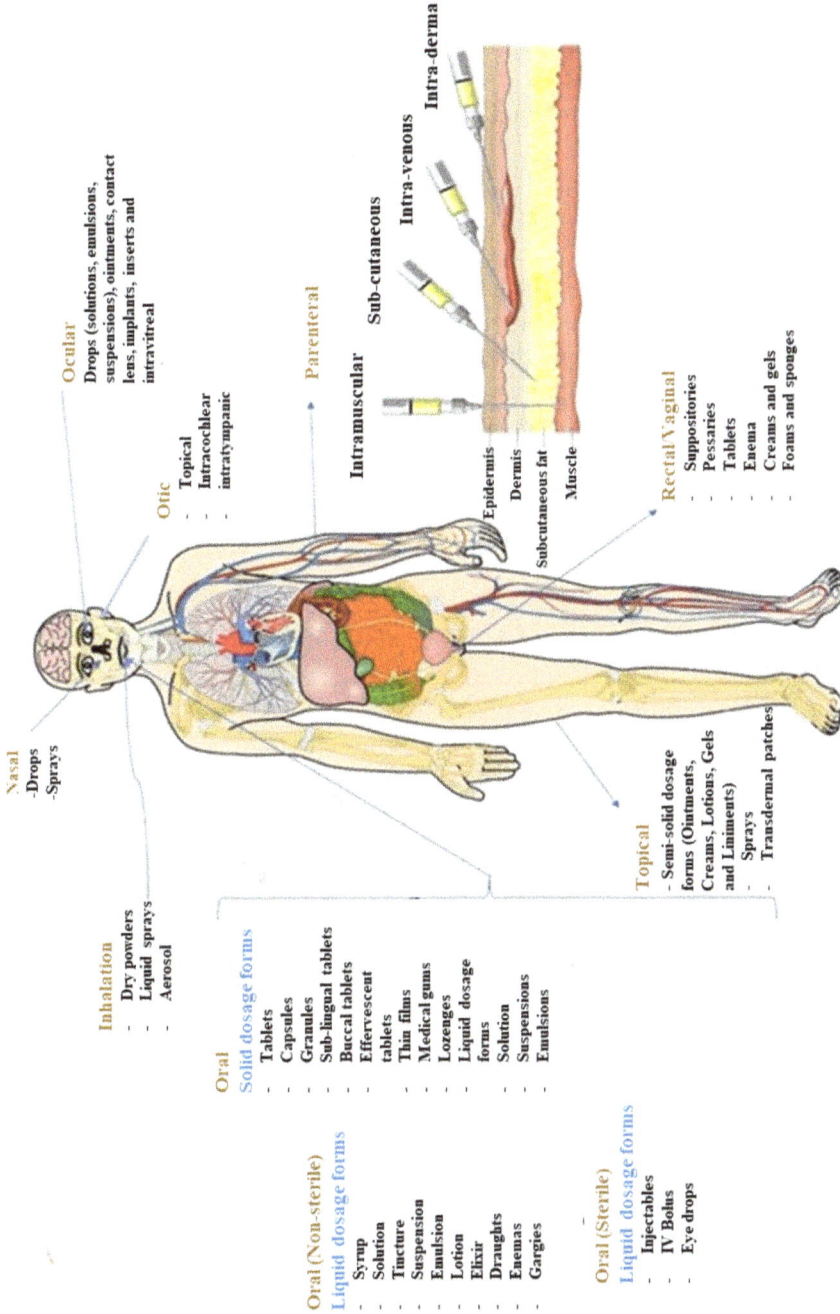

**Figure 13.1:** Type of drug formulations and routes of their administration (reprinted with permission from MDPI [2]).

the target site [2, 3, 11] through the use of DDS. The conventional DDS are based on drug formulations such as tablets, capsules, powders, syrups, ointments, lotion, and droplets, which are inspired by the drug solubility, half-life, permeability, pharmacokinetics, and route of administration. Typical routes for drug administration are shown in Figure 13.1, which takes into account the characteristic behaviour of the drug within the body and the drug target. Oral administration can subject the drugs to degradation by digestive enzymes and the low pH in the stomach, resulting in less amount of drugs reaching the diseased tissues [12, 13], while the parenteral administration route can have 100% bioavailability and drug targeting [2]. The efficiency and potency of a drug is highly dependent on its bioavailability, as well as the drug's ability to escape biodegradation by digestive enzymes or at acidic pH, and early drug clearance [1]. The conventional DDS are able to deliver the drugs at disease sites; however, most, if not all, of these systems have several disadvantages listed in Table 13.1, such as bystander toxicity, early drug release, early drug clearance, and poor solubility that limit their application as drug delivery agents [3].

**Table 13.1:** Attributes and limitations of the conventional DDS.

| DDS route of administration | Application | Advantages | Limitations | References |
|---|---|---|---|---|
| Oral | Drug administration by tablets | Cost-effective Ease of large-scale production Long shelf life Ease of administration | Degradation by digestive enzymes and gastric juices Effect of the drug may sometimes be too slow for emergencies Low bioavailability, some of drugs absorbed in non-targeted sites Not useful for unconscious patients | [12, 13] |
| Injectable | Used for drugs with short half-life | Can be used for unconscious and comatose patients Concentration of drug needed is low Rapid absorption of drug No risk of degradation of drug | Pain at the injection site Risk of embolism | [14] [13] |

**Table 13.1** (continued)

| DDS route of administration | Application | Advantages | Limitations | References |
|---|---|---|---|---|
| Transdermal | A topical DDS using patches and through a percutaneous absorption Drugs are released to the blood stream in a controlled manner | Frequency of dosing is reduced Bioavailability is improved Flexibility in terminating drug administration Useful for potent drugs | Skin irritation on the site of application Limited number of drugs that can be used due to skin permeability | [15] [12] |
| Pulmonary | Drugs administered through mouth or nose inhalation | Requires small doses Onset effect Adverse effects are less severe Drugs are delivered to target organ | Performed by health professionals Airways should be accessible Can cause irritation of airways | [16] |
| Drug carrier | Incorporating and encapsulating drugs or therapeutic agents | Enhance the efficiency, bioavailability, and efficiency of drugs Reduces frequency of administration Simplified administration ensures patient compliance. Minimizes drug side effects | Possible toxicity Difficulty scaling-up production | [17, 18] |

The attributes that had made DDS to be successful drug delivery agents are also highlighted in Table 13.1; however, for the drugs to successfully execute their therapeutic effects they must also overcome unfavourable physiological barriers. Development of new drugs is time consuming and expensive; hence, the use of DDS can improve the efficacy of pre-existing drugs. Drug carriers are engineered for targeted and controlled drug release in the target cells to enhance drug potency, safety, bioavailability [18], and improved patient compliance. An appropriate drug carrier should be biocompatible, biodegradable, and highly soluble, and the most important feature is their ability to transport the drug in a targeted manner. There are many strategies reported for drug delivery; however, drug delivery must work synergistically with the drug release profile for enhanced therapeutic effect. Nanomaterials with special focus on CDs as discussed in this chapter emerged as a viable approach for targeted drug delivery.

## 13.2.1 Nanocarriers in drug delivery

The use of nanocarriers as DDS demonstrates good prospects, and can be used to enhance the efficacy and reduce adverse effects of the drugs [19]. Nanotechnology has alleviated the challenges associated with conventional DDS as well as the drugs by developing "smart delivery systems". The theory behind these systems is that they are easily manipulated by physical, chemical, and biological stimuli; and when subjected to these stimuli, they can be programmed to release drugs in a regulated manner [4]. The inspiration for introducing nanomaterials to DDS is due to their surface-area-to-volume ratio which enables nanomaterials to adsorb and carry multiple bioactive compounds [20]. Organic and inorganic nanocarriers, such as liposomes, micelles, dendrimers, quantum dots (QDs) and CDs, among others, have emerged as promising drug delivery agents. After conjugation or encapsulation of the drugs on the nanomaterials, they will be delivered to the target site in a controlled manner [5] via passive or active targeting. An ideal nanocarrier is expected to have the following properties: mechanical stability, high drug loading capacity, biocompatibility, low cytotoxicity, and a long in vivo residence time [6]. The prolonged circulation or half-life of NP-based DDS in the body provides an added advantage for diagnostic and therapeutic agents for targeted drug delivery and accumulation at a diseased site. The NPs can move freely in the body due to their small sizes [7].

Various types of organic and inorganic nanomaterials have been used in cancer [21] and diabetes [22] as drug delivery vehicles or as treatments. These include metal and soft NPs such as QDs, liposomes, and micelles. QDs are semiconductor nanocrystals typically with a diameter of 2–10 nm; they possess a size tunable absorption and emission bands due to the quantum confinement effects [23]. The size of the QD conjugates can increase to 5–20 nm when functionalized with some biomaterials, and these sizes can passively penetrate solid tissues and be cleared by renal filtration. Due to long-term toxicity and degradation presented by the heavy metals in the core of the QDs, the application of QDs is limited to cell-based and small animal studies. Conjugating QDs with non-toxic compounds such as silicon and carbon can also mask the toxicity of the QDs and enhance their biocompatibility [24].

Liposomes and micelles are among the soft NPs that have been widely used as drug delivery agents, due in part to their biocompatible and biodegradable nature. Liposomes are made up of double phospholipids layers and steroids and are already used in pharmaceuticals as DDS [25]. Amphotericin and daunorubicin are examples of the drugs that were transported by liposomes [26]. The membrane structure of liposomes is analogous to the cell membranes, making them better DDS. They have also been proven to stabilize therapeutic compounds, and improve drug biodistribution. Positively charged liposomes are currently being used in gene therapy as DNA DDS [15]. As with other foreign particles that enter the body, liposomes encounter immunogenicity, opsonization, and reticuloendothelial system defences [27] which can possibly lead to their physical instability [15].

Micelles are made of surfactant amphiphilic block polymers that are spontane-ously self-arranged into spherical aggregates [28]. They have hydrophobic cores which can be loaded with hydrophobic drugs such as docetaxel and camptothecin. This core assimilation with the drugs results in an enhanced drug bioavailability and stability [25]. Micelles are non-toxic, have high drug-loading capacity and a faster rate of clearance, and this makes them suitable for intravenous drug delivery [29]. Despite these advantages, the stability of micelles is reduced when they encounter changes in environments and when the concentration is low, they can dissociate [30].

Both metal and soft nanomaterials are, however, plagued by some challenges which limit their application in in vivo drug delivery. Inorganic or metal NP-based systems are non-biodegradable and their accumulation in vital organs such as the heart, kidneys, liver, and spleen could induce cytotoxicity [31]. The organic NPs are not reproducible and their loading efficiency cannot be ascertained, which then re-sults in reduced drug efficacy [32]. Moreover, the polymeric NPs and liposomes can induce toxic effects on healthy tissues and trigger immune responses. Thus, many of the NP-based drug delivery technologies are restricted by the impact they have on bi-ological processes resulting in poor biocompatibility, bystander toxicity, non-specific adsorption, and formation of protein coronas [33]. To be effective, a nanocarrier must not only facilitate systemic and intracellular drug delivery but must also retain the drug's original features with high precision and sensitivity. This chapter focuses only on the CDs as the drug delivery agents.

## 13.3 CDs as potential drug nanocarriers

CDs are the newest and among the sought-after nanomaterials which meet the needs for targeted drug delivery. They are the most promising drug delivery vehicles, mainly due to their bio-friendly properties. They are considered to be less toxic compared to other nanomaterials with similar properties such as fluorescent metal-based QDs [9, 34]. And unlike others, CDs possess unique and distinct mechanisms of actions to their cargoes, able to retain their original properties and mode of actions; and therefore regarded as capable nanocarriers for delivering cargoes to the target sites. CDs can be used to regu-late and monitor their own distribution, dispersion, and clearance from various parts of the body, concurrently doing the same for the molecules they carry [35]. Through their tunable PL properties, CDs together with their cargoes can be monitored in real time [9]. CDs were shown to enhance the drug solubility and uptake [36] through either passive or active targeting. Drug release from CDs usually occurs via desorption, diffusion, or tethered to be released through external or internal stimuli [10, 36–38]. The size and charge of the CDs influence drug permeation, where smaller sizes are able to pass through cellular barriers including the blood–brain barrier (BBB) [39, 40].

## 13.4 CD drug loading

CDs have a large surface-area-to-volume ratio providing more room to functionalize them with various biomolecules; these can be targeting moieties, diagnostic and therapeutic agents; individually or all combined depending on the downstream application. This is made possible by the carbon sources used in the synthesis of CDs; they contain a lot of functional groups such as $NH_2$, OH, or COOH which are then conferred on their surface creating multiple binding sites on the CDs. These functional groups make it easy to use different chemistries to load biomolecules onto the CDs, which may include drugs or contrast agents [3, 10]. Several chemical interactions were explored to load bioactive molecules on the CDs as shown in Figure 13.2A. The most commonly used chemistries include electrostatic interaction, covalent bonding, pi–pi conjugation, and the 1-ethyl-3-(3-dimethyl-aminopropyl) carbodiimide (EDC) and *N*-hydroxysuccinimide (NHS) covalent crosslinking. Surface functionalization using EDC and NHS chemistry is a common practice for biomolecules containing either carboxyl (COOH) or amine ($NH_2$) groups to covalently conjugate drugs containing these functional groups (Figure 13.2B). The COOH group on one of the biomolecules (CDs or the drug) allows for carboxyl-to-amine covalent bonding with the biomolecules that have $NH_2$ group [37].

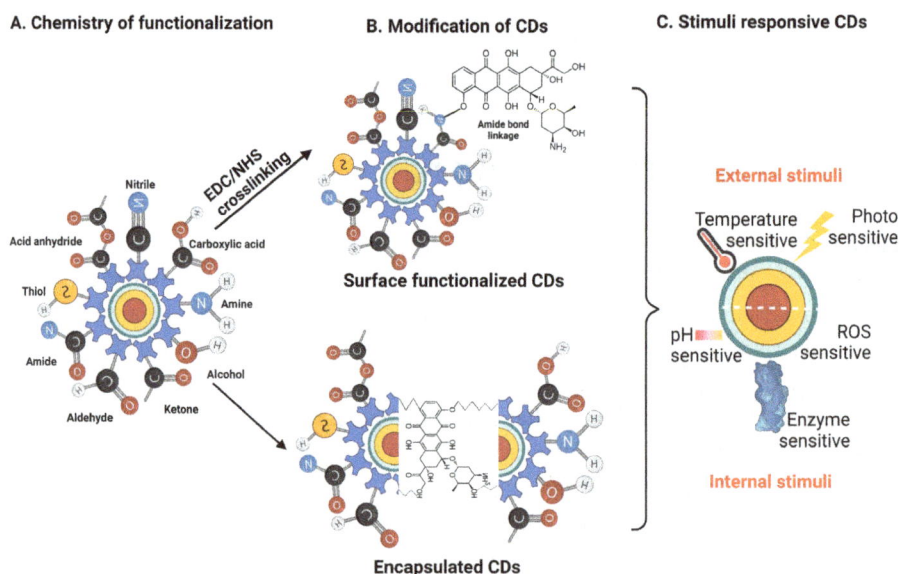

**Figure 13.2:** Different chemistries (A) for modification of CDs through surface functionalization and/or encapsulation (B) to develop stimuli-responsive CD–drug conjugates.

CDs are flexible and drugs can be loaded either through surface functionalization or via encapsulation or both (Figure 13.2B) to obtain CDs for a desired function. Additionally, the CDs can be designed to release the drugs following an encounter with either external or internal stimuli as highlighted in Figure 13.2C [37] such as cellular enzymes [38], environmental pH [10], light [36], or temperature. CD–drug conjugate formed by non-covalent bond between the COOH and $NH_2$ groups is usually exploited for drug release at the diseased tissues as it was shown to be sensitive to changes in pH [3]. Of importance, a promising drug nanocarrier must be able to reach the target area to facilitate drug release and function. Although CDs have exhibited many excellent drug loading and targeting properties, not all CD formulations are able to target the bones and the BBB. Their targeting is often improved by attaching targeting moieties that are specific for certain disease biomarkers [41, 42] to reduce drug side effects, and in the process enhance drug efficacy [42].

## 13.5 CDs as a potential DDS

The potential role of CDs as drug delivery vehicles is increasingly being acknowledged in medicine, and they emerge as the best candidates for disease diagnosis and therapy. CDs have shown many attributes that qualifies them as putative drug delivery agents even in difficult to reach areas. Cancer and neurological diseases are among many diseases that stand to benefit from CDs as delivery vehicles, as they are characterized by a highly complex and heavily barricaded microenvironment [43–45], and BBB [39, 40]. A prerequisite for any successful disease therapy is for the drugs to reach the target organs or tissues, intact and in sufficient dosage. However, for cancer, factors such as variable hypoxia, intratumoural pressure gradient, and abnormal vasculature within the tumours make it difficult to accomplish cancer targeting, thus reducing the drug efficacy and, at times, causing drug resistance [43–46].

The BBB is another physiological barrier for drug delivery [46] in the brain tissues for the treatment of brain cancer and other neurological diseases, as it hinders the delivery of drugs from the blood into the central nervous system (CNS) [40, 47]. In fact, the BBB, blood–brain tumour barrier, and the weak enhanced permeability and retention effect (EPR) for brain tumour treatment exclude almost 100% of the drugs from targeting the brain tissues [48, 49]. CDs have shown ability to open and cross through tight junctions of the BBB due to their unique physicochemical properties. CDs as drug carriers can then be used to transport and deliver drugs into the CNS and also to increase the BBB permeability as shown in Figure 13.3 [50]. There is a growing body of literature that recognizes CDs to have the ability to cross the BBB in vitro [51] and in vivo [51, 52]. The CDs were able to target the glioma cells and tissue in rats [51] and the BBB in zebrafish and rats [52] without the aid of targeting molecules. Their targeting abilities are associated with the type of molecules covering their surface, consistent with this hypothesis, CDs synthesized using glucose (GluCDs) as a precursor were shown to target and be internalized by cells expressing the glucose transporters. Evidently so, the

GluCDs were only localized in glucose transporter-positive yeast strain and not in the glucose-deficient strains. The GluCDs were able to reach and target the BBB after injection through the heart of the zebrafish and intravenous tail vein in the rats. In the rats, the GluCDs conjugated with the fluorescein (GluCD-F) were located in the neurons found in the brainstem, the ventral horn, medial grey, and dorsal grey of the cervical spinal cord [52]. The CDs also proved to be target-specific, CD-Asp synthesized using equimolar concentrations of D-glucose and L-aspartic acid, were able to target the brain C6 glioma cells and the glioma tissues in the rat brain within 15 min after intravenous injection through the tail vein. None of the CDs synthesized using the individual materials or those that substituted L-aspartic acid with an amino acid that differ with a single methyl group (L-glutamic acid) were able to replicate these effects or show glioma targeting or specificity [51]. Using passive diffusion transport, Yellow-CDs (Y-CDs) were able to cross the BBB of the zebrafish. Their uptake inhibited the overexpression of the human amyloid precursor and the β-amyloid (Aβ) proteins that are responsible for development of Alzheimer's disease [49]. The ability of the GluCD-F to cross the BBB with their cargoes and accumulate in the brainstem and cervical spinal cord, and glioma targeting by CD-Asp without any targeting ligands, indicated the feasibility of CD-based DDS for the CNS-related diseases such as neurodegenerative disease, traumatic injury [52], and brain tumours [51]. CDs were successfully used as delivery vehicles for chemotherapeutic drugs [9], antipsychotic drug (haloperidol, HaLO) [53], antibiotics [54, 55], as well as mineral supplements. Ferric ammonium citrate (FAC), an iron supplement, was successfully loaded in the CDs (CD-FAC) and improved the biocompatibility of the FAC on liver (HepG2) cells up to 1,000 μg mL$^{-1}$. The free FAC reduced cell viability at ≥100 μg mL$^{-1}$. These suggested that the CDs could do a better job in delivering the iron supplement safely to the cells and enhanced its cellular uptake [56].

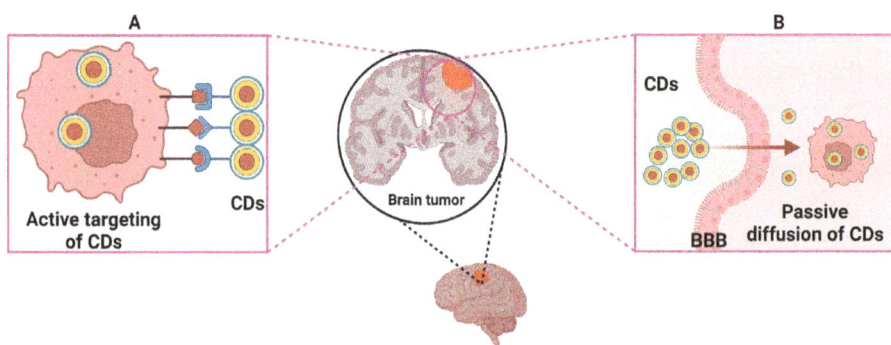

**Figure 13.3:** Passive (A) and active (B) translocation of CDs in the BBB for drug delivery and imaging of the brain tumours.

Passive targeting of CDs is a desirable trait for drug delivery agents; however, not all CDs possess this property [51] and at times those that have lack target specificity. To increase their specificity and localization, targeting moieties such as aptamers, antibodies, and nanobodies can be easily attached to the surface of CDs. Active targeting has emerged as a powerful platform for CD-based DDS for the treatment of diseases, including the CNS-associated diseases [57] and cancer. Transferrin [40] and folic acid [49] receptors are among the biomarkers that were identified as targets for the BBB and cancer. And when the targeting ligands for these receptors were covalently conjugated to CDs, the CDs were able to target and pass through the BBB of zebrafish via receptor-mediated endocytosis and accumulated in the brain [40].

The CDs are also compatible with other nanosystems and can be used to create CD–nanohybrid DDS with synergistic, but superior, activities than the individual nanomaterials [38, 58]. Mathew and co-workers produced a chitosan (CS)/CD-based nanocomposite, wherein dopamine was later entrapped in the matrix to form a dopamine@CS/CDs nanocomposite. The originality of their technology was influenced by the use of a non-toxic carrier to successfully deliver drugs to all diseases; in particular, the CDs permitted consistent dopamine release in neurodegenerative diseases. The chitosan/CD nanocomposite allowed for efficient transportation of dopamine and prolonged drug release [58]. Mesoporous silica nanoparticles (MSNs) were developed into MSNs-CD nanocomposite, with the CDs used as imaging agents [38].

## 13.6 CDs as a trackable drug delivery agents

CDs are a subset of nanomaterials [59] with multifunctional capabilities and used as an alternative drug delivery vehicle that enables real-time imaging [3]. CDs have been studied extensively since their first discovery in 2004; mainly due to their photoluminescent, strong absorption properties, photostability, resistance to photobleaching, low toxicity, environmental-friendliness, biocompatibility, and facile preparation, amongst others [60].

CDs stand out as delivery agents due to their PL and fluorescent properties which could be simultaneously used for both drug delivery and bioimaging. The CDs can help follow or track the movements and localization of the CDs [3, 9] and indirectly its cargoes in real time [3]. CDs with a size distribution between 1 and 4 nm could be detected in the kidneys with a strong fluorescent signal after 30 min, and the intensity of the signal was reduced over time. Negligible amounts of CDs were also observed in the liver and spleen after 30 min of treatment, and the signal was completely undetected at the other time points. Their presence in the kidneys and bladder signifies that the CDs were excreted through glomerular filtration, and most of them were removed from the system within 3 hr. The fluorescent intensity of these CDs was maintained for 2 weeks on the samples that were stored at room temperature and 4 °C. Unlike other nanofluorescent agents such as QDs and carbon nanotubes (CNTs); CDs are compatible with various biological media, stable in acidic and basic media at pH range of 3–9, and are resistant to photobleaching [9].Thus, CDs can be used as alternatives to the traditional and semiconductor fluorophores (QDs), all of which suffer from toxicity, high cost and photobleaching [34]. Wide application of CDs has been recorded in biomedical fields, not limited to imaging, drug delivery, and theranostics [50]. These applications are inspired by the exceptional properties of CDs shown in Table 13.2.

**Table 13.2:** Properties that play a role in bioimaging.

| Parameter | CDs | QDs (semiconductors) | CNTs | Nanodiamonds | Graphene | Fluorophores |
|---|---|---|---|---|---|---|
| Fluorescence | High | High | High | Optically transparent | No fluorescence (originally) | Low |
| Solubility | High | Poor | Poor | Poor | Poor | High |
| Photostability | High | High | High | High | Poor | Varies |
| Toxicity | Low | High | High | Low | Low | High |
| Cost | Low | Low | Low | Low | Low | Low |
| Quantum yield | High | High | Low | Low | Low (no modification) | Moderate |
| References | [61] | [61, 62] | [61–63] | [64, 65] | [61, 64] | [64, 66] |

**Note:** Most nanomaterials on the table must be modified to improve their properties, therefore recorded as "poor".

Amongst the different properties of CDs responsible for their recognition in drug delivery, their emission qualities are their most enticing attributes that can aid in tracking the CDs when used in vivo. CDs absorb light over a wide UV spectrum and emit light with high intensity in a narrow spectral range [67]. CDs have high tunable PL and excitation capabilities that are exploited for a variety of applications. The optical properties of CDs are responsible for their trackable ability when used in biomedical application as either drug delivery or therapeutic agents. The excellent PL properties of CDs and high fluorescence intensity have been exploited as imaging-guided nanocarriers for the delivery of drugs [68]. For instance, CDs functionalized with an anticancer drug conjugated to a photo trigger agent (3-bromopropoxy)-2-quinolylmethyl chlorambucil; Qucbl) demonstrated an adjustable emission range of 325–550 nm and a broad absorption range of 350–450 nm. The emission spectrum of Qucbl-CDs spans from visible to NIR wavelengths and was not associated with the photo trigger (quinoline moiety) as it does not have this kind of variable emission spectra. It was then concluded that the tunable emission observed with the Qucbl-CDs must be caused by the CDs [53]. Thus, surface passivation or modification of CDs can alter the absorption properties of the CDs, which are responsible for the fluorescent and the trackable ability of the CDs. Green-emitting CDs (G-CDs) were used to track delivery and localization of a cancer treatment in a mouse model of liver cancer. CDs generally have a high fluorescence intensity and durable photostability, which were exploited in the G-CDs to monitor doxorubicin (DOX) and proved that CDs provide a smart DDS that allows for both traceability and controlled drug release [3]. The following properties highlight the mechanisms behind the trackable ability of the CDs:

## 13.6.1 Absorption

CDs have a unique absorption spectrum with absorption peaks typically in the range between 230 and 320 nm that extend into the visible region as an absorption tail. The $\pi-\pi^*$ transition of C–C bonds, the $n-\pi^*$ transition of C = O bonds, and/or others may be responsible for various absorption shoulders. Although this is typically the case, the UV absorption peaks may vary significantly depending on the method that was employed to synthesize the CDs [69]. The hollow CDs (HCDs) demonstrated bright PL and were used for DOX loading and release. The HCD-DOX with inherent fluorescence served as a prototype of anticancer drug carrier with a dual emission. The HCD-DOX were trackable by a noticeable peak around 490 nm, which was a typical absorption wavelength of DOX. The HCD-DOX remained stable for numerous weeks without clumping and generated a red fluorescence when exposed to UV light [70].

## 13.6.2 Photoluminescence

CDs demonstrated excitation-dependent radiation, which was explored for the PL mechanism and adapting CDs for biomedical applications. The PL features of the CDs are the most relevant [69, 71] for monitoring the drug uptake and response in real time. The PL of CDs can be triggered by photoexcitation once internalized by the cells. The CD photons or electromagnetic radiation absorption will then result in the emission of an electromagnetic wave in the visible light wavelength range [72]. When compared to the emission of organic fluorophores, the emission peaks of CDs were frequently wide, with substantial Stokes shifts. This phenomenon could be due to a wide range of CD sizes and their surface chemistry, distinct emissive traps, or other factors. Each CD contains a variety of emissive locations, and while the optical properties of small-sized CDs are suspected for their PL properties, their precise mechanism is not well understood [73].

### 13.6.3 Up-conversion PL (UCPL)

Up-conversion fluorescence (UCPL) is the process of converting longer wavelength light into higher energy photons. As a result, the UCPL property of CDs can be linked to multi-photon excitation, which results in the emission of light at shorter wavelengths and higher energy than the excitation wavelength due to the simultaneous absorption of two or more photons [72]. The longer excitation wavelengths and diverse labelling with distinct emission wavelengths reduce background autofluorescence, and photon tissue permeability improves because of UCPL, making it an important function in bio-imaging [70]. UCPL is a common fluorescence that is excited by flow from the second dispersion in a single-phase spectrophotometer and can be eliminated by using a passing filter in the excitation route of a conventional fluorescent spectrophotometer. The UCPL of CDs opens new possibilities for cell imaging with two-photon luminescence microscopy [74]. CDs' UCPL was evaluated in the dopamine@CS/CD nanocomposite and confirmed that the PL property observed in the nanohybrid was based solely on the CDs distinguishing attribute. The dopamine@CS/CD nanocomposite was stimulated at 510 nm with an emission peak generated at 550 nm. As a result of the excitation and emission in the visible spectra, the CD up-conversion fluorescence property has been shown to aid in bioimaging, suggesting that CDs can be employed as a bioimaging tracer for tracking delivery of CDs and their cargoes in neurodegenerative diseases [58] and other diseases.

## 13.7 Drug release from CDs

CD–drug conjugates can be tailored based on the disease environment, to help dislodge the drugs from the CDs and activate their functions when they reach the target site [9]. The release of the drug from CDs can be regulated by one of the following mechanisms: diffusion of the loaded drug [75] and stimuli-responsive drug release [76]. The latter concept raises concerns, as like any other carbon-based nanomaterials, the cargoes (drugs) loaded in the CDs may diffuse when placed in an aqueous environment and only a fraction of the drug will reach the targeted tissues [77]. This clashes with the desired release profile of the therapeutic drugs, hence the design of CDs for controlled drug release using stimuli-responsive strategies.

In these strategies, the ability of the drug to be released is driven by exerting a stimulus upon the nanocarrier at the target site to unload their cargo. Stimuli-responsive DDS can prevent premature drug release, which is a common issue with traditional DDS [78]. The stimuli or the triggers are classified into endogenous (internal) and exogenous (external) stimuli (Figure 13.4). The endogenous stimuli are localized and contained either within the diseased cells or its microenvironment [79]; examples include pH, cellular chemical (particularly oxidation–reduction, redox), and enzymatic reactions. The pH disparity between the pathological tissue's microenvironment when compared to the physiological pH in the normal cells is the most widely explored endogenous stimuli for drug release. The tumour's acidic extracellular ambiance and low pH conditions in the endosomal and lysosomal compartments were shown to cause drug release when subjected to the pathological site. Similarly, a change in glutathione (GSH) levels associated with disease development can also trigger drug release by cleaving the disulphide bonds used in the conjugation of the drug to the nanocarrier, causing redox susceptibility [79]. Furthermore, the enzyme-susceptible nanocarriers can result in flexible drug-release kinetics, which can be caused by disease-induced expression of enzymes such as intracellular proteases or kinases, among others. Aside from these, increased adenosine 5′-triphosphate (ATP) and reactive oxygen species (ROS) levels in cancer cells can promote an ATP- and ROS-responsive drug release [80].

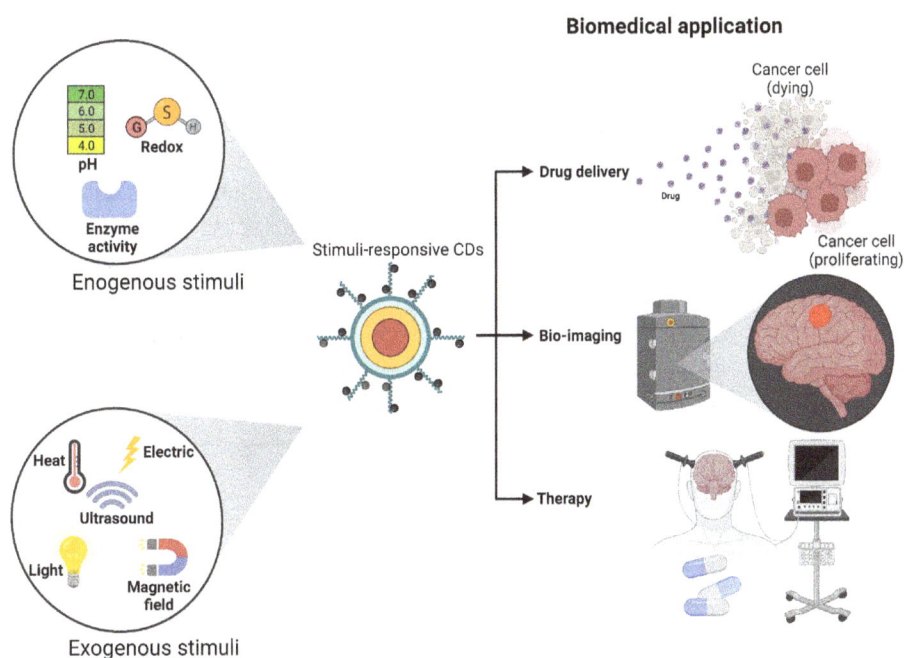

**Figure 13.4:** Drug release from stimuli-responsive CDs based on endogenous and exogenous stimuli.

Exogenous stimuli have more clinical applications due to the ability to control the accumulation of the nanocarriers in specified diseased sites or tissues and the precision with which the load is released when modulated by an external source. This type of release is independent of the biological environment or circumstances, allowing for adaptive activation of the drug release [79]. There is a common concept that some tissues such as tumours and infections are associated with elevated temperatures when compared to normal cells. However, due to the thermo-influence of various cellular activities and the extracellular environment, such temperature changes inside cells are usually on a small scale and transient, making them difficult to measure. As a result, developing customized thermo-responsive CDs carriers based on these internal temperature differences remains uncertain. Along these lines, heat-susceptible carriers can be created by using an external heat source [81]. Exogenous stimuli, such as photo- and ultrasound-responsive NPs that destabilize when exposed to a specific wavelength of light, were also explored. Many studies have emphasized the delivery of drugs via dual- or multi-responsive nanocarriers, with the combination of stimuli working synergistically and methodically for efficient drug release, such as pH/redox-, pH/enzyme-, and pH/light-responsive drug carriers [79].

## 13.7.1 Stimuli-responsive drug release from CDs

### 13.7.1.1 pH stimuli-responsive drug release

pH discrepancy is prevalent in many parts of the human body as summarized in Table 13.3, and is critical to CDs' functionality and often explored for controlled drug release. This distinction, particularly the lower pH levels in tumour cells under certain conditions such as increased glycolic pathway and lactic acid production, has been used to develop pH-sensitive CDs. These pH susceptibilities were processed through the disruption of certain bonds that the CDs are equipped with, resulting in the release of the drug that is normally protected under physiological environment. Evidently, DOX release from CD–DOX was higher at lower pH (pH 5.0, 82%) compared to pH 6.8 and 7.4 [10].

**Table 13.3:** pH pertaining to different human body parts.

| Body part | pH |
|-----------|-------|
| Stomach | 1–2 |
| Colon | 7 |
| Lysosomes | 4.5–5 |
| Endosomes | 5.5–6 |
| Cytosol | 7.4 |

pH sensing is considered a highly beneficial property of CDs. CD–DOX pH-responsive nanoconjugate resulted in approximately 12% of the DOX released from the CDs in vitro at pH 7.2, while 4% of the drug was released at pH 5.8. Using a pH that mimics the tumour environment as a trigger for drug release ensures that the drug reaches the target site in its intact form and that the drug effects are confined to the tumour site [82]. In a similar strategy, DOX loaded in HCDs demonstrated an improved DOX release profile from HCD-based nanocarriers at pH 5.0 which mimics the acidic environment of intracellular lysosomes when compared to the physiological environment (pH 7.4). The HCD nanocarriers only released 4% of the drug at pH 7.4, while 70% of the drug was released at pH 5.0 [70]. The CDs produced from *Daucus carota* subsp. *sativus* (carrot) roots used hydrogen bonding to couple mitomycin, a chemotherapy drug used to treat cancers; breakage of the hydrogen bonding at the mildly acidic tumour extracellular microenvironment (pH ~ 6.80) caused the release of mitomycin [83].

CDs@MSN-DOX nanohybrid also demonstrated the ability to increase DOX release under slightly acidic conditions (pH 5.0) at 37 °C. The CDs@MSN-DOX nanohybrid was internalized by HeLa cells after 2 h and was found primarily in the lysosomes, where the low pH (pH 5.0) triggered DOX release from the nanocarrier, resulting in cell death [84]. CD–DOX resulting from electrostatic interaction between positively and negatively charged functional groups on CDs and DOX also dissociated as the pH value decreased. The carboxyl groups on the surface of CDs were gradually protonated, and the electrostatic interaction between DOX and CDs was gradually weakened, increasing the rate and amount of DOX released [71]. The surface charge of the CDs can be controlled using pH. This behaviour has been reported to alter drug release at different pH levels. pH-responsive CDs were produced by modifying their surface with zwitterionic molecules, which have a positive charge at low or acidic pH and change to a negative charge when exposed to high or basic pH. As a result of the CDs changing conformation in the acidic environment of the tumour cells (pH 6.5–6.9), the therapeutic agents were released [38, 85]. Notably, the CDs-MSN nanohybrid loaded with DOX and capped with a zwitterionic antibiofouling layer allowed DOX release when subjected to a pH of the tumour extracellular fluid. In this way, DOX release was controlled and occurred only at pH 6.8 after enzymatic removal of polycaprolactone (PCL); the MSN gate keeper prevented early drug release. This was evident because nothing happened for 5 days, and when esterase was added on day six, 90% of the drug was released within 3 days. To prove that the DOX release was dependent on both pH and enzyme, when the MSN-CDs were subjected to pH 7.4 in the presence of the enzyme or pH 6.8 buffer without the enzyme, ~ 15% of the drug was released after 28 days [38].

In addition to chemotherapy, CDs can be used to deliver and release other therapeutic agents such as antibiotics [54, 55] and antipsychotic drug [53]. CD-coated alginate beads (CA-CDs) were produced by electrostatic interactions between the positively charged CDs and negatively charged alginate, and used as nanocarriers for tetracycline (TC) and TC associated with β-cyclodextrin. After 96 h at pH 1, the CA-CDs released 70% of TC, indicating their applicability in the gastrointestinal tract [54]. In another study, CDs loaded with the broad-spectrum antibiotic ciprofloxacin (Cipro@CDs) could release the antibiotic under physiological conditions (pH 7.4); the release profile was sustained for 24 h. Cipro@CDs demonstrated time-dependent and controlled release of Cipro as well as potent antibacterial activity against both gram-positive and gram-negative strains [55]. Furthermore, CDs were used to deliver HaLO; a constant HaLO release from the nanocarrier was achieved for more than 40 h under physiological conditions (pH 7.4). The HaLO release profile from CDs was found to have an initial burst in the first 4–5 h, followed by a sustained release for up to 40 h [53].

### 13.7.1.2 Redox stimuli-responsive drug release

Normal cells have a significantly lower redox potential than tumour cells, and this difference exists between the intracellular and extracellular environments. The cellular redox environment is known to be primarily regulated by GSH levels. GSH is a thiol compound found in high concentrations, ranging between 5 and 10 mM, in the cytoplasm of mammalian cells. GSH deficiency always increases susceptibility to oxidative stress, whereas GSH excess generally increases antioxidant capacity and oxidative stress resistance [79]. As a result, intracellular GSH may be a promising stimulus for triggering drug release from CDs.

A redox-responsive MSN capped by fluorescent CDs showed controlled DOX release by exploiting their sensitivity towards the intracellular GSH. The CDs that were electrostatically anchored onto the surface of the MSNs blocked the pores of the MSNs and prevented the leakage of DOX. When GSH was added to the physiological environment, the integrity of the system was disrupted due to the breakage and detachment of the disulphide bonds, resulting in the rapid release of DOX [86].

### 13.7.1.3 Temperature stimuli-responsive drug release

CDs are often used with other systems to create thermo-responsive nanohybrid carriers; most times, thermosensitive polymers such as poly(N-isopropyl acrylamide) and poly(methyl vinyl ether) [2] and other nanosystems [87] are used in conjugation with CDs to prevent drug leakage and ensure controlled release. A thermo-responsive CD-hydrogel nanocarrier for diclofenac demonstrated controlled drug release mediated by an external source. Drug release relied on the thermo-susceptible hydrogels after topical administration in the eyes of the rabbits. CDs fluorescent in various ocular tissues (cornea, crystalline lens, sclera, and conjunctiva) was an indication that diclofenac was distributed in these tissues [87]. Thus, in situ or external temperature variation from other systems, and not directly from the CDs, can also be exploited in the development of thermo-responsive CD-based DDS.

### 13.7.1.4 Light stimuli-responsive drug release

Light-responsive DDS that operate at different wavelengths have been reported. UV light has poor tissue penetration, so it is generally replaced by NIR wavelengths, which not only ensure better tissue penetration but are also safe for in vivo use [88].

Owing to the intrinsic photothermal conversion ability of CDs, NIR laser irradiation could also be used to enhance drug release from the nanocarriers. The increase in temperature in the local tissues will help with the detachment of the drug. CD–DOX exhibited strong absorbance in the NIR region with higher photothermal conversion efficiency. The NIR light-stimulated DOX release, allowing for drug accumulation in the cancer cells, and increasing the drug's effectiveness in killing the cancer cells. Wang et al. created a CD--nanohybrid loaded with DOX; the nanohybrid was composed of magnetic iron oxide ($Fe_3O_4$) nanocrystals and fluorescent CDs, with the $Fe_3O_4$ nanocrystals clustered in the core and the CDs implanted in the porous carbon shell. DOX release profile from $Fe_3O_4$@CDs was monitored in the dark under NIR irradiation. The kinetics of DOX release from the nanocarrier proceeded slowly and almost steadily. Irradiation with NIR light significantly increased the rate of DOX release from the hybrid nanocarrier [89].

Another photo-responsive system, the nitrogen-doped and PEG200-coated CDs initiated delivery of DOX in human breast cancer (MCF-7) cells via NIR two-photon excitation. Under NIR laser irradiation, DOX was successfully released from the CDs, subsequently contributing to enhanced cell death. These findings implied that NIR could be used to control the rate of DOX release from the CD nanocarriers [90]. Other light sources have been investigated in addition to NIR irradiation. Qucbl-CDs demonstrated light-controlled drug release mediated by W light ($\geq$365 nm, Hg-vapour lamp, 120 mW cm$^{-2}$) and He–Ne laser (5 mW cm$^{-2}$). Approximately 73% and 20% of Qucbl was released after 30 min of irradiation with W light and He–Ne laser, respectively [91].

### 13.7.1.5 Multiple stimuli-responsive CDs for drug release

To create potent formulations, multiple stimuli-responsive agents have been used to trigger drug release. These triggers can be a combination of endogenous stimuli, exogenous stimuli, or a combination of both endogenous and exogenous stimuli. This is currently being investigated to improve the efficiency of these systems [92]. pH- and enzyme-responsive CD-MSN-DOX hybrid relied on the tumour pH to remove the zwitterionic layer that prevented formation of protein corona on the surface, and the esterases to degrade the PCL and release DOX from the MSNs. The low pH was the first and vital trigger which then allowed esterase an access to the PCL and controlled DOX release. On the contrary, the enzyme at high pH conditions failed to degrade the PCL and no drug release was observed, thus indicating a highly selective drug release performance to tumour tissue [38].

## 13.8 Biocompatibility and cytotoxicity of CDs

CDs are smaller in size, with a diameter less than 10 nm and can easily interact with various cellular organelles and induce cellular damage. They are presumed to be biocompatible and have reduced toxicity compared to other QDs synthesized from semiconducting materials [93] and toxic metal compounds, because they are synthesized from carbon, one of the building blocks of nature [56, 94, 95]. Thus, CDs are biocompatible and safer for the living organisms and the environment. Several in vitro and in vivo studies have been done to evaluate the toxicity of CDs using mammalian (normal vs cancerous) cell lines, as well as animal models with remarkable results. The biocompatibility and safety of CDs are dependent on factors such as the methods of synthesis, surface composition, and surface charge. CDs synthesized through hydrothermal methods were shown to be non-toxic to normal breast (MCF-10A) cell line compared to calcination- and microwave-based CDs. The latter showed time-dependent cytotoxicity effects, while the hydrothermal synthesized CDs remained non-toxic up to 72 h at concentrations $\leq 4$ mg mL$^{-1}$ [96]. CDs synthesized by the hydrothermal treatment of folic acid and kappa-carrageenan (FKC-CDs) had negligible haemolytic effect of 2.71% (haemolysis values <10% are in the accepted range) at 600 µg mL$^{-1}$, and thus have good overall blood biocompatibility. The FKC-CDs up to 600 µg mL$^{-1}$ showed no visible reduction in the viability of HeLa and normal fibroblast cells, suggesting that the FKC-CDs have no cytotoxic and no anticancer activities [97]. CDs synthesized by the hydrothermal treatment of $_p$-phenylenediamine with various metal ions as catalysts produced non-toxic CDs: pPCDs, Ag–pPCDs, Cu–pPCDs, Pt–pPCDs, Fe–pPCDs, Pd–pPCDs, and Ni–pPCDs. The metal ions were undetected in the final products, and the CDs were considered as biocompatible as they did not contain metal elements which tend to be cytotoxic. Out of all the CDs, the Ni–pPCDs showed better physicochemical and PL properties than the rest, with cell viabilities ~ 80% at $\leq 50$ µg mL$^{-1}$ of Ni–pPCDs in both normal (AT II) and lung cancer (A549) cells. In a mouse model, the Ni–pPCDs accumulated mostly in the tumour site and had no notable accumulation in the major organs such as the heart, liver, spleen, lungs, and kidneys [98]; further certifying the biocompatibility of the CDs. Many other studies also demonstrated the non-cytotoxic nature of CDs synthesized through different methods in both normal and cancer cell lines [93, 99–101]. The unloaded CDs did not show any toxicity on both normal (HL-7702, H9C2, HUVEC) and cancer (HeLa, HepG2, MCF-7) cells at concentrations up to 100 µg mL$^{-1}$ from 0–96 h [3]. Similarly, CDs were shown to have no cytotoxic effects on the mouse fibroblast (L929) and MCF-7 cells at concentrations up to 0.5 mg per mlat 24 and 48 h [10].

### 13.8.1 Selective toxicity of drug-loaded CDs

Due to their non-cytotoxic and biocompatible nature, CDs were used in drug delivery applications to improve the performance and the therapeutic index of certain drugs.

Anticancer drugs such as DOX are non-selective in their action towards both cancer and normal cells. However, when they are loaded into CDs their cytotoxicity to normal cells was significantly reduced, while their anticancer activity was enhanced. To demonstrate this effect, the DOX-loaded CDs (CD–DOX) compared to unloaded CDs had no cytotoxic effect in L929 cells at ≤600 µg mL$^{-1}$ for 24–72 h of treatment. The unloaded CDs had negligible cytotoxicity towards L929 cells, while the free DOX and the CD–DOX at 1.5 µM DOX showed differential activities. The free DOX significantly reduced cell viability in the L929 cells compared to the treatment with the CD–DOX at all the time points. This suggested that the cytotoxic effect of DOX was reduced when it was loaded into the CDs and could potentially mean that the side effects exhibited by free DOX onto normal cells can be minimized when the drug is loaded on CDs. Also, in the cancerous ACC-2 cells, CD–DOX exhibited a higher antitumor activity than free DOX at all the time points. This effect was more pronounced in the 72 h treatment resulting in 5% ACC-2 cell viability after treatment with the CD–DOX. The antitumor effect as well as the minimal cytotoxicity effect was attributed to the slow release of DOX from the CD–DOX [95]. Similarly, in another study, CD–DOX at ≤500 µg mL$^{-1}$ showed no cytotoxic effect in the L929 cells at all time points. Moreover, no notable cytotoxicity or anticancer activity was observed in both the L929 and MCF-7 cells treated with the unloaded CDs, at all investigated concentrations and time points. On the contrary, free DOX and CD–DOX reduced the viability of the MCF-7 cells. The reduction was more pronounced in the CD–DOX treatments compared to the free DOX, with the IC$_{50}$ values of 0.983 and 0.939 µg mL$^{-1}$ of free DOX at 24 and 48 h, respectively, which were further reduced to be 0.652 and 0.356 µg mL$^{-1}$ at 24 and 48 h for CD–DOX, respectively [10].

The cytotoxicity of CD–DOX was reported to be time- and concentration-dependent, and showed similar or superior effects to free DOX. Their effects were significant from 0.0625 µg mL$^{-1}$ at 24 h and increased to 0.15625 µg mL$^{-1}$ at 48 h for the CDs-DOX in the cancer cells. Both the unloaded CDs and CD–DOX were shown to be equally taken in by the normal and cancer cells, and the CDs were able to offer some protection to the non-cancer cells. Their cytotoxicity was significantly reduced than that of free DOX [3]. The cytotoxicity of the CDs is also dependent on their biodistribution. Large amounts of CD–DOX intravenously injected in HepG2 tumour-bearing mice were located by the CDs fluorescence in the tumour and the kidneys after 24 h, with negligible amounts in the liver, spleen, heart, and lungs. This was an indication that the nanomaterials have some degree of biocompatibility and could be safe to use in vivo [3]. The CDs might not have long-term toxic effects on mice, as the histological analysis of vital tissues at the baseline, 7 and 21 days showed no changes in the morphology of these tissues [9].

The selectivity of drug-loaded CDs was also reported for other drugs such as epirubicin, temozolomide [102], mitoxantrone (MTO) [103], coptisine [94], and FAC [56]. In these four studies, it was evident that unloaded CDs had no cytotoxicity against the test cell lines, normal and diseased alike. The drug-loaded CDs also demonstrated some selective toxicity towards diseased cells [94, 102, 103]; with the exception of CD–FAC [56] and CD–temozolomide [102] which were non-toxic to both normal and diseased cells. Moreover, the drug-loaded CDs showed a significant cytotoxic effect on the cells than the free drugs in a dose-dependent manner [84, 93, 94]. In vivo studies showed no apparent changes in the size of the tumour as well as the weight of the mice treated with both free coptisine and CD-coptisine, indicating their biocompatibility and non-toxic effect. Furthermore, the CD-coptisine only accumulated in the tumour site as opposed to other organs such as the spleen, heart, kidneys, and liver, further confirming the biocompatibility and non-toxic nature of the CD-coptisine [94]. In a study by Hettiarachchi et al., multi-drug (epirubicin and temozolomide)-loaded CDs with or without targeting moiety (transferrin) compared to single drug-loaded-CDs with or without transferrin and the free drugs had enhanced cytotoxicity towards several cancerous (SJGBM2, CHLA266, CHLA200, and U87) cells lines. Of the single drug-loaded-CDs, only the CD-epirubicin reduced cell viabilities to 17% and 30% in all cells at 10 μM while the CD-temozolomide had no notable cytotoxic effect at all tested concentrations in all the cancer cells. A targeting ligand (transferrin) enhanced the efficacy of the drug-loaded CDs even at lower drug concentrations [102].

From these studies, it is quite evident that decorating CDs with various biomolecules can improve the selectivity of the CDs as well as the performance and delivery of the loaded drug [104]. Functionalization of CD–DOX with 4-carboxybenzylboronic acid (CBBA; DOX/CBBA-CDs) reduced the cell viability in a concentration-dependent manner in HeLa cells, while the unloaded CDs and CBBS-CDs were non-cytotoxic up to 400 μg mL$^{-1}$ [71]. Furthermore, the use CDs with other DDS was shown to improve the delivery, performance, biocompatibility, and efficacy of the drugs. This was demonstrated by a CDs nanohybrid containing hyaluronic acid, disulphide bonds, and hMSNs (hMSN-SS-CD$_{PEI}$@HA) for the delivery of DOX. The hMSN-SH and hMSN-SS-CD$_{PEI}$@HA had no notable cytotoxic effect on both normal mouse fibroblast (NIH-3T3) and A549 cells. The hMSN-SS-CD$_{PEI}$@ showed selectivity cytotoxicity towards the cancer cells [105]. A number of studies can attest to the non-toxicity and potential biocompatibility of the unloaded CDs, and their ability to enhance drug effects.

## 13.9 Conclusion

CDs hold promise as an effective DDS, mainly due to the fact that they are made from biocompatible carbon precursors. Moreover, their exceptional physicochemical properties encourage their dual bio-applications for construction of smart diagnostic and therapeutic agents. CDs have been timeously proven to be non-toxic and biocompatible towards cells and animals. These desirable properties make CDs excellent DDS. They can transport their cargoes without exhibiting any bystander effects on neighbouring cells. Their flexibility to encapsulate or load drugs on their surface is especially appealing and can be exploited for multiplexing. Additionally, CDs can be tailored to be stimuli-responsive and only release their cargoes in specified targets for more control over the drug kinetics. Thus, CDs as DDS can minimize bystander effects on the normal cells while inhibiting the growth and killing the targeted cells. All these events can be monitored in real time due to their PL and fluorescent properties. However, much work remains to be done to translate this technology into clinically viable systems.

# References

[1]   Svenson S. Carrier-Based Drug Delivery, 2004, 2–23. https://doi.org/10.1021/BK-2004-0879.CH001.

[2]   Adepu S, Ramakrishna S, Costa-Pinto R, Oliveira AL. Controlled drug delivery systems: current status and future directions. Mol. 2021, 26, 5905. https://doi.org/10.3390/MOLECULES26195905.

[3]   Zeng Q, Shao D, He X, Ren Z, Ji W, Shan C, Qu S, Li J, Chen L, Li Q. Carbon dots as a trackable drug delivery carrier for localized cancer therapy: In vivo. J Mater Chem B. 2016, 4, 5119–5126. https://doi.org/10.1039/C6TB01259K.

[4]   Liu D, Yang F, Xiong F, Gu N. The smart drug delivery system and its clinical potential. Theranostics. 2016, 6, 1306–1323. https://doi.org/10.7150/THNO.14858.

[5]   Amer Ridha A, Pakravan P, Hemati Azandaryani A, Zhaleh H, Carbon dots; the smallest photoresponsive structure of carbon in advanced drug targeting. J Drug Deliv Sci Technol, 2020, 55. https://doi.org/10.1016/J.JDDST.2019.101408.

[6]   Zhao MX, Zhu BJ. The research and applications of quantum dots as nano-carriers for targeted drug delivery and cancer therapy. Nanoscale Res Lett. 2016, 11, 1–9. https://doi.org/10.1186/s11671-016-1394-9.

[7]   Phillips MA, Gran ML, Peppas NA. Targeted nanodelivery of drugs and diagnostics. Nano Today. 2010, 5, 143–159. https://doi.org/10.1016/j.nantod.2010.03.003.

[8]   Xia J, Kawamura Y, Suehiro T, Chen Y, Sato K. Carbon dots have antitumor action as monotherapy or combination therapy. Drug Discov Ther. 2019, 13, 114–117. https://doi.org/10.5582/DDT.2019.01013.

[9]   Sun Y, Zheng S, Liu L, Kong Y, Zhang A, Xu K, Han C. The cost-effective preparation of green fluorescent carbon dots for bioimaging and enhanced intracellular drug delivery. Nanoscale Res Lett. 2020, 15, 1–9. https://doi.org/https://doi.org/10.1186/s11671-020-3288-0.

[10]  Kong T, Hao L, Wei Y, Cai X, Zhu B. Doxorubicin conjugated carbon dots as a drug delivery system for human breast cancer therapy. Cell Prolif. 2018, 51. https://doi.org/10.1111/CPR.12488.

[11]  Sultana A, Zare M, Thomas V, Kumar TSS, Ramakrishna S. Nano-based drug delivery systems: Conventional drug delivery routes, recent developments and future prospects. Med Drug Discov. 2022, 15, 100134. https://doi.org/10.1016/J.MEDIDD.2022.100134.

[12]  Homayun B, Lin X, Choi H-J. pharmaceutics challenges and recent progress in oral drug delivery systems for biopharmaceuticals. 2019. doi: https://doi.org/10.3390/pharmaceutics11030129.

[13]  Kerz T, Paret G, Herff H. Routes of drug administration. Card Arrest Sci Pract Resusc Med. 2007, 614–638. https://doi.org/10.1017/CBO9780511544828.035.

[14]  Norouzi M, Nazari B, Miller DW. Injectable hydrogel-based drug delivery systems for local cancer therapy. Drug Discov Today. 2016, 21, 1835–1849. https://doi.org/10.1016/J.DRUDIS.2016.07.006.

[15]  Escobar-Chavez J, Diaz-Torres R, Rodriguez-Cruz IM, Domínguez-Delgado S-M, Angeles-Anguiano M-C. Nanocarriers for transdermal drug delivery. Res Reports Transdermal Drug Deliv. 2012, 3. https://doi.org/10.2147/rrtd.s32621.

[16]  Rau JL, Faarc R. The inhalation of drugs: Advantages and problems introduction: the inhalation of drugs for respiratory disease. Respir Care. 2005, 50, 367–382.

[17]    Saleem K, Khursheed Z, Hano C, Anjum I, Anjum S, Applications of nanomaterials in leishmaniasis: A focus on recent advances and challenges, Nanomaterials. 2019, 9. https://doi.org/10.3390/NANO9121749.

[18]    Felice B, Prabhakaran MP, Rodríguez AP, Ramakrishna S. Drug delivery vehicles on a nano-engineering perspective. Mater Sci Eng C: Mater Biol Appl. 2014, 41, 178–195. https://doi.org/10.1016/J.MSEC.2014.04.049.

[19]    Bartosova L, Bajgar J. Transdermal drug delivery in vitro using diffusion cells. Curr Med Chem. 2012, 19, 4671–4677. https://doi.org/10.2174/092986712803306358.

[20]    De Jong WH, Borm PJA. Drug delivery and nanoparticles: Applications and hazards. Int J Nanomedicine. 2008, 3, 133. https://doi.org/10.2147/IJN.S596.

[21]    Yao Y, Zhou Y, Liu L, Xu Y, Chen Q, Wang Y, Wu S, Deng Y, Zhang J, Shao A. Nanoparticle-based drug delivery in cancer therapy and its role in overcoming drug resistance. Front Mol Biosci. 2020, 7, 193. https://doi.org/10.3389/FMOLB.2020.00193.

[22]    Souto EB, Souto SB, Campos JR, Severino P, Pashirova TN, Zakharova LY, Silva AM, Durazzo A, Lucarini M, Izzo AA, Santini A. Nanoparticle delivery systems in the treatment of diabetes complications. Molecules. 2019, 24, 4209. https://doi.org/10.3390/MOLECULES24234209.

[23]    Munasinghe E, Aththapaththu M, Jayarathne L. Magnetic and quantum dot nanoparticles for drug delivery and diagnostic systems. Colloid Sci Pharm Nanotechnol. 2020. https://doi.org/10.5772/INTECHOPEN.88611.

[24]    Qi L, Gao X. Emerging application of quantum dots for drug delivery and therapy. Expert Opin Drug Deliv. 2008, 5, 263–267. https://doi.org/10.1517/17425247.5.3.263.

[25]    Patra JK, Das G, Fraceto LF, Campos EVR, Rodriguez-Torres MDP, Acosta-Torres LS, Diaz-Torres LA, Grillo R, Swamy MK, Sharma S, Habtemariam S, Shin HS. Nano based drug delivery systems: Recent developments and future prospects. J Nanobiotechnol. 2018, 16, 71. https://doi.org/10.1186/s12951-018-0392-8.

[26]    Wu W, Arsene AL. Drug carriers: Classification, administration, release profiles, and industrial approach. Process. 2021, 9, 470. https://doi.org/10.3390/PR9030470.

[27]    Tan A, Wang Z, Lust R, Hua S, Sercombe L, Veerati T, Moheimani F, Wu SY, Sood AK. Advances and challenges of liposome assisted drug delivery. Front Pharmacol. 2015, Www.Frontiersin.Org, 6, 286. https://doi.org/10.3389/fphar.2015.00286.

[28]    Rossi F, Sharon M, Irudayaraj J, Vega-Vásquez P, Mosier NS. Nanoscale drug delivery systems: From medicine to agriculture. Front Bioeng Biotechnol. 2020, Www.Frontiersin.Org, 8, 79. https://doi.org/10.3389/fbioe.2020.00079.

[29]    Mourya VK. Polymeric micelles: General considerations and their applications. Indian J Pharm Res Educ. 2011, 45, https://doi.org/Nillz.

[30]    Lu Y, Zhang E, Yang J, Cao Z. Strategies to improve micelle stability for drug delivery. Nano Res. 2018, 11, 4985–4998. https://doi.org/10.1007/s12274-018-2152-3.

[31]    Sibuyi NRS, Moabelo KL, Fadaka AO, Meyer S, Onani MO, Madiehe AM, Meyer M. Multifunctional gold nanoparticles for improved diagnostic and therapeutic applications: A review. Nanoscale Res Lett. 2021, 161(16). 1–27. https://doi.org/10.1186/S11671-021-03632-W.

[32]    Jahangirian H, Lemraski EG, Webster TJ, Rafiee-Moghaddam R, Abdollahi Y. A review of drug delivery systems based on nanotechnology and green chemistry: Green nanomedicine. Int J Nanomedicine. 2017, 12, 2957–2978. https://doi.org/10.2147/IJN.S127683.

[33]    Inglut CT, Sorrin AJ, Kuruppu T, Vig S, Cicalo J, Ahmad H, Huang HC. Immunological and toxicological considerations for the design of liposomes. Nanomaterials. 2020, 10. https://doi.org/10.3390/NANO10020190.

[34]    Chu KW, Lee SL, Chang CJ, Liu L. Recent progress of carbon dot precursors and photocatalysis applications. Polym. 2019, 11, 689. https://doi.org/10.3390/POLYM11040689.

[35]   Probst CE, Zrazhevskiy P, Bagalkot V, Gao X. Quantum dots as a platform for nanoparticle drug delivery vehicle design. Adv Drug Deliv Rev. 2013, 65, 703–718. https://doi.org/10.1016/j.addr.2012.09.036.

[36]   Aguilar Cosme JR, Bryant HE, Claeyssens F. Carbon dot-protoporphyrin IX conjugates for improved drug delivery and bioimaging. PLoS One. 2019, 14, e0220210. https://doi.org/10.1371/JOURNAL. PONE.0220210.

[37]   Haque Adrita S, Nujhat Tasnim K, Hyun Ryu J, Md Sharker S, Ki Choi S. Nanotheranostic carbon dots as an emerging platform for cancer therapy. n.d. https://doi.org/10.3390/jnt1010006.

[38]   Liu Z, Chen X, Zhang X, Gooding JJ, Zhou Y. Carbon-quantum-dots-loaded mesoporous silica nanocarriers with pH-switchable zwitterionic surface and enzyme-responsive pore-cap for targeted imaging and drug delivery to tumor. Adv Healthc Mater. 2016, 5, 1401–1407. https://doi.org/ 10.1002/ADHM.201600002.

[39]   Fasipe O. Recent advances and current trend in the pharmacotherapy of obesity. Arch Med Heal Sci. 2018, 6, 99. https://doi.org/10.4103/amhs.amhs_30_18.

[40]   Li S, Peng Z, Dallman J, Baker J, Othman AM, Blackwelder PL, Leblanc RM. Crossing the blood–brain–barrier with transferrin conjugated carbon dots: A zebrafish model study. Coll Surf B: Biointerfac. 2016, 145, 251–256. https://doi.org/10.1016/J.COLSURFB.2016.05.007.

[41]   Zhou Y, Mintz KJ, Cheng L, Chen J, Ferreira BCLB, Hettiarachchi SD, Liyanage PY, Seven ES, Miloserdov N, Pandey RR, Quiroga B, Blackwelder PL, Chusuei CC, Li S, Peng Z, Leblanc RM. Direct conjugation of distinct carbon dots as Lego-like building blocks for the assembly of versatile drug nanocarriers. J Colloid Interface Sci. 2020, 576, 412–425. https://doi.org/10.1016/J.JCIS.2020.05.005.

[42]   Sharma SK, Micic M, Li S, Hoar B, Paudyal S, Zahran EM, Leblanc RM, Conjugation of carbon dots with β-galactosidase enzyme: surface chemistry and use in biosensing, Mol 2019, 24, 3275. https://doi.org/10.3390/MOLECULES24183275.

[43]   Zhou Y, Chen X, Cao J, Gao H. Overcoming the biological barriers in the tumor microenvironment for improving drug delivery and efficacy. J Mater Chem B. 2020, 8, 6765–6781. https://doi.org/ 10.1039/D0TB00649A.

[44]   Sriraman SK, Aryasomayajula B, Torchilin VP. Barriers to drug delivery in solid tumors. Tissue Barriers. 2014, 2, e29528-1-e29528-10. https://doi.org/10.4161/TISB.29528.

[45]   Baghban R, Roshangar L, Jahanban-Esfahlan R, Seidi K, Ebrahimi-Kalan A, Jaymand M, Kolahian S, Javaheri T, Zare P. Tumor microenvironment complexity and therapeutic implications at a glance. Cell Commun Signal. 2020, 18(1), 1–19. https://doi.org/10.1186/S12964-020-0530-4.

[46]   Blakeley J. Drug delivery to brain tumors. Curr Neurol Neurosci Rep. 2008, 8, 235. https://doi.org/ 10.1007/S11910-008-0036-8.

[47]   Arvanitis CD, Ferraro GB, Jain RK. The blood–brain barrier and blood–tumour barrier in brain tumours and metastases. Nat Rev Cancer. 2019, 201(20), 26–41. https://doi.org/10.1038/s41568- 019-0205-x.

[48]   Calabrese G, De Luca G, Nocito G, Rizzo MG, Lombardo SP, Chisari G, Forte S, Sciuto EL, Conoci S. Carbon dots: An innovative tool for drug delivery in brain tumors. Int J Mol Sci. 2021, 22. https://doi.org/10.3390/IJMS222111783.

[49]   Zhou Y, Liyanage PY, Devadoss D, Rios Guevara LR, Cheng L, Graham RM, Chand HS, Al-Youbi AO, Bashammakh AS, El-Shahawi MS, Leblanc RM. Nontoxic amphiphilic carbon dots as promising drug nanocarriers across the blood–brain barrier and inhibitors of β-amyloid. Nanoscale. 2019, 11, 22387–22397. https://doi.org/10.1039/C9NR08194A.

[50]   Ashrafizadeh M, Mohammadinejad R, Kailasa SK, Ahmadi Z, Afshar EG, Pardakhty A, Carbon dots as versatile nanoarchitectures for the treatment of neurological disorders and their theranostic applications: A review. Adv Colloid Interface Sci. 2020, 278. https://doi.org/10.1016/J. CIS.2020.102123.

[51] Zheng M, Ruan S, Liu S, Sun T, Qu D, Zhao H, Xie Z, Gao H, Jing X, Sun Z. Self-targeting fluorescent carbon dots for diagnosis of brain cancer cells. ACS Nano. 2015, 9, 11455–11461. https://doi.org/10.1021/ACSNANO.5B05575/ASSET/IMAGES/LARGE/NN-2015-05575B_0007.JPEG.

[52] Seven ES, Seven YB, Zhou Y, Poudel-Sharma S, Diaz-Rucco JJ, Cilingir EK, Mitchell GS, Van Dyken JD, Leblanc RM. Crossing the blood–brain barrier with carbon dots: Uptake mechanism and in vivo cargo delivery. Nanoscale Adv. 2021, 3, 3942–3953. https://doi.org/10.1039/D1NA00145K.

[53] Karthik S, Saha B, Ghosh SK, Pradeep Singh ND. Photoresponsive quinoline tethered fluorescent carbon dots for regulated anticancer drug delivery. Chem Commun. 2013, 49, 10471–10473. https://doi.org/10.1039/C3CC46078A.

[54] Gogoi N, Chowdhury D. Novel carbon dot coated alginate beads with superior stability, swelling and pH responsive drug delivery. J Mater Chem B. 2014, 2, 4089–4099. https://doi.org/10.1039/C3TB21835J.

[55] Thakur M, Pandey S, Mewada A, Patil V, Khade M, Goshi E, Sharon M. Antibiotic conjugated fluorescent carbon dots as a theranostic agent for controlled drug release, bioimaging, and enhanced antimicrobial activity. J Drug Deliv. 2014, 2014, 1–9. https://doi.org/10.1155/2014/282193.

[56] Shi N, Sun K, Zhang Z, Zhao J, Geng L, Lei Y. Amino-modified carbon dots as a functional platform for drug delivery: Load-release mechanism and cytotoxicity. J Ind Eng Chem. 2021, 101, 372–378. https://doi.org/10.1016/j.jiec.2021.05.046.

[57] Zhang W, Sigdel G, Mintz KJ, Seven ES, Zhou Y, Wang C, Leblanc RM. Carbon dots: A future blood–brain barrier penetrating nanomedicine and drug nanocarrier. Int J Nanomedicine. 2021, 16, 5003. https://doi.org/10.2147/IJN.S318732.

[58] Mathew SA, Praveena P, Dhanavel S, Manikandan R, Senthilkumar S, Stephen A. Luminescent chitosan/carbon dots as an effective nano-drug carrier for neurodegenerative diseases. RSC Adv. 2020, 10, 24386–24396. https://doi.org/10.1039/D0RA04599C.

[59] Yao B, Huang H, Liu Y, Kang Z. Carbon dots: A small conundrum. Trends Chem. 2019, 1, 235–246. https://doi.org/10.1016/J.TRECHM.2019.02.003.

[60] Xia C, Zhu S, Feng T, Yang M, Yang B, Xia C, Feng T, Yang M, Yang B, Zhu S. Evolution and synthesis of carbon dots: From carbon dots to carbonized polymer dots. Adv Sci. 2019, 6, 1901316. https://doi.org/10.1002/ADVS.201901316.

[61] Xiao D, Qi H, Teng Y, Pierre D, Kutoka PT, Liu D. Advances and challenges of fluorescent nanomaterials for synthesis and biomedical applications. Nanoscale Res Lett. 2021, 16, 1–23. https://doi.org/https://doi.org/10.1186/s11671-021-03613-z.

[62] Rani R, Sethi K, Singh G. Nanomaterials and their applications in bioimaging. Nanotechnol Life Sci. 2019, 429–450. https://doi.org/10.1007/978-3-030-16379-2_15.

[63] Kobayashi N, Izumi H, Morimoto Y. Review of toxicity studies of carbon nanotubes. J Occup Health. 2017, 59, 394–407. https://doi.org/10.1539/JOH.17-0089-RA.

[64] Alkahtani MH, Alghannam F, Jiang L, Almethen A, Rampersaud AA, Brick R, Gomes CL, Scully MO, Hemmer PR, Fluorescent nanodiamonds: Past, present, and future. Nanophotonics. 2018, 7, 1423–1453. https://doi.org/10.1515/NANOPH-2018-0025/ASSET/GRAPHIC/J_NANOPH-2018-0025_FIG_006.JPG.

[65] Qin JX, Yang XG, Lv CF, Li YZ, Liu KK, Zang JH, Yang X, Dong L, Shan CX. Nanodiamonds: Synthesis, properties, and applications in nanomedicine. Mater Des. 2021, 210, 110091. https://doi.org/10.1016/J.MATDES.2021.110091.

[66] Resch-Genger U, Grabolle M, Cavaliere-Jaricot S, Nitschke R, Nann T. Quantum dots versus organic dyes as fluorescent labels. Nat Methods. 2008, 59(5), 763–775. https://doi.org/10.1038/nmeth.1248.

[67] Wang J, Liu G, Leung K, Loffroy R, Lu P-X, Wáng YJ. Opportunities and challenges of fluorescent carbon dots in translational optical imaging. Curr Pharm Des. 2015, 21, 5401–5416. https://doi.org/10.2174/1381612821666150917093232.

[68]    Feng T, Ai X, Ong H, Zhao Y. Dual-responsive carbon dots for tumor extracellular microenvironment triggered targeting and enhanced anticancer drug delivery. ACS Appl Mater Interfaces. 2016, 8, 18732–18740. https://doi.org/10.1021/ACSAMI.6B06695/SUPPL_FILE/AM6B06695_SI_001.PDF.

[69]    Nair A, Haponiuk JT, Thomas S, Gopi S. Natural carbon-based quantum dots and their applications in drug delivery: A review. Biomed Pharmacother. 2020, 132. https://doi.org/10.1016/J. BIOPHA.2020.110834.

[70]    Wang Q, Huang X, Long Y, Wang X, Zhang H, Zhu R, Liang L, Teng P, Zheng H. Hollow luminescent carbon dots for drug delivery. Carbon N Y. 2013, 59, 192–199. https://doi.org/10.1016/J. CARBON.2013.03.009.

[71]    Duan Q, Ma Y, Che M, Zhang B, Zhang Y, Li Y, Zhang W, Sang S. Fluorescent carbon dots as carriers for intracellular doxorubicin delivery and track. J Drug Deliv Sci Technol. 2019, 49, 527–533. https://doi.org/10.1016/j.jddst.2018.12.015.

[72]    Behboudi H, Mehdipour G, Safari N, Pourmadadi M, Saei A, Omidi M, Tayebi L, Rahmandoust M. Carbon quantum dots in nanobiotechnology. Adv Struct Mater. 2019, 104, 145–179. https://doi. org/10.1007/978-3-030-10834-2_6.

[73]    Mintz KJ, Zhou Y, Leblanc RM. Recent development of carbon quantum dots regarding their optical properties, photoluminescence mechanism, and core structure. Nanoscale. 2019, 11, 4634–4652. https://doi.org/10.1039/C8NR10059D.

[74]    Wen X, Yu P, Toh YR, Ma X, Tang J. On the upconversion fluorescence in carbon nanodots and graphene quantum dots. Chem Commun. 2014, 50, 4703–4706. https://doi.org/10.1039/C4CC01213E.

[75]    Tangboriboon N. Carbon and carbon nanotube drug delivery and its characterization, properties, and applications. Nanocarriers Drug Deliv Nanosci Nanotechnol Drug Deliv. 2018, 451–467. https://doi.org/10.1016/B978-0-12-814033-8.00015-1.

[76]    Salve R, Gajbhiye KR, Babu RJ, Gajbhiye V. Carbon nanomaterial-based stimuli-responsive drug delivery strategies. Stimuli-Responsive Nanocarriers. 2022, 367–392. https://doi.org/10.1016/ B978-0-12-824456-2.00006-0.

[77]    Gisbert-Garzarán M, Manzano M, Vallet-Regí M. bioengineering pH-responsive mesoporous silica and carbon nanoparticles for drug delivery. Bioeng. 2017, 4(1). https://doi.org/10.3390/ bioengineering4010003.

[78]    Pham SH, Choi Y, Choi J. pharmaceutics stimuli-responsive nanomaterials for application in antitumor therapy and drug delivery. Pharmaceutics. 2020, 12, 630. https://doi.org/10.3390/ pharmaceutics12070630.

[79]    Karimi M, Ghasemi A, Sahandi Zangabad P, Rahighi R, Moosavi Basri SM, Mirshekari H, Amiri M, Shafaei Pishabad Z, Aslani A, Bozorgomid M, Ghosh D, Beyzavi A, Vaseghi A, Aref AR, Haghani L, Bahrami S, Hamblin MR. Smart micro/nanoparticles in stimulus-responsive drug/gene delivery systems. Chem Soc Rev. 2016, 45, 1457–1501. https://doi.org/10.1039/C5CS00798D.

[80]    Liou G-Y, Storz P. Reactive oxygen species in cancer. Free Radic Res. 2010, 44, 479–496. https://doi.org/10.3109/10715761003667554.

[81]    Karimi M, Sahandi Zangabad P, Ghasemi A, Amiri M, Bahrami M, Malekzad H, Ghahramanzadeh Asl H, Mahdieh Z, Bozorgomid M, Ghasemi A, Reza Rahmani Taji Boyuk M, Hamblin MR. Temperature-responsive smart nanocarriers for delivery of therapeutic agents: Applications and recent advances HHS public access. ACS Appl Mater Interfaces. 2016, 8, 21107–21133. https://doi.org/10.1021/acsami.6b00371.

[82]    Mewada A, Pandey S, Thakur M, Jadhav D, Sharon M. Swarming carbon dots for folic acid mediated delivery of doxorubicin and biological imaging. J Mater Chem B. 2014, 2, 698–705. https://doi.org/ 10.1039/C3TB21436B.

[83]    D'souza SL, Chettiar SS, Koduru JR, Kailasa SK. Synthesis of fluorescent carbon dots using Daucus carota subsp. sativus roots for mitomycin drug delivery. Optik (Stuttg). 2018, 158, 893–900. https://doi.org/10.1016/J.IJLEO.2017.12.200.

[84] Zhou L, Li Z, Liu Z, Ren J, Qu X. Luminescent carbon dot-gated nanovehicles for ph-triggered intracellular controlled release and imaging. Langmuir. 2013, 29, 6396–6403. https://doi.org/10.1021/LA400479N/SUPPL_FILE/LA400479N_SI_001.PDF.

[85] Kang EB, Lee JE, Mazrad ZAI, In I, Jeong JH, Park SY. PH-Responsible fluorescent carbon nanoparticles for tumor selective theranostics: Via pH-turn on/off fluorescence and photothermal effect in vivo and in vitro. Nanoscale. 2018, 10, 2512–2523. https://doi.org/10.1039/C7NR07900A.

[86] Zhang Y, Han L, Zhang Y, Chang YQ, Chen XW, He RH, Shu Y, Wang JH. Glutathione-mediated mesoporous carbon as a drug delivery nanocarrier with carbon dots as a cap and fluorescent tracer. Nanot. 2016, 27, 355102. https://doi.org/10.1088/0957-4484/27/35/355102.

[87] Wang L, Pan H, Gu D, Sun H, Chen K, Tan G, Pan W. A novel carbon dots/thermo-sensitive in situ gel for a composite ocular drug delivery system: Characterization, ex-vivo imaging and in vivo evaluation. Int J Mol Sci. 2021, 22. https://doi.org/10.3390/IJMS22189934/S1.

[88] Wu S, Butt HJ. Near-infrared photochemistry at interfaces based on upconverting nanoparticles. Phys Chem Chem Phys. 2017, 19, 23585–23596. https://doi.org/10.1039/C7CP01838J.

[89] Wang H, Shen J, Li Y, Wei Z, Cao G, Gai Z, Hong K, Banerjee P, Zhou S. Magnetic iron oxide–fluorescent carbon dots integrated nanoparticles for dual-modal imaging, near-infrared light-responsive drug carrier and photothermal therapy. Biomater Sci. 2014, 2, 915–923. https://doi.org/10.1039/C3BM60297D.

[90] Ardekani SM, Dehghani A, Hassan M, Kianinia M, Aharonovich I, Gomes VG. Two-photon excitation triggers combined chemo-photothermal therapy via doped carbon nanohybrid dots for effective breast cancer treatment. Chem Eng J. 2017, 330, 651–662. https://doi.org/10.1016/J.CEJ.2017.07.165.

[91] Karthik S, Puvvada N, Kumar BNP, Rajput S, Pathak A, Mandal M, Singh NDP. Photoresponsive coumarin-tethered multifunctional magnetic nanoparticles for release of anticancer drug. ACS Appl Mater Interfaces. 2013, 5, 5232–5238. https://doi.org/10.1021/AM401059K.

[92] Jia R, Teng L, Gao L, Su T, Fu L, Qiu Z, Bi Y. Advances in multiple stimuli-responsive drug-delivery systems for cancer therapy. Int J Nanomedicine. 2021, 16, 1525–1551. https://doi.org/10.2147/IJN.S293427.

[93] Moradi S, Sadrjavadi K, Farhadian N, Hosseinzadeh L, Shahlaei M, Easy synthesis, characterization and cell cytotoxicity of green nano carbon dots using hydrothermal carbonization of Gum Tragacanth and chitosan bio-polymers for bioimaging. J Mol Liq. 2018, 259, 284–290. https://doi.org/10.1016/j.molliq.2018.03.054.

[94] Ren W, Nan F, Li S, Yang S, Ge J, Zhao Z. Red emissive carbon dots prepared from polymers as an efficient nanocarrier for coptisine delivery in vivo and in vitro. Chem Med Chem. 2021, 16, 646–653. https://doi.org/10.1002/cmdc.202000420.

[95] Yuan Y, Guo B, Hao L, Liu N, Lin Y, Guo W, Li X, Gu B, Doxorubicin-loaded environmentally friendly carbon dots as a novel drug delivery system for nucleus targeted cancer therapy. Coll Surf B: Biointerfac. 2017, 159, 349–359. https://doi.org/10.1016/j.colsurfb.2017.07.030.

[96] Vale N, Silva S, Duarte D, Crista DMA, Pinto da Silva L, Esteves da Silva JCG. Normal breast epithelial MCF-10A cells to evaluate the safety of carbon dots. RSC Med Chem. 2021, 12, 245. https://doi.org/10.1039/D0MD00317D.

[97] Das P, Ganguly S, Agarwal T, Maity P, Ghosh S, Choudhary S, Gangopadhyay S, Maiti TK, Dhara S, Banerjee S, Das NC. Heteroatom doped blue luminescent carbon dots as a nano-probe for targeted cell labeling and anticancer drug delivery vehicle. Mater Chem Phys. 2019, 237, 121860. https://doi.org/10.1016/j.matchemphys.2019.121860.

[98] Hua XW, Bao YW, Zeng J, Wu FG. Nucleolus-targeted red emissive carbon dots with polarity-sensitive and excitation-independent fluorescence emission: high-resolution cell imaging and in vivo tracking. ACS Appl Mater Interfaces. 2019, 11, 32647–32658. https://doi.org/10.1021/acsami.9b09590.

[99]   Qing W, Chen K, Yang Y, Wang Y, Liu X. Cu2+-doped carbon dots as fluorescence probe for specific recognition of Cr(VI) and its antimicrobial activity. Microchem J. 2020, 152, 104262. https://doi.org/ 10.1016/j.microc.2019.104262.

[100]  Arul V, Sethuraman MG. Facile green synthesis of fluorescent N-doped carbon dots from Actinidia deliciosa and their catalytic activity and cytotoxicity applications. Opt Mater (Amst). 2018, 78, 181–190. https://doi.org/10.1016/j.optmat.2018.02.029.

[101]  Du J, Yang Y, Shao T, Qi S, Zhang P, Zhuo S, Zhu C. Yellow emission carbon dots for highly selective and sensitive OFF-ON sensing of ferric and pyrophosphate ions in living cells. J Colloid Interface Sci. 2021, 587, 376–384. https://doi.org/10.1016/j.jcis.2020.11.108.

[102]  Hettiarachchi SD, Graham RM, Mintz KJ, Zhou Y, Vanni S, Peng Z, Leblanc RM. Triple conjugated carbon dots as a nano-drug delivery model for glioblastoma brain tumors. Nanoscale. 2019, 11, 6192–6205. https://doi.org/10.1039/C8NR08970A.

[103]  Wen X, Zhao Z, Zhai S, Wang X, Li Y. Stable nitrogen and sulfur co-doped carbon dots for selective folate sensing, in vivo imaging and drug delivery. Diam Relat Mater. 2020, 105, 107791. https://doi. org/10.1016/j.diamond.2020.107791.

[104]  Fahmi MZ, Haris A, Permana AJ, Nor Wibowo DL, Purwanto B, Nikmah YL, Idris A. Bamboo leaf-based carbon dots for efficient tumor imaging and therapy. RSC Adv. 2018, 8, 38376–38383. https://doi.org/10.1039/c8ra07944g.

[105]  Zhao Q, Wang S, Yang Y, Li X, Di D, Zhang C, Jiang T, Wang S. Hyaluronic acid and carbon dots-gated hollow mesoporous silica for redox and enzyme-triggered targeted drug delivery and bioimaging. Mater Sci Eng C. 2017, 78, 475–484. https://doi.org/10.1016/j.msec.2017.04.059.

N.B. Iroha*, C.O. Ezenwaka, C.N. Opara and F.E. Abeng

# Chapter 14
# Carbon dots in protein and nucleic acid delivery

**Abstract:** Proteins play significant roles in biology, like gene expression regulation, cellular signalling pathways, enzyme catalysis, and fine balance maintenance between programmed death and cell survival. The two main kinds of nucleic acids, DNA and RNA, are mainly responsible for the synthesis of protein in a cell. Simple and general strategies of protein and nucleic acid delivery within a cell have been developed by assembling carbon dots along with proteins and nucleic acids. In comparison with free nucleic acids or proteins, this system of delivery defends the proteins and nucleic acids against enzymatic hydrolysis and delivers EGFP effectively into HeLa cells. Carbon dots of recent have stimulated immense awareness owing to some of their outstanding characteristics, like uniform distribution, biocompatibility, high quantum yield, and fluorescence. These properties are the reason carbon dots are used in therapeutic delivery, bioimaging, theranostics and optogenetics. This chapter presents the current application of carbon dots nanocomposite in plants and animals proteins and nucleic acids delivery.

**Keywords:** Proteins, carbon dots, deoxyribonucleic acid, ribonucleic acid, therapeutic delivery

## 14.1 Introduction

The rapid progress in the synthetic preparation of carbon dots (CDs) offers one of a kind opportunity for investigation into their potentiality as new carriers for protein and nucleic acid delivery. The observation and discovery of CDs were made by Xu et al. [1] when they tried to separate from carbon soot, single-walled carbon nanotubes utilizing the arc-discharge method. CDs are nanocarriers that are spherical in shape and with diameter less than 10 nm [2]. The surface of CDs is made up of a carbon core compressing of various functional groups (e.g. $-NH_2$, $-COOH$, and $-OH$) which presents high biological activity and excellent solubility, enabling them to form conjugates with inorganic

---

*Corresponding authors: N.B. Iroha, Department of Chemistry, Federal University Otuoke, Bayelsa State, Nigeria, e-mail: irohanb@fuotuoke.edu.ng
C.O. Ezenwaka, Department of Biology, Federal University Otuoke, Bayelsa State, Nigeria
C.N. Opara, Department of Microbiology, Federal University Otuoke, Bayelsa State, Nigeria
F.E. Abeng, Department of Chemistry, Cross River University of Technology, Calabar, Nigeria

https://doi.org/10.1515/9783110799958-014

and other organic substances [3]. On account of unique properties of CDs such as easy synthesis and modification, extraordinarily small size, low cytotoxicity, excellent water retention, and high degree of oxidation [4], they have been usefully applied in cell imaging and fluorescent labelling [5], heavy metal detection [6], bacterial labelling by charge-selective interaction [7], and drug delivery in cancer-targeted treatment [8].

Protein delivery involves the description of methods used to introduce proteins into cells. Proteins can be utilized in the production of high specific drugs and the methods of delivery must sustain the activity and bioavailability of proteins and should be non-toxic. In view of CDs' promising applications in nanomedicine, their safety has been a source of concern and has recently drawn increased attention [9], and studies on luminescent CD cytotoxicity have been detailed. Studies performed in vitro have shown that CDs are generally safe for several cell lines. Diverse other studies performed in vivo demonstrated that though CDs may be detected in different organs, the accumulated quantity was very low. No reported cases of clinical symptoms, toxicity, weight drops, or death [10]. The delivery of nucleic acid has been commonly undertaken by making use of cationic nanoparticles. These nanoparticles successfully condense the nucleic acid on account of electrostatic interactions. The CDs nanocarriers have the ability to escape rapidly the endo-lysosomal chamber on account of charge interplay with endo-lysosomal membrane.

RNA interference (RNAi) initiation through topical application of interfering RNA has potentials of being used for functional genomics of plant, crop protection and crop improvement, but the effective delivery of RNAi effectors into cells have been a major obstacle in developing this technology. Again, the effectiveness of RNAi in eliciting gene silencing response is highly dependent on the delivery of the double-stranded RNA (dsRNA) molecules into the target cell. This delivery has been a challenge in developing RNAi-based therapies, but the advent of CDs has greatly improved protein and nucleic acid delivery. In DNA delivery, carbon nanoparticles have been used to prevent most of the limitations related to conventional methods of gene delivery, including transformation complexity, host specificity, and tissue and cell damage resulting from external forces [11]. As reported by Dou et al. [12], quaternized CD gene transfection capability was discovered to be about $10^4$ times more effective than plain DNA delivery.

## 14.2 Carbon dots in the delivery of protein and nucleic acid into plant cells

In their recent work, Delgado-Martín et al. [13] used glucose as nucleation source to derive CDs and they were passivated with branched polyethylenimine (PEI) for the purpose of developing dsRNA nanocomposites. They characterized the CDs fully by utilizing transmission electron microscopy, hydrodynamic analyses, X-ray photoelectron spectroscopy, and Fourier-transform infrared spectroscopy. Their findings indicated that the

CDs possessed positive charges, excellent conductivity and electrophoretic mobility, and ideal in getting dsRNA nanocomposites. Cucumber plants leaves received the naked dsRNA and dsRNA coated with the CDs by spraying. The delivery of dsRNA to the leaves was better with CD-coated dsRNA, with 50-fold more dsRNA delivery than naked dsRNA. Moreover, certain short interfering RNAs (siRNAs) obtained from the sprinked dsRNAs were more abundant with CD-coated dsRNA. In distal leaves, the systemic dsRNAs determined showed an enhancement in the concentration when it was delivered as CD-coated nanocomposite. Likewise, in distal leaves were found more abundant systemic siRNAs when sprayed with the CD-coated nanocomposite. The results show that CDs derived by hydrothermal syntheses are satisfactory for the delivery of dsRNA in RNAi plant applications. Schwartz et al. [14] illustrated a technique of delivering siRNA into the model plant *Nicotiana benthamiana* and *Solanum lycopersicum* (tomato) using a class of tiny nanoparticles known as CDs. In both plant species, low-pressure spray utilization of the CD formulations with spreading surfactants gave rise to strong green fluorescent protein (GFP) transgenes silencing. The efficiency of delivery of CD formulations was further illustrated by the silencing of endogenous genes encoding two magnesium chelatase subunits, an enzyme required for chlorophyll synthesis. The visible phenotypes observed with the CD-facilitated delivery were authenticated by measuring notable reductions in the protein levels and/or target gene transcript. Figure 14.1 [14] shows the high levels of activity for siRNA delivery and GFP silencing in tomato (Figure 14.1A) and GFP and CHLH silencing in *N. benthamiana* (Figure 14.1B and C).

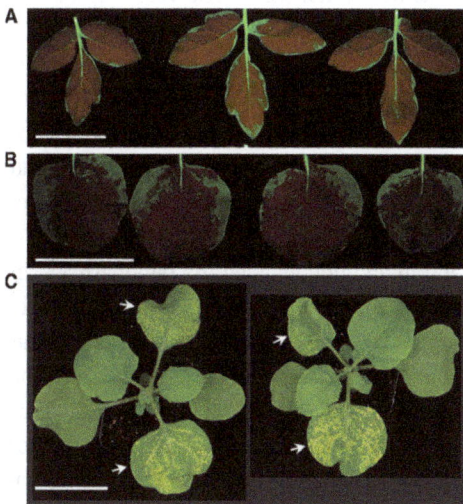

**Figure 14.1:** Gene silencing efficacy of CD-5 K-bPEI: (A) GFP silencing in HP375 tomato line, (B) GFP silencing in *N. benthamiana* 16C line, and (C) silencing of CHLH in *N. benthamiana* with arrows used to indicate the application leaves (reprinted with permission from [14]).

## 14.3 Carbon dots in the delivery of protein and nucleic acid into animal cells

CD-based nanomaterials' emergence offers great possibilities for extensive biomedical applications like cellular labelling and fluorescence imaging. They possess distinctive properties that manifest great ability in various applications. Wang et al. [15], in their report, illustrated a new photonic CD-based nanocarrier by utilizing low-molar-weight amphiphilic PEI (alkyl-PEI2k) in surface passivation. The water-dispersible alkyl-PEI2k-CD nanocarrier that was obtained possesses fluorescence properties, mono-dispersity, and good stability. In addition, the alkyl-PEI2k-CD nanocarrier possesses a remarkably low toxicity and exceptional gene transfection result in vivo and in vitro. The authors opined that considering the good gene delivery efficiency, fluorescence performance, and low cytotoxicity of alkyl-PEI2k-CD, it could serve as a new imaging-trackable favourable gene-delivery nanocarrier for optical molecular imaging and gene therapy. The application of alkyl-PEI2k-CD/siRNA complexes confirmed the decreased gene expression as the luciferase expression level was remarkably reduced, while the gene-silence effect was rarely seen in the other control groups.

The growing demands for better gene carrier achievement have led to the suggestion that a multifunctional vector may hugely clarify gene delivery for disease therapeutics. However, the non-viral vectors currently available are deficient of self-tracking abilities. Cao et al. [16] produced new dual-functional cationic CDs by one-step, microwave-assisted pyrolysis of glucose and arginine, being used as both a non-viral gene vector and a self-imaging agent for chondrogenesis from fibroblasts. The cationic CDs have the ability of condensing the model-gene plasmid SOX9 (pSOX9) forming extremely small nanoparticles ($10^{-30}$ nm), which possesses various good properties, like tunable fluorescence, high solubility, high yield, outstanding biocompatibility, and low cytotoxicity. Their results also reveal that the mean transit time assay showed that CDs/pSOX9 nanoparticles had minute cytotoxicity against mouse embryonic fibroblasts (MEFs) when compared with Lipofectamine 2000 and PEI (25 kDa). The CDs/pSOX9 nanoparticles precisely delivered the pSOX9 into MEFs with remarkably high efficiency and in addition enabled the nanoparticles' intracellular tracking. In addition, obvious chondrogenic differentiation was shown by the CDs/pSOX9 nanoparticle-mediated transfection of MEFs. The findings fully revealed that their prepared CDs could be presented as a dual-functional reagent paradigmatic example for both effective non-viral gene delivery and self-imaging. Pierrat et al. [17] fabricated cationic CDs by bPEI25k and citric acid pyrolysis under microwave radiations. The method they adopted produced 20–30% yield of different nanoparticles by simply modifying the reaction parameters. The researchers paid attention to the reaction products purification to make certain the residual starting polyamine is satisfactorily eliminated. They measured the intrinsic properties (surface charge, size, quantum yield, and photoluminescence) of particles and determined their effectiveness in forming stable complexes with nucleic acid. They also investigated the ability of the fabricated CDs in delivering siRNA or plasmid

DNA (pDNA) to different cell lines and compared it with that of bPEI25k. Similar efficiency of pDNA in vitro transfection for these CDs were observed when compared with the parent PEI, including their cytotoxicity. The lower cytotoxicity of CD/siRNA complexes in comparison with bPEI25k/siRNA complexes had pronounced outcome on the two carriers' gene silencing efficiency. However, the results are not entirely in agreement with some other results earlier reported on similar nanoparticles, which reveal that CDs' toxicity depends strongly on their fabrication protocol. The carriers were finally assessed for in vivo gene delivery via non-invasive pulmonary pathway in mice. Intense transgene expressions were achieved in the lungs that were comparable to those achieved with GL67A standard formulation, but were accompanied with remarkably minimized toxicity. Confocal microscopy was used to investigate the dissociation and cellular trafficking of the molecular assemblies of the CD/nucleic acid. The cell lines A549 were incubated at various time points (1, 4, and 24 h) with CD-7/DNA-Cy5 (Figure 14.2). Excitation wavelengths ($\lambda_{ext}$) of 488, 405, and 635 nm were used in measuring the images. Blue CD fluorescence was hugely seen in the cells after 1 h incubation with cellular uptake increasing with time. Notably, the red-punctuated DNA-Cy5 fluorescence could be found only outside the cells. These observations indicate that the CD-7/DNA-Cy5 complexes disassembled rapidly once inside the cells.

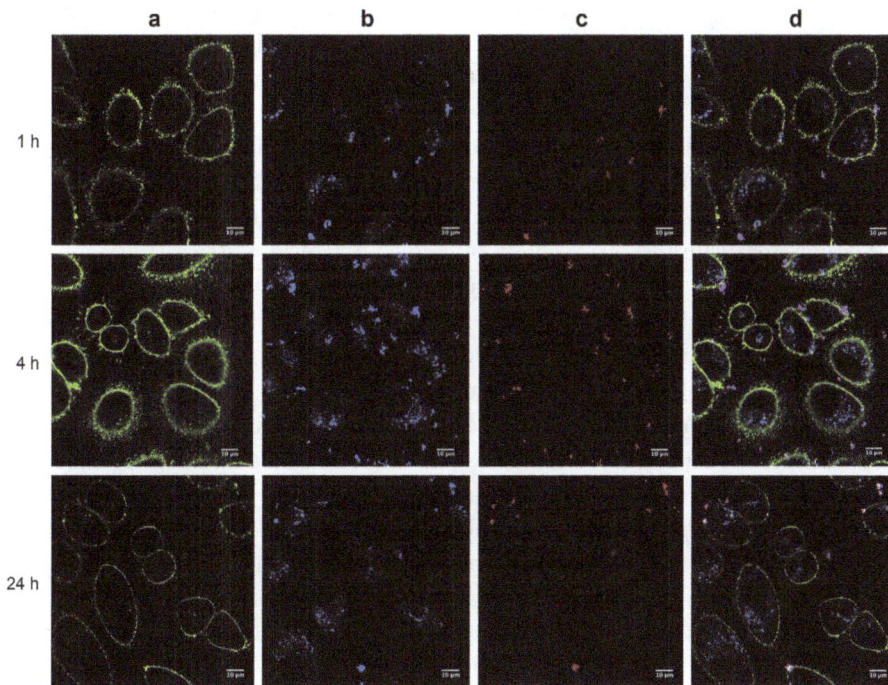

**Figure 14.2:** Confocal microscopy images of the delivery of CD-7/DNA-Cy5 complexes into A549 cells. Cells incubated at various time points (1, 4, and 24 h): (a) $\lambda_{ext}$:488 nm; (b) $\lambda_{ext}$:405 nm; (c) $\lambda_{ext}$:635 nm; and (d) merged images (reprinted with permission from [17]. Copyright 2022 Elsevier).

The remodelling of induced cardiomyocytes (iCMs) is of specific importance in regenerative medicine. It, however, remains a major challenge to assemble a safe and effective process of delivering gene and to effect the remodelling of iCMs for therapeutic implementation in heart trauma. Yang et al. [18] reported branched polyethyleneimine (BP)-coated nitrogen-enriched CDs (BP-NCDs) as effective nanocarriers filled with microRNAs-combo (BP-NCDs/MC) for cardiac remodelling. They prepared and characterized the BP-NCD nanocarriers using different analytical techniques. From their results, the BP-NCD nanocarriers indicated minor cytotoxicity, long-term microRNAs impression, and good microRNA-combo binding affinity. Another important observation is leading BP-NCDs/MC nano-complexes to effective direct remodelling of fibroblasts into iCMs in the absence of genomic integration and bringing about efficient cardiac function recovery following myocardial infarction. Their study offered a new approach in providing efficient and safe microRNA-delivery nano-platforms predicated on CDs for favourable disease therapy and cardiac regeneration. In another research, Wang et al. [19] developed tetraethylene pentamine CDs (TEPA-CDs) as nanocarriers of siRNA based on TEPA and glucose using bottom-up method. They prepared the CDs by one-step oil bath heating. The synthesized TEPA-CDs have a zeta potential of 19.17 mV, an average diameter of 8.6 nm and with positively charged decoration. Their result reveals that TEPA-CDs could precipitate siRNA into firm complexes without detectable premature release. Based on cellular uptake analysis, the Cy3-labelled siRNA was taken up by HeLa cells. Their study showed that TEPA-CDs could be useful as siRNA nanocarrier to tumour cells.

Branched PEI-based CDs (PCD) were developed by Hu et al. [20] from PEI by modified hydrothermal and oxidation reactions. The produced PCD from composition and structure analysis was found to possess a diameter of 3–4 nm and a 0.30 nm lattice spacing graphitic structure. The PCD with a 54.3% quantum yield possesses a bright photoluminescence indicating its usefulness shows in cell imaging. In addition, the excellent biocompatibility exhibited by the PCD shows it can be used for gene delivery. The specific nanostructure of the prepared PCD and its photoluminescence property indicate ability to be used in gene delivery and bioimaging. Wu et al. [21] devised an integrated theranostic, multi-functionalized nano-agent formulated on folate-conjugated reducible PEI-passivated CDs (fc-rPEICdots). The developed nanoagents discharge conspicuous blue photoluminescence below excitation of 360 nm and can enclose multiple siRNAs followed by delivering them in intracellular surroundings. The fc-rPEI-Cdots from in vitro cell culture investigation is a material that is very biocompatible and a good carrier of siRNA gene to deliver to targeted lung cancer therapy. Moreover, fc-rPEI-Cdots/pooled siRNAs could get accumulated through endocytosis in lung cancer cells, leading to enhanced anti-cancer effect and gene silencing. In combination of stimulus-responsive characteristics, CD bioimaging, active targeting motif, and gene silencing strategy, these fc-rPEI-Cdots may present a functional tool that will of benefit to clinicians in adjusting therapeutic strategy.

Liu et al. [22] fabricated PEI-functionalized CDs (CD-PEI) using microwave pyrolysis of a mixture of branched PEI25k and glycerol where carbon nanoparticles formation and the surface passivation were simultaneously accomplished. The nitrogen-rich nature of PEI ensured its important role in CD-PEI, enhancing the fluorescence by surface passivation. The PEI also functioned as a polyelectrolyte to condense DNA. It was revealed that the fabricated CD-PEI is water soluble and emits bright stable multicolour fluorescence which relies on excitation wavelength. The cytotoxicity and capability of CD-PEI is regulated by pyrolysis time probably due to PEI destruction during CD formation. The obtained CD-PEI exhibited comparable or higher gene expression of pDNA in COS-7 cells, lower toxicity, and HepG2 cells comparable to control PEI25k. The CD-PEIs is intriguingly internalized into tunable fluorescent emission shown by cells under differing excitation wavelength, indicating the efficacy of CD-PEI application in bioimaging and gene delivery.

## 14.4 Carbon dots in the delivery of protein and nucleic acid into bacterial cells

Manipulating an organism genetically is a vital facet of functional genomics research and recombinant DNA technology. These technologies depend on using bacterial cells to mass produce the required metabolites and proteins. Bacterial cells are equally utilized in amplifying the recombinant DNA prior to inserting the cells into the targeted organism. However, to successfully deliver the recombinant DNA into the bacterial cell poses a serious challenge as some bacteria are not compliant to these methods. Artificial transformation via electroporation and heat shock has been the commonly used technique of DNA delivery into bacteria. These methods involve tedious steps and needs sophisticated instruments in preparing the competent cells (Table 14.1). CD nanoparticles have been well utilized in therapeutics to deliver drugs into animal cells. They have gained popularity

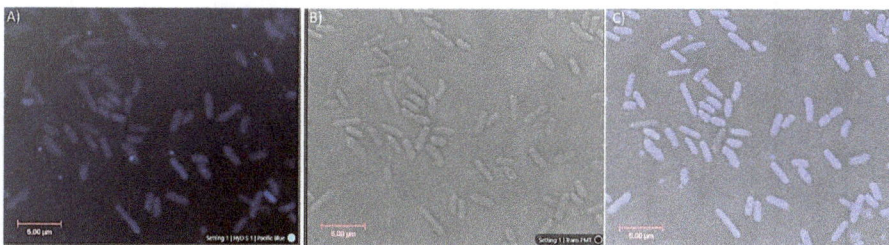

**Figure 14.3:** (a) Cells incubated with citric acid/β-alanine CDs for 15 min under confocal fluorescence microscope; (b) cells incubated with citric acid/β-alanine CDs for 15 min under differential interference contrast (DIC) microscopy; and (c) merge visualization of CD expression and live cells under confocal microscopy (reprinted with permission from [23]).

**Table 14.1:** Summary of studies on the use of carbon dots in proteins and nucleic acid delivery.

| Carbon dots | Precursors | Size (nm) | Synthesis method | Cargo | Cell lines | Findings | Ref. |
|---|---|---|---|---|---|---|---|
| CD-dsRNA | Glucose, branched PEI | 4–5 | Hydrothermal synthesis | dsRNA | 16C | Indicated that CD-dsRNA is appropriate for dsRNA foliar delivery in RNAi plant applications | [13] |
| CD-5 K-bPEI | Branched PEIs (25-kDa, 5-kDa bPEI) | <10 | Monowave | siRNA | 16C GFP, HP375 | Demonstrated strong GFP transgenes silencing and efficient siRNA delivery | [14] |
| Alkyl-PEI2k-CD | Amphiphilic PEI | 10–15 | Laser ablation | siRNA and DNA | 4T1-luc | Confirmed decreased gene expression as the luciferase expression level was remarkably reduced and could serve as a favourable gene delivery nanocarrier for gene therapy | [15] |
| CDs/pSOX9 | Glucose, arginine | 10⁻³⁰ | Microwave-assisted pyrolysis | pSOX9 | MEFs | Precisely delivered pSOX9 into mouse embryonic fibroblasts (MEFs) with remarkably high efficiency | [16] |
| CD/siRNA | Citric acid, bPEI25k | 12–13.2 | Microwave radiation | siRNA and pDNA) | A549 | Reported to be efficient in delivering pDNA and siRNA in A549 cells for lung disease treatment | [17] |
| BP-NCDs | Branched PEI, nitrogen | 42.1 | Hydrothermal | miRNA | C57BL/6 | Provided efficient and safe microRNA-delivery nano-platforms predicated on CDs for favourable disease therapy and cardiac regeneration | [18] |
| TEPA-CDs | Tetraethylene pentamine, glucose | <10 | Hydrothermal reaction | siRNA | HeLa | TEPA-CDs could precipitate siRNA into stable complexes and is useful as siRNA nanocarrier into tumour cells | [19] |
| PCD | PEI | 3–4 | Hydrothermal | DNA | MCF-7; 293T | Exhibited excellent biocompatibility which shows PCD can be used for gene delivery | [20] |
| fc-rPEICdots | PEI, glycerol | 143.1 | Microwave pyrolysis | siRNA (EGFR and cyclin B1) | H460; 3T3 | Good carrier of siRNA gene to deliver to targeted lung cancer therapy | [21] |

| | | | | | | | |
|---|---|---|---|---|---|---|---|
| CD-PEI | Branched PEI, glycerol | 7–12 | Microwave pyrolysis | pDNA | COS-7, HepG2 | Exhibited high gene expression of pDNA in COS-7 cells showing the efficacy of CD-PEI in gene delivery | [22] |
| CA/β-alanine CDs | Citric acid, β-alanine | 20–40 | Microwave | pDNA | E. coli cells | Demonstrated that CA/β-alanine CDs can be successfully used for foreign DNA delivery into E. coli cells of up to 10 kb | [23] |
| CD-PDMA-PMPD | Citric acid | 50 | Microwave | DNA | COS-7 | Provided a platform for serum-resistant gene delivery and imaging and displayed higher transfection efficiency | [24] |
| Positive charge CD | FA and PEI | – | Hydrothermal | pDNA | HeLa, 293T | Demonstrated efficient pDNA transfected into cells and exhibited photoluminescent property for gene therapy and cancer diagnosis | [25] |

as new DNA nanocarriers in animal and plant sciences. Pandey et al. [23] reported the synthesis of CDs from β-alanine and citric acid and their usage in delivering DNA into *E. coli* cells. The authors fabricated CDs utilizing microwave supported synthesis. Plasmids conveying ampicillin resistance and red fluorescent protein reporter genes were transported to bacterial cells and additionally established utilizing polymerase chain reaction. From the reports, it was discovered that CDs can be successfully used for foreign DNA delivery into *E. coli* of up to 10 kb. The use of citric acid/β-alanine CDs as nanocarriers of DNA into *E. coli* cells was demonstrated, and their limitations in terms of the capacity of pDNA they are able to carry were identified. The application of CDs in remote DNA delivery into bacterial cells is a new method and has a potential to transform resistant organism for which no definitive DNA delivery systems have been identified. Figure 14.3 [23] shows the confocal microscopy (magnification: ×63) of CD interaction with *E. coli* cells, and the CDs were found trapped around the surface of the bacterial cells.

## 14.5 Conclusions

In this chapter, the applications of CDs in protein and nucleic acid delivery into animal, plants, and bacteria cells were reviewed. CDs were found to be excellent in delivering siRNA, DNA, and other gene cargos into different targeted cell lines. The CDs are known to exhibit excellent biocompatibility stable fluorescence, high quantum yield, and negligible toxicity. The interesting properties of CDs make them safe and efficient nucleic acid and protein nanocarriers. The potent fluorescence and good stability of CDs indicate that the delivery of protein and nucleic acid efficiency could be determined easily by fluorescence, which exhibits a great capacity as delivery agent with distinctive property.

## References

[1]   Xu X, Ray R, Gu Y, Ploehn HJ, Gearheart L, Raker K, Scrivens WA. Electrophoretic analysis and purification of fluorescent single-walled carbon nanotube fragments. J Am Chem Soc. 2004, 126, 12736–12737.

[2]   Mishra V, Patil A, Thankur S, Prashant K. Carbon dots: Emerging theranostic nanoarchitectures. Drug Discov Today. 2018, 23, 1219–1232.

[3]   Hu SL, Niu KY, Sun J, Yang J, Zhao NQ, Du XW. One-step synthesis of fluorescent carbon nanoparticles by laser irradiation. J Mater Chem. 2009, 19, 484–488.

[4]   Lim SY, Shen W, Gao Z. Carbon quantum dots and their applications. Chem Soc Rev. 2015, 44, 362–381.

[5]   Luo PG, Yang F, Yang ST, Sonkar SK, Yang L, Broglie JJ, Liu Y, Sun YP. Carbon-based quantum dots for fluorescenceimaging of cells and tissues. RSC Adv. 2014, 4, 10791–10807.

[6]   Gao X, Du C, Zhuang Z, Chen W. Carbon quantum dot-based nanoprobes for metal ion detection. J Mater Chem C. 2016, 4, 6927–6945.
[7]   Hua XW, Bao YW, Wang HY, Chen Z, Wu FG. Bacteria-derived fluorescent carbon dots for microbial live/dead differentiation. Nanoscale. 2017, 9, 2150–2161.
[8]   Zheng M, Liu S, Li J, Qu D, Zhao H, Guan X, Hu X, Xie Z, Jing X, Sun Z. Integrating oxaliplatin with highly luminescent carbon dots: An unprecedented theranostic agent for personalized medicine. Adv Mater. 2014, 26, 3554–3560.
[9]   Hong W, Liu Y, Li MH, Xing YX, Chen T, Fu YH, et al. In vivo toxicology of carbon dots by (1)H NMR-based metabolomics. Toxicol Res. 2018, 7, 834–847.
[10]  Chong Y, Ma Y, Shen H, Tu X, Zhou X, Xu J, et al. The in vitro and in vivo toxicity of graphene quantum dots. Biomaterials. 2014, 35, 5041–5048.
[11]  Demirer GS, Zhang H, Goh NS, González-Grandío E, Landry MP. Carbon nanotube-mediated DNA delivery without transgene integration in intact plants. Nat Protoc. 2019, 14, 2954–2971.
[12]  Dou Q, Fang X, Jiang S, Chee PL, Leed T, Loh XJ. Multi-functional fluorescent carbon dots with antibacterial and gene delivery properties. RSC Adv. 2015, 5, 46817–46822.
[13]  Delgado-Martín J, Delgado-Olidén A, Velasco L. Carbon dots boost dsRNA delivery in plants and increase local and systemic siRNA production. Int J Mol Sci. 2022, 23, 5338.
[14]  Schwartz SH, Hendrix B, Hoffer P, Sanders RA, Zheng W. Carbon dots for efficient small interfering RNA delivery and gene silencing in plants. Plant Physiol. 2020, 184, 647–657.
[15]  Wang L, Wang X, Bhirde A, Cao J, Zeng Y, Huang X, Sun Y, Liu G, Chen X. Carbon dots based two-photon visible nanocarriers for safe and highly efficient delivery of siRNA and DNA. Adv Healthc Mater. 2014, 3(8), 1203–1209.
[16]  Cao X, Wang J, Deng W, Chen J, Wang Y, Zhou J, Du P, Xu W, Wang Q, Wang Q, Yu Q, Spector M, Yu J, Xu X. Photoluminescent cationic carbon dots as efficient non-viral delivery of plasmid SOX9 and chondrogenesis of fibroblasts. Sci Rep. 2018, 8, 7057.
[17]  Pierrat P, Wang R, Kereselidze D, Lux M, Didier P, Antoine Kichler A, Pons F, Lebeau L. Efficient in vitro and in vivo pulmonary delivery of nucleic acid by carbon dot-based nanocarriers. Biomaterials. 2015, 51, 290–302.
[18]  Yang L, Xue S, Du M, Lian F. Highly efficient microRNA delivery using functionalized carbon dots for enhanced conversion of fibroblasts to cardiomyocytes. Int J Nanomed. 2021, 16, 3741–3754.
[19]  Wang J, Liu S, Chang Y, Fang L, Han K, Li M. High efficient delivery of siRNA into tumor cells by positively charged carbon dots. J Macromol Sci Part A. 2019, doi: https://doi.org/10.1080/10601325.2018.1526043.
[20]  Hu L, Sun Y, Li S, Wang X, Hu K, Wang L, Liang XJ, Wu Y. Multifunctional carbon dots with high quantum yield for imaging and gene delivery. Carbon. 2014, 67, 508–513.
[21]  Wu YF, Wu HC, Kuan CH, Lin CJ, Wang LW, Chang CW, Wang TW. Multi-functionalized carbon dots as theranostic nanoagent for gene delivery in lung cancer therapy. Sci Rep. 2016, 6, 21170.
[22]  Liu C, Zhang P, Zhai X, Tian F, Li W, Yang J, Liu Y, Wang H, Wang W, Liu W. Nano-carrier for gene delivery and bioimaging based on carbon dots with PEI-passivation enhanced fluorescence. Biomaterials. 2012, 33, 3604–3613.
[23]  Pandey A, Devkota A, Sigdel A, Yadegari Z, Dumenyo K, Taheri A. Citric acid/β-alanine carbon dots as a novel tool for delivery of plasmid DNA into E. coli cells. Sci Rep. 2021, 11, 23964. doi: https://doi.org/10.1038/s41598-021-03437-y.
[24]  Cheng L, Li Y, Zhai X, Xu B, Cao Z, Liu W. Polycation-b-polyzwitterion copolymer grafted luminescent carbon dots as a multifunctional platform for serum-resistant gene delivery and bioimaging. ACS Appl Mat Interf. 2014, 6, 20487–20497.
[25]  Yang X, Wang Y, Shen X, Su C, Yang J, Piao M, Lin Q. One-step synthesis of photoluminescent carbon dots with excitation-independent emission for selective bioimaging and gene delivery. J Colloid Interface Sci. 2017, 492, 1–7.

# Index

https://doi.org/10.1515/9783110799958-015

www.ingramcontent.com/pod-product-compliance
Lightning Source LLC
Chambersburg PA
CBHW080916220326
41598CB00034B/5585

9 783110 799927